The Geometric Induction
of Bone Formation

T0136212

The Geometric Induction of Bone Formation

Ugo Ripamonti
Director: Bone Research Laboratory
Visiting Professor, School of Clinical
Medicine – Internal Medicine
Faculty of Health Sciences
University of the Witwatersrand, Johannesburg,
South Africa

CRC Press
Taylor & Francis Group
Boca Raton London New York

CRC Press is an imprint of the
Taylor & Francis Group, an **informa** business

Cover: The spontaneous induction of chondrogenesis within the macroporous spaces of a coral-derived bioreactor *solo* harvested on day 21 after intramuscular implantation in the shark *Carcharhinus obscurus*. Lack of vascular invasion followed by chondrolysis is responsible for the lack of bone formation after evolutionary expression and synthesis of powerful inhibitors of angiogenesis that block *osteogenesis in angiogenesis* (see text).

First edition published 2021
by CRC Press
6000 Broken Sound Parkway NW, Suite 300, Boca Raton, FL 33487-2742

and by CRC Press
2 Park Square, Milton Park, Abingdon, Oxon, OX14 4RN

© 2021 Taylor & Francis Group, LLC

CRC Press is an imprint of Taylor & Francis Group, LLC

ISBN: 9780367195786 (hbk)
ISBN: 9780367682606 (pbk)
ISBN: 9780429203299 (ebk)

Typeset in Times
by Deanta Global Publishing Services, Chennai, India

Dedication

This contribution to The Geometric Induction of Bone Formation *is dedicated to my daughter Daniella Bella for the creativity to write. Daniella may one day understand in more detail what it is that her papà has been doing working often late across the nights and across the planet – writing lovely "stories" about the "concavity: the shape of life".*

Contents

Acknowledgements

Our discovery by serendipity of the so-defined spontaneous and/or intrinsic induction of bone formation by coral-derived macroporous bioreactors was the beginning of a fascinating and intriguing scientific ride across cell differentiation, de-differentiation, tissue induction and morphogenesis, morphogens and stem cells, angiogenesis and organogenesis. This ride has highlighted to no end the scientific and intellectual journey of the writer and editor of this book on the geometric induction of bone formation, beyond morphogens and stem cells. Many thanks to the Chacma baboon *Papio ursinus* for continuously providing across several years unique microenvironments for the heterotopic intramuscular induction of bone formation without the exogenous applications of the osteogenic soluble molecular signals of the transforming growth factor-β (TGF-β) supergene family. My Unit and I would like to thank Edwin Clayton Shors, Interpore International, and later Mike Ponticiello, Zimmer Biomet, for the continuous supply of coral-derived macroporous bioreactors for implantation in several *Papio ursinus* across the decades. A special thanks to Barbara van den Heever, Senior Research Technologist firstly at the Dental Research Institute, later at the Bone Research Laboratory, who greatly helped shape the Bone Research Laboratory into a fully fledged Unit of the SA Medical Research Council and the University of the Witwatersrand, Johannesburg. Barbara started the precision cutting of undecalcified sections using Leica sledge microtomes with carbide-tungsten knifes often cut at 2–3 μm thick. Ruqayya Parak, Senior Research Technologist at the Bone Research Unit, has outstandingly added the third dimension of undecalcified histology by cutting impeccable sections on the Exakt precision cutting and grinding diamond saw. Barbara and Ruqayya's research presented decalcified and undecalcified histological material now spread across more than 250 contributions around the planet, including three CRC Press volumes reporting the induction of bone formation. To both, special recognition is offered, particularly for having been able to handle the time constraints of continuously preparing impeccable undecalcified whole mounted and stained sections and to deal with the author and Director of the Unit far too often asking what next and when the sections would be finally stained and ready for analyses and digital photography. Laura Roden (née Yeates) significantly contributed to our studies on the geometric induction of bone formation by continuously offering lucid and clear scientific thinking and analyses highlighted by impeccable purification schemes of naturally derived highly purified osteoinductive proteins extracted and purified to apparent homogeneity after gel filtration chromatography from demineralized and chaotropically extracted baboon bone matrices. The molecular biology team headed by Raquel Duarte, School of Clinical Medicine, Internal Medicine, mechanistically resolved the molecular key of the spontaneous induction of bone formation. Special thanks to Raquel Duarte, Roland Manfred Klar, Caroline Dickens and Therese Dix-Peek for several joint experiments, many discussion hours and for finally resolving the spontaneous induction of bone formation with gene expression analyses regulating the heterotopic induction of bone formation by coral-derived macroporous bioreactors when implanted in the *rectus abdominis* muscle of *Papio ursinus*, where there is no bone. Special thanks to Hari A. Reddi, formerly of the NIH, Bone Cell Biology Section, who mentored the author and editor of this volume on fundamental manuscripts firstly reporting the effect of the geometry of the inductor on the induction of bone formation. To Hari, now Distinguished Professor at the University of California, Davis, School of Medicine, Sacramento, and to his mentor, Prof Charles Huggins, the author is particularly indebted. While this CRC Press volume was being composed and printed, the author and editor were awarded the highest honour given by the global biomaterials community of being elected to the status of "Fellow of Biomaterials Science and Engineering" (FBSE). Tony Dakyns, a neighbour and friend close by my Unit at Club Nautique, Broederstroom, kindly provided the password for his Internet access so that I could work close to his Unit in the mobile office of my car. Tony also provided a heater that helped a lot in writing and revising manuscripts in the cold and bitter high veld winter nights; to him, special recognition is offered. My grateful thanks to C.R. Crumbly PhD, Senior Acquisition Editor, Life Sciences, and Ana Lucia Eberhart, Editorial Assistant, CRC Press, Boca Raton, US, for coping with an often-recalcitrant editor and writer and for providing the platform for constructing a written vehicle for the geometric induction of bone formation.

At Club Nautique, Broederstroom,
North West Province, South Africa

Author

Ugo Ripamonti is a leader in the field of tissue induction and differentiation in postnatal osteogenesis. His principal achievement has been the demonstration of the induction of bone formation in non-human primate models. Through an imaginative multifaceted combination of heterotopic implantation, self-inducing biomaterial matrices, soluble molecular signals and non-human primates of the species *Papio ursinus*, he has systematically studied the induction of bone formation in non-human primates as a prerequisite for therapeutic application in humans.

As a research scientist attached to the Medical Research Council (MRC) at the Dental Research Institute of the University of the Witwatersrand, Dr Ripamonti made a seminal contribution by showing the induction of bone formation within the macroporous spaces of the biomimetic matrices when implanted in the *rectus abdominis* muscle where there is no bone. In several important published papers he defined the mechanistic insights of the intrinsic induction in non-human primates *Papio ursinus*.

During the late 1980s–early 1990s Dr Ripamonti, then a visiting scientist at the Bone Cell Biology Section of the National Institute of Dental Research, National Institutes of Health, Bethesda, Maryland, joined forces with NIH Chief Hari Reddi, PhD, to research the induction of bone formation with naturally derived bone morphogenetic proteins from baboon bone matrices and implanted in *Papio ursinus*. In 1994, he was appointed Professor at the Medical School University of the Witwatersrand leading the newly founded Bone Research Laboratory of the MRC and the hosting university.

Professor Ugo Ripamonti has been active in international affairs through his studies, with more than 200 publications and more than 120 invited lectures at International Symposia having also organized a series of workshops at several International Conferences including World Biomaterials Congresses and the International Conferences on Bone Morphogenetic Proteins. He is a member of the International Organizing Committee since the 1st Conference in 1994 and he has been invited to deliver the 2006 Marshall Urist Awarded Lecture. These remarkable different achievements are also due to Dr Ripamonti's eclectic, multiform and diverse, but focused, clinical and research training starting with a medical doctor degree (*cum laude*), a diploma in Odontology and Stomatology (*cum laude*), a diploma in Maxillo-Facial Surgery (*cum laude*), a Masters' in Periodontology and Oral Medicine and a doctoral thesis on the induction of bone formation in the non-human primate *Papio ursinus*.

In summary, Prof Ripamonti has been a pioneer of the redundancy of molecular signals initiating the induction of bone formation in non-human and human primates, and he is a leading international scholar in the field of regenerative medicine and bone tissue engineering. His incisive experiments in *Papio ursinus* are considered the benchmark in bone induction and because of the impact of his research on bone tissue engineering, he has been invited three times to present new data at the Keystone Symposia on Molecular and Cellular Biology highlighting new challenges for regenerative medicine. He has been inducted to the College of Fellow Biomaterials Science and Engineering (FBSE). His work has significance for understanding skeletal development as well as providing necessary methods for bone tissue engineering.

List of Contributors

Ugo Ripamonti
Bone Research Laboratory
School of Clinical Medicine - Internal
    Medicine
Faculty of Health Sciences
University of the Witwatersrand
Johannesburg, South Africa

Hari A Reddi
Center for Tissue Regeneration and
    Repair
University of California–Davis
School of Medicine
Sacramento, California, US

Laura Roden
School of Life Sciences
Coventry University
Coventry, CV1 2DS, UK
*Formerly, the Dental Research
    Institute, and the Bone Research
    Unit, the MRC/University of the
    Witwatersrand, Johannesburg*

Raquel Duarte
Department of Internal Medicine
School of Clinical Medicine
Faculty of Health Sciences
University of the Witwatersrand
Johannesburg, South Africa

Introduction

Ugo Ripamonti and Hari A. Reddi

THE GEOMETRIC INDUCTION OF BONE FORMATION

Central to self-inductive biomaterials are interactions of cell–biomaterials (Williams 2011; Ripamonti 2006). These include cell attachment, spreading, proliferation, differentiation and tissue induction (Ripamonti et al. 1993; Ripamonti 2006). Extensive research work has led to our deep understanding of the role of biomaterial surfaces in controlling the induction of cell differentiation (Williams 2011).

Perhaps two titles put this introduction in a clear biological perspective, i.e. the title of McNamara et al. in the *Journal of Tissue Engineering*: "Nanotopographical control of stem cell differentiation" (2010), and Liu et al. in *Biomaterials*: "Subcellular geometry on micropillars regulates stem cell differentiation" (2016). More revealing however is the *Cell* paper of Discher's laboratory titled "Matrix elasticity directs stem cell differentiation" (Engler et al. 2006).

Geometry appears thus to be a critical characteristic imprinted into biomaterials which may or may not differentiate cellular elements resting on specific geometric topographies. The geometric guided tissue induction does result in specific genetic and morphogenetic pathways initiating in the induction of selected morphogenetic processes including the induction of bone formation (Fig. 0.1) as well as the geometric control of capillary architecture (Sun et al. 2014). Geometric topographies set into motion gene expression pathways, resulting in the induction of specific morphologies as predicted by micro surface topographies or geometries (Klar et al. 2013; Ripamonti 2017; Ripamonti 2018) (Fig. 0.1).

The aim of these short introductory notes to this CRC Press Volume on *The Geometric Induction of Bone Formation* is to concisely review the past and present of a biological scenario of cell differentiation and tissue induction operated by the geometric configuration of extracellular matrices as substrata that *per se* initiate self-inductive phenomena (Ripamonti et al. 1999) (Fig. 0.2).

Firstly, however, we have to briefly review and introduce a series of recent incisive communications that set the molecular pathways of cellular functions regulated by the geometry of the substratum (Engler et al. 2006; Discher et al. 2005; Discher et al. 2009; Buxboim and Discher 2010; Buxboim et al. 2010). These communications report the mechanical regulation of cell functions by geometrically modulated substrata.

Exquisitely novel, Liu et al. (2016) report that mechanical regulation on micropillars also affects subcellular nuclear geometry which further regulates stem cell differentiation and the induction of tissue patterning and differentiation (Liu et al. 2016). The results report that the geometry of cell nuclei is an intracellular signal modulator for stem cells' lineage commitment and for topographically induced regulation of cell differentiation (Liu et al. 2016).

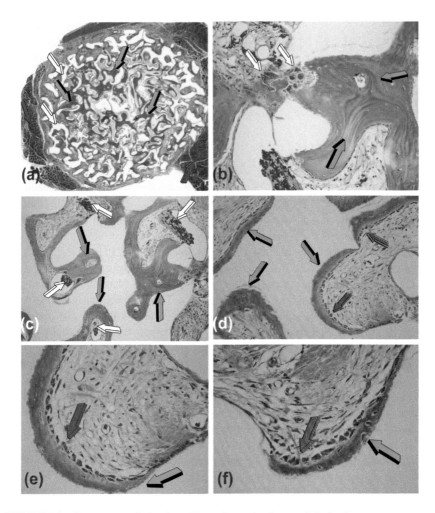

FIGURE 0.1 Spontaneous induction of bone formation by coral-derived macroporous constructs implanted in the rectus abdominis muscle of the adult Chacma baboon *Papio ursinus*, where there is no bone, and harvested on day 90 after heterotopic implantation. (a) Low-power macro photograph of a coral-derived construct (*white* arrows) implanted in the *rectus abdominis* muscle (*magenta* arrows) showing the induction of bone formation across the macroporous spaces (*light blue* arrows). The coral-derived substratum (*white* arrows) is shown after demineralization of the calcium-carbonate constructs. (b) High-power view showing the induction of lamellar osteonic bone (*light blue* arrows) that formed by day 90 after implantation within the macroporous spaces. (c) Remodelling with the induction of lamellar osteonic bone (*light blue* arrows) supported by capillary sprouting and invasion (*white* arrows). (d,e,f) Preferential induction of bone formation within concavities of the calcium phosphate-based coral-derived bioreactors induced and tightly attached to the concavity of the calcium phosphate-based bioreactors (*light blue* arrows). Newly formed bone (*light blue* arrows) is covered by contiguous osteoblasts (*magenta* arrows) with osteocytes embedded within the newly formed matrix (*light blue* arrow in f). Decalcified paraffin embedded sections cut at 4 μm stained with Goldner's trichrome.

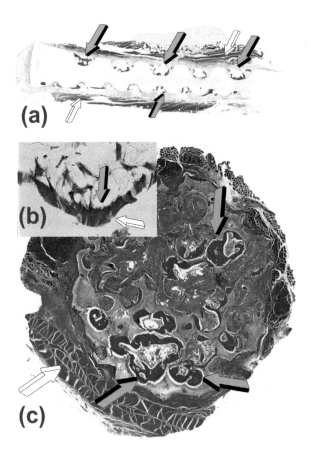

FIGURE 0.2 "The geometric induction of bone formation" (Ripamonti et al. 1999) is defined by repetitive sequences of concavities with specific radii of curvature prepared in macroporous sintered crystalline hydroxyapatites or in solid discs of crystalline sintered bioreactors when implanted in heterotopic site of the *rectus abdominis* muscle of the non-human primate *Papio ursinus* (Ripamonti et al.1999; Ripamonti and Kirkbride 2001; Klar et al. 2013; Ripamonti 2004; Ripamonti 2017). (a) Longitudinal section of a crystalline sintered hydroxyapatite disc showing that the induction of bone formation by day 90 after intramuscular implantation only initiates within the concavities of the substratum (*light blue* arrows) (Ripamonti et al. 1999; Ripamonti and Kirkbride 2001; Ripamonti 2004; Ripamonti 2017). (b) *In vitro* studies showed the organization, alignment and orientation of pre-osteoblastic cells (*light blue* arrow) within concavities of the coral-derived biomimetic construct (*white* arrow). (Ripamonti 2012; Ripamonti et al. 2012a; Ripamonti et al. 2012b). (c) Low-power macro photographic image showing the spontaneous induction of bone formation (*light blue* arrows) within the macroporous spaces of the crystalline sintered construct heterotopically implanted in the *rectus abdominis* muscle (*white* arrows) (Ripamonti et al. 1999). Decalcified paraffin embedded sections cut at 4 μm stained with Goldner's trichrome. Original magnification (a) ×30 and (b) ×20.

The incisive work of Discher's laboratories shows how microenvironments are critical in directing stem cell specification (Engler et al. 2006). As such, stem cells are shown to commit to specific phenotypes *via* tissue-level elasticity (Engler et al. 2006). Discher's statements in his *Cell* paper (Engler et al. 2006) are perhaps the most molecularly and intellectually fascinating aspect of biomimetic science generated so far, that is: "soft matrices that mimic brain are neurogenic"; in contrast, "comparatively rigid matrices that mimic collagenous bone prove to be osteogenic" (Engler et al. 2006).

Mechanical regulation on micropillars also affects subcellular nuclear geometry which further regulates stem cell differentiation (Liu et al. 2016). Again, as the laboratories of Discher stated, substrata of different stiffness are recognized by stem cells which "feel the difference" (Buxboim et al. 2010; Buxboim and Discher 2010) between soft and hard substrata (Discher et al. 2005; Engler et al. 2006).

The realization that stem cells "feel the difference" (Buxboim et al. 2010; Buxboim and Discher 2010) has indicated that matrices upon which cells grow and differentiate could be tailored to functionalize biomaterials' surfaces (McNamara et al. 2010; McNamara et al. 2011). Cell/matrix functionalization results in the induction of cellular differentiation with the expression of the osteogenic phenotype. The reported data on how to functionalize titania surface topographies are a step forward in biomaterials science ultimately constructing biomaterial surfaces intrinsically capable of inducing the osteogenic phenotype in pre-clinical and clinical contexts (McNamara et al. 2011).

Of interest, Liu et al. (2016) showed that subcellular geometry on selected micropillar arrays regulates stem cell differentiation (Liu et al. 2016). Piercing-like nuclear deformation and altered nuclear geometry of stem cells resting on micropillar arrays showed that cells cultured on 4.6 μm patterned surfaces express more than twofold *Runx2* and *Alkaline phosphatase* (*ALP*) than other dimensions used in the micropillar study (Liu et al. 2016). It is noteworthy to quote again that

> nuclear deformation as a result of topographical cues set by the geometry of the substratum is the ultimate and clear-cut proof that the geometry of the substratum regulates cell differentiation and the expression of the osteogenic phenotype both *in vitro* and *in vivo*.
>
> **(Ripamonti 2017)**

Geometry, geometric configurations, nanotopographic surface alterations and modifications, sintered surface microstructures, pits, micro-concavities and concavities are all geometric cues that define geometric configurations endowed with the striking capacity to induce cellular differentiation and tissue induction (Ripamonti et al. 1999; Ripamonti 2004; Ripamonti 2017; Ripamonti et al. 2017; Ahn et al. 2014; Barradas et al. 2011; Bettinger et al. 2009; Brunette 1988; Clark et al. 1987; Clark et al. 1990; Costa-Rodriguez et al. 2012; Curran et al. 2006; Curtis and Wilkinson 1997; Gittens et al. 2011; Kiang et al. 2013; Killian et al. 2010; Lamers et al. 2010; Li et al. 2008; Habibovic et al. 2005; Metawarayuth et al. 2016; Miyoshi and Adachi 2014; Vlacic-Zischke et al. 2011; Watari et al. 2012; Wilkinson et al. 2011; Zhang et al. 2014; Zhang et al. 2017a; Zhang et al. 2017b; Zhang et al. 2018); Danoux et al. 2016; Fiedler et al. 2013; Kim et al. 2012.

As we have previously stated, "The geometric induction of bone formation", that titled an experimental study by sintered crystalline hydroxyapatites with concavities prepared on both planar surfaces (Fig. 0.2a) (Ripamonti et al. 1999), stems from classic last-century reports that unequivocally showed that the geometry of the inductor tightly controls the induction of bone formation (Fig. 0.3) (Reddi and Huggins 1973).

Demineralized incisors harvested and prepared from adult rats were transplanted in symmetrical subcutaneous contralateral sites of allogeneic animals (Reddi and Huggins 1973). Harvested specimens showed the induction of chondrogenesis in the apical part of the pulp chamber by day 7 after transplantation (Reddi and Huggins 1973). On day 28 there was the induction of vascularized bone formation within the central and coronal part of the incisors' endodontic space (Fig. 0.3a). A vascularized ossicle (Fig. 0.3a O) with hemopoietic marrow formed against the cartilage still present within the apical part of the canal (Fig. 0.3a *magenta* arrow).

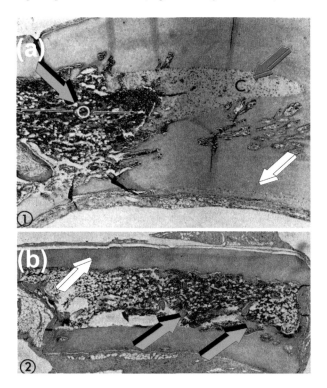

FIGURE 0.3 Demineralized continuously growing rat mandibular incisors were demineralized and implanted in the subcutaneous space of allogeneic rats (Reddi and Huggins 1973). Close incisors, with the tooth apex (*white* arrow), showed the induction of cartilage in the close pulpal space (*magenta* arrow in (a). The remaining pulpal space showed the induction of an ossicle (O) with bone and bone marrow *light blue* arrows O). (b) Implantation of demineralized tooth tubes without the apices showed bone induction throughout the open endodontic space on day 28 with lack of cartilage induction. Histological sections courtesy of A.H. Reddi (Reddi and Huggins 1973).

In additional incisor transplants, the apex of the demineralized incisors was cut to form a demineralized incisor tube with open ends (Reddi and Huggins 1973). As it is stated, "the results were utterly different" (Fig. 0.3b) (Reddi and Huggins 1973). On day 14, the induction of bone formation was seen at both ends of the demineralized dentinal cylinders. Of note, islands of cartilage separated the osteogenetic fronts forming at both open ends of the demineralized dentinal tubes (Reddi and Huggins 1973). On day 28, there was the induction of trabecular bone throughout the open demineralized dentinal tube (Fig. 0.3b *light blue* arrows) without evidence of cartilage remnants (Reddi and Huggins 1973). The communication concluded that the morphological temporal sequence of the induction of bone formation from fibroblast invasion, chondrogenesis and the induction of bone with later bone marrow formation "was profoundly influenced by the geometry of the transformant", i.e. the geometry of the transplanted demineralized dentinal matrices (Reddi and Huggins 1973; Reddi 1974).

Further experiments by A.H. Reddi, then at the National Institutes of Health (NIH), Bethesda, showed that subcutaneous implantation of coarse demineralized bone matrix (particle size 74–420 μm) in the subcutaneous space of allogeneic rats resulted in the induction of endochondral bone differentiation (Fig. 0.4a) (Sampath and Reddi, 1984). In contrast, subcutaneous implantation of demineralized bone with fine particle size of 44–74 μm did not result in the induction of bone formation (Fig. 0.4b) (Sampath and Reddi, 1984).

The critical role of the geometry of the inductor was investigated by solubilizing putative osteogenic proteins from fine demineralized bone matrices (particle size of 44–74 μm) after dissociative extraction in 4 M guanidine HCl (Sampath and Reddi 1981; Sampath and Reddi 1983; Sampath and Reddi 1984). Extracts from fine demineralized particles were reconstituted with the inactive collagenous demineralized bone matrix residue of coarse powders (particle size 74–420 μm) (Sampath and Reddi 1984). The reconstitution of coarse powder matrices with fine powder extracts restored the biological activity of the transplanted and reconstituted matrices with the induction of endochondral bone formation (Fig. 0.4c) (Sampath and Reddi 1984).

Of note, partial purification of fine particle extracts on Sepharose CL-6B gel filtration chromatography showed that the reconstitution of column fractions of 20,000–50,000 mol. wt. with inactive coarse residue induced reproducible and prominent endochondral bone formation (Fig. 0.4d) (Sampath and Reddi 1984). Solubilized extracellular matrix components of coarse bone matrix (particle size of 74–420 μm) were reconstituted with fine matrix inactive residue (particle size 74–420 μm) and bioassayed for endochondral bone formation at subcutaneous bilateral sites in rodents (Sampath and Reddi 1984). The results showed that demineralized extracted collagenous matrices of coarse powders did not support the induction of bone formation when reconstituted with fine matrix (particle size 44–74 μm) (Sampath and Reddi 1984). On the other hand, the reconstitution of fine matrix extracts with coarse demineralized matrices (particle size 74–420 μm) restored the biological activity with the induction of endochondral bone formation (Fig. 0.4c) (Sampath and Reddi 1984).

The above experiments in heterotopic subcutaneous sites of rodents show how the geometry of the inductors blocks the initiation of bone formation by demineralized

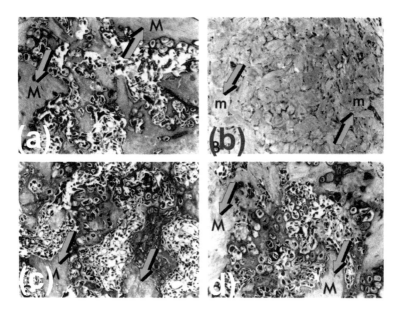

FIGURE 0.4   The effect of geometry of particle size of allogeneic demineralized bone matrices after implantation in the subcutaneous space of the rat (Sampath and Reddi 1984). (a) Induction of endochondral bone formation by coarse matrix particles (M *light blue* arrows, 74–420 μm). (b) Lack of bone differentiation and endochondral bone formation after implantation of fine matrix particles (m *light blue* arrows, 44–74 μm). Coarse matrix particles (74–420 μm) were reconstituted with guanidinium extracted (c) and partially purified fine matrices and implanted subcutaneously in the rodent bioassay (Sampath and Reddi 1984). Reconstitution of fine matrix extracts restored the biological activity of dissociatively extracted coarse matrix. (d) Partial purification of fine matrix extracts when reconstituted with dissociatively extracted coarse matrix restored the biological activity of the whole demineralized bone matrix, particle size 74–420 μm. Images courtesy of A.H. Reddi (Sampath and Reddi 1984) from the *J. Cell Biology* 1984.

bone matrices, and that the induction of bone formation is dependent on the geometry of the inductor which overrules the morphogenetic drive of bone morphogenetic proteins embedded within the demineralized matrices or reconstituted with the insoluble inactive collagenous matrices.

The studies concluded that the extracellular matrices of bone, both of coarse and fine particle size, contain osteoinductive components necessary for the induction of bone formation as previously reported by the dissociative extraction and reconstitution of solubilized proteinaceous components from the intact, demineralized and guanidine HCl extracted bone matrices (Sampath and Reddi 1981; Sampath et al. 1983). The results showed the critical role of the geometry of particle size in controlling the induction of bone formation in heterotopic sites of the rodent (Reddi 1974; Sampath and Reddi 1984).

The critical role of geometry in the induction of bone formation was later shown by a series of experiments in heterotopic subcutaneous and intramuscular sites of both rodents and the Chacma baboon *Papio ursinus* (Ripamonti et al. 1992a; van Eeden and Ripamonti 1994), respectively.

Coral-derived bioreactors with two geometric configurations were implanted in heterotopic subcutaneous sites of Long–Evans rats (Ripamonti et al. 1992a). Configurations were particulate granular vs. discs, 4 mm in diameter, of coral-derived calcium phosphate-based macroporous bioreactors (Ripamonti et al. 1992a). Bioreactors were preloaded with doses of bovine highly purified naturally derived osteogenic fractions after sequential hydroxyapatite Ultrogel adsorption and heparin-Sepharose affinity chromatography columns as described (Luyten et al. 1999). The 500 mM heparin-Sepharose eluates were concentrated and exchanged with 4 M guanidine HCl buffer and loaded onto tandem Sephacryl S-200 gel filtration columns eluted in running buffer (Luyten et al. 1999). Proteins eluates with an apparent molecular mass of 28–42 kDa after gel filtration chromatography purified greater than 50,000 fold with respect to the initial crude urea extract (Luyten et al. 1999) were exchanged with 5 mM hydrochloric acid (HCl) using Amicon ultrafiltration membranes with a 10,000 Dalton molecular weight cut off (Ripamonti et al. 1992a). Macroporous substrata were placed in sterile tissue culture dishes in a laminar flow cabinet, and 50 µg osteogenic fractions in 50 µl 5 mM HCl acid were loaded onto both particulate granular and solid discs of calcium phosphate-based bioreactors, lyophilized and implanted subcutaneously in anaesthetized Long–Evans rats (Ripamonti et al. 1992a).

Tissue specimens were harvested on days 7, 11 and 21 days after heterotopic implantation, fixed in Bouin's fluid, dehydrated through a graded series of ethanol infiltrated and embedded in JB4 resin. Undecalcified sections, cut at 2 µm, were stained with toluidine blue (Fig. 0.5) (Ripamonti et al. 1992a).

Results showed firstly that the biological activity of highly purified osteogenic proteins could be restored and delivered by a substratum other than the insoluble collagenous bone matrix or inactive residue so far used to test the biological activity of both naturally derived and recombinantly produced bone morphogenetic proteins (BMPs ((Ripamonti et al. 1992a).

Above all, however, the study showed the critical role of geometry in regulating the biological activity and the induction of bone by highly purified osteogenic fractions (Luyten et al. 1999; Ripamonti et al. 1992b). On day 7, preloaded constructs showed the induction of chondrogenesis in close relationship with the calcium phosphate-based bioreactor (Figs. 0.5a,b) (Ripamonti et al. 2007). Chondrocytes and cartilaginous matrix were tightly attached to the coral-derived bioreactor (Fig. 0.5c) with the development of hypertrophic chondrocytes (Fig. 0.5c *magenta* arrow). It was noteworthy that the coral-derived bioreactor almost equally delivered the biological activity of the highly purified osteogenic fractions when compared to the induction of chondrogenesis with hypertrophic chondrocytes when delivered by insoluble collagenous bone matrix residues (Ripamonti et al. 1992a; Ripamonti et al. 199b) (Fig. 0.5d *white* arrow).

In contrast, coral-derived granular particulate constructs even when preloaded with highly purified osteogenic fractions showed lack of cartilage induction on day 7 and lack of bone differentiation on day 11 after heterotopic implantation (Ripamonti et al. 1992a).

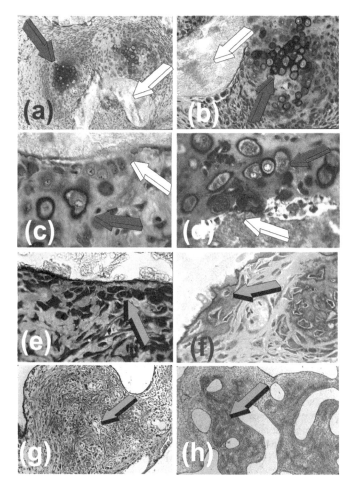

FIGURE 0.5   The critical role of the geometry of porous hydroxyapatite delivery system in
the induction of bone by highly purified naturally derived bone morphogenetic protein fractions. Fifty µg highly purified protein fractions in 50 µl 5 mM hydrochloric acid were loaded
onto coral-derived constructs in disc or granular particulate configuration (Ripamonti et al.
1992a). The effect of the geometry of particle size on the induction of bone in coral-derived
discs implanted subcutaneously in the rodent bioassay and harvested on day 7 after implantation (Ripamonti et al. 1992a). (a,b) Induction of cartilage anlages (*magenta* arrows) within the
macroporous spaces close to the coral-derived constructs (*white* arrows). (c,d) Differentiating
chondroblastic cells at the hydroxyapatite interface (*white* arrows) with the induction of hypertrophic chondrocytes (*magenta* arrows). (e,f) Differentiation of osteoblasts at the hydroxyapatite
interface (*light blue* arrows) with deposition of bone matrix with entrapped osteocytes (*light
blue* arrows). (g) Differentiation of bone on day 7 after heterotopic implantation with construction of the early trabeculae around blood vessels (*light blue* arrow). (h) Further induction of
bone formation 11 days after heterotopic implantation with newly formed trabeculae surrounding invading morphogenetic vessels (*light blue* arrow). Tissue specimens were fixed in Bouin's
fluid, dehydrated through a grade series of ethanol, infiltrated and embedded in JB4 resin.
Sections, cut at 2 µm, were stained in toluidine blue in 30% ethanol (Ripamonti et al. 1992a).

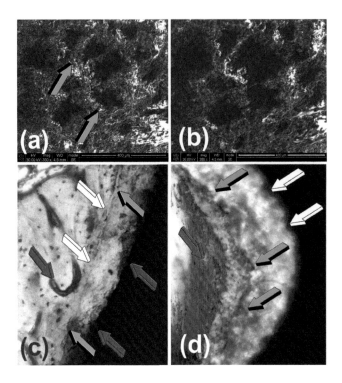

FIGURE 0.6 The induction of bone formation by geometrically constructed titanium implants with a series of concavities constructed along the titanium surfaces coated by crystalline plasma-sprayed sintered hydroxyapatites (Ripamonti and Kirkbride 2001; Ripamonti et al. 2012a; Ripamonti et al. 2012b; Ripamonti et al. 2013). Scanning electron microphotographs (SEM) defining the landscape of the experimental geometric constructs with repetitive concavities coated by crystalline plasma sprayed hydroxyapatites (Ripamonti and Kirkbride 2001; Ripamonti et al 2012a; Ripamonti et al. 2012b; Ripamonti et al. 2013). (a,b) SEM high-power views within concavities of plasma sprayed titania showing nanotopographical surface modifications characterized by micro-lacunae, pits and concavities (*light blue* arrows in a) highly suitable for stem cell differentiation into the osteoblastic phenotype. (c) Short-term intramuscular implantation (3 months) indicates that titanium geometric constructs coated by sintered crystalline plasma sprayed hydroxyapatite (*light blue* arrows) are ideal bioreactors to achieve optimal osteointegration (*light blue* and *white* arrows) with bone formation tightly attached the hydroxyapatite coating (*light blue* and *white* arrows) with the mineralized matrix permeated by capillary sprouting and invasion (*red* arrow). (d) Long-term intramuscular implantation (31 months) shows that titanium bioreactors coated by sintered crystalline plasma sprayed hydroxyapatite are spontaneously osteoinductive within the concavities of the bioreactors, *per se* initiating the induction of bone formation even without the exogenous applications of the osteogenic proteins of the TGF-β supergene family. The hydroxyapatite coating is tightly bound to the titanium body (*white* arrows) with the induction of bone formation and mineralization initiating within the concavity of the plasma sprayed hydroxyapatite (*light blue* arrows). On the surface of mineralized bone (*light blue* arrows), there is the induction of newly formed matrix as yet to be mineralized or osteoid (*orange* arrow). Exakt diamond saw cutting and grinding equipment with sections stained by Goldner's trichrome (Ripamonti et al. 2012a).

On day 11 after heterotopic implantation, coral-derived constructs in disc configuration combined with highly purified osteogenic fractions showed the induction of bone formation across the macroporous spaces. The induction of bone was tightly attached to the calcium phosphate-based bioreactors (Figs. 0.5e,f *light blue* arrows) with the induction of trabeculations across the macroporous spaces (Figs. 0.5g,h *light blue* arrows) (Ripamonti et al. 1992a). Granular particulate coral-derived constructs showed a lack of bone differentiation when pre-treated with highly purified osteogenic fractions (Ripamonti et al. 1992a).

How can the "geometric induction of bone formation" (Ripamonti et al. 1999) be translated in pre-clinical and clinical contexts? This question is addressed in Chapter 7 of this CRC Volume which describes the development of titanium implants carved with a series of concavities coated by crystalline sintered hydroxyapatite that *per se* are endowed with the unique prerogative of inducing new bone formation when implanted in intramuscular *rectus abdominis* sites, where there is no bone.

The described titanium implants coated by crystalline sintered hydroxyapatite so far represent the only titanium-based bioreactor that *per se* imitates the induction of bone formation even without the exogenous application of the osteogenetic soluble molecular signals of the transforming growth factor-β supergene family (Ripamonti and Kirkbride 2001; Ripamonti et al. 2012a; Ripamonti et al. 2012b; Ripamonti 2012; Ripamonti et al. 2013).

ACKNOWLEDGEMENTS

We thank the several students, laboratory technologists, scientists and visiting professors who have significantly contributed to the several discoveries highlighted in the manuscript both at the Bone Cell Biology Section, National Institutes of Health and at the Bone Research Laboratory at the University of the Witwatersrand, Johannesburg. We wish to thank N.S. Cunningham, S.-S. Ma, T.K. Sampath, W. Paralkar and F.P. Luyten at the NIH and Barbara van den Heever, Laura Yeates, Ruqayya Parak, Manolis Heliotis, Carlo Ferretti, Roland Klar, Therese Dix-Peek, Brenda Milner and Raquel Duarte at the Bone Research Laboratory in South Africa. Our studies on the geometric induction of bone formation were supported by grants from the University of the Witwatersrand, Johannesburg, the South African National Research Foundation and the NIH, Bethesda, including a joint NIH grant to A.H. Reddi and U. Ripamonti on craniofacial bone regeneration in primates (DE 10712-01).

REFERENCES

Ahn, E.H.; Kim, Y.; Kshitiz An, S.S.; Afzal, J.; Lee, S.; Kwak, M.; Suh, K.-Y.; Kim, D.-H.; Levchenko, A. Spatial Control of Adult Stem Cell Fate Using Nanotopographic Cues. *Biomaterials* 2014, *35*(8), 2401–10.

Barradas, A.M.C.; Yuan, H.; van Blitterswijk, C.A.; Habibovic, P. Osteoinductive Biomaterials: Current Knowledge of Properties, Experimental Models and Biological Mechanisms. *Eur. Cell. Mater.* 2011, *21*, 407–29.

Bettinger, C.; Langer, R.; Borenstein, J. Engineering Substrate Topography at the Micro- and Nanoscale to Control Cell Function. *Angew. Chem. Int. Ed. Engl.* 2009, *48*(30), 5406–15.

Brunette, D.M. The Effects of Implant Surface Topography on the Behavior of Cells. *Int. J. Oral Maxillofac. Implants* 1988, *3*(4), 231–46.

Buxboim, A.; Discher, D.E. Stem Cells Feel the Difference. *Nat. Methods* 2010, *7*(5), 695–7.

Buxboim, A.; Ivanovska, I.L.; Discher, D.E. Matrix Elasticity, Cytoskeletal Forces and Physics of the Nucleus: How Deeply Do Cells "Feel" Outside and In? *J. Cell Sci.* 2010, *123*(3), 297–308.

Clark, P.; Connolly, P.; Curtis, A.S.G.; Dow, J.A.T.; Wilkinson, D.W. Topographical Control of Cell Behaviour I. Simple Step Cues. *Development* 1987, *108*, 635–44.

Clark, P.; Connolly, P.; Curtis, A.S.G.; Dow, J.A.T.; Wilkinson, D.W. Topographical Control of Cell Behaviour: II. Multiple Grooved Substrata. *Development* 1990, *99*, 493–48.

Costa-Rodriguez, J.; Fernandes, A.; Lopes, M.A.; Fernandes, M.H. Hydroxyapatite Surface Roughness: Complex Modulation of the Osteoclastogenesis of Human Precursor Cells. *Acta Biomater.* 2012, *8*(3), 1137–45.

Curran, J.M.; Chen, R.; Hunt, J.A. The Guidance of Human Mesenchymal Stem Cell Differentiation In Vitro by Controlled Modifications to the Cell Substrate. *Biomaterials* 2006, *27*(27), 4783–93.

Curtis, A.C.; Wilkinson, C. Topographical Control of Cells. *Biomaterials* 1997, *18*(24), 1573–83.

Danoux, C.; Pereira, D.; Döbelin, N.; Stähli, C.; Barralet, J.; van Blitterswijk, C.; Habibovic, P. The Effects of Crystal Phase and Particle Morphology of Calcium Phosphates on Proliferation and Differentiation of Human Mesenchymal Stromal Cells. *Adv. Health Mater.* 2016, *5*(14), 1775–85, doi:10.1002/adhm.201600184.

Discher, D.E.; Janmey, P.; Wang, Y.-L. Tissue Cells Feel and Respond to the Stiffness of Their Substrate. *Science* 2005, *310*(5751), 1139–43.

Discher, D.E.; Mooney, D.J.; Zandstra, P.W. Growth Factors, Matrices, and Forces Combine and Control Stem Cells. *Science* 2009, *324*(5935), 1673–77.

Engler, A.J.; Sen, S.; Sweeney, H.L.; Discher, D.E. Matrix Elasticity Directs Stem Cell Lineage Specification. *Cell* 2006, *126*(4), 677–89.

Fiedler, J.; Özdemir, B.; Bartholomä, J.; Pletti, A.; Brenner, R.E.; Ziemann, P. The Effect of Substrate Surface Nanotopogrphy on the Behavior of Multipotent Mesenchymal Stromal Cells and Osteoblasts. *Biomaterials* 2013, *34*(35), 8851–59.

Gittens, R.A.; McLachlan, T.; Olivares-Navarrete, R.; Cai, Y.; Berner, S.; Tannenbaum, R.; Schwartz, Z.; Sandhage, K.H.; Boyan, B.D. The Effects of Combined Micron-/Submicron-Scale Surface Roughness and Nanoscale Features on Cell Proliferation and Differentiation. *Biomaterials* 2011, *32*(13), 3395–403.

Habibovic, P.; Yuan, H. Van der Valk, C.M.; Meijer, G.; van Blitterswijk, C.A.; de Groot, K. 3D Microenvironment as Essential Element for Osteoinduction by Biomaterials. *Biomaterials* 2005, *26*, 3565–3575.

Kiang, J.D.; Wen, J.H.; del Alamo, J.C.; Engler, A.J. Dynamic and Reversible Surface Topography Influences Cell Morphology. *J. Biomed. Mater. Res. A* 2013, *101*(8), 2313–21.

Kilian, K.A.; Bugarija, B.; Lahn, B.T.; Mrksich, M. Geometric Cues for Directing the Differentiation of Mesenchymal Stem Cells. *Proc. Natl. Acad. Sci. U.S.A.* 2010, *107*(11), 4872–877.

Kim, D.-H.; Provenzano, P.P.; Smith, C.L.; Levchenko, A. Matrix Nanotopography as a Regulator of Cell Function. *J. Cell. Biol.* 2012, *197*(3), 351–60.

Klar, R.M.; Duarte, R.; Dix-Peek, T.; Dickens, C.; Ferretti, C.; Ripamonti, U. Calcium Ions and Osteoclastogenesis Initiate the Induction of Bone Formation by Coral-Derived Macroporous Constructs. *J. Cell. Mol. Med.* 2013, *17*(11), 1444–57.

Lamers, E.; van Horssen, R.; te Riet, J.; van Delft, F.C.; Luttge, R.; Walboomers, X.F.; Jansen, J.A. The Influence of Nanoscale Topographical Cues on Initial Osteoblast Morphology and Migration. *Eur. Cell Mater.* 2010, *20*, 329–43.

Li, X.; van Blitterswijk, C.A.; Feng, Q.; Cui, F.; Watari, F. The Effect of Calcium Phosphate Microstructure on Bone-Related Cells *In Vitro*. *Biomaterials* 2008, *29*(23), 3306–16.

Liu, X.; Liu, R.; Cao, B.; Ye, K.; Li, S.; Gu, Y.; Pan, Z.; Ding, J. Subcellular Cell Geometry on Micropillars Regulates Stem Cell Differentiation. *Biomaterials* 2016, *111*, 27–39.

Luyten, F.P.; Cunningham, N.S.; Ma, S.; Muthukumaran, N.; Hammonds, R.G.; Nevins, W.B.; Wood, W.I.; Reddi, A.H. Purification and Partial Amino Acid Sequence of Osteogenin, a Protein Initiating Bone Differentiation. *J. Biol. Chem.* 1999, *264*(23), 13377–80.

Metavarayuth, K.; Sitasuwan, P.; Zhao, X.; Lin, Y.; Wang, Q. Influence of Surface Topographical Cues on the Differentiation of Mesenchymal Stem Cells In Vitro. *ACS Biomater. Sci. Eng.* 2016, *2*(2), 142–51.

McNamara, L.E.; McMurray, R.J.; Biggs, M.J.P.; Kantawong, F.; Oreffo, M.J.; Dalby, M.J. Nanotopographical Control of Stem Cell Differentiation. *J. Tissue Eng.* 2010, doi:10.4061/2010/120623.

McNamara, L.E.; Sjöström, T.; Burgess, K.E.V.; Kim, J.J.W.; Liu, E.; Gordonov, S.; Moghe, P.V.; Meek, R.M.D.; Oreffo, O.C.; Su, B.; Dalby, M.J. Skeletal Stem Cell Physiology on Functionally Distinct Titania Topographies. *Biomaterials* 2011, *32*, 7403–10.

Miyoshi, H.; Adachi, T. Topography Design Concept of a Tissue Engineering Scaffold for Controlling Cell Function and Fate through Actin Cytoskeletal Modulation. *Tissue Eng.* 2014, *20*(6), 609–27.

Reddi, A.H.; Huggins, C.B. Influence of Geometry of Transplanted Tooth and Bone on Transformation of Fibroblasts. *Proc. Soc. Exp. Biol. Med.* 1973, *143*(3), 634–37.

Reddi, A.H. Bone Matrix in the Solid State: Geometric Influence on Differentiation of Fibroblasts. *Adv. Biol. Med. Phys.* 1974, *15*, 1–18.

Ripamonti, U.; Ma, S.; van den Heever, B.; Reddi, A.H. The Critical Role of Geometry of Porous Hydroxyapatite Delivery System in Induction of Bone by Osteogenin, a Bone Morphogenetic Protein. *Matrix* 1992a, *12*(3), 202–12.

Ripamonti, U.; Ma, S.; Cunningham, N.; Yeates, L.; Reddi, A.H. Initiation of Bone Regeneration in Adult Baboons by Osteogenin, a Bone Morphogenetic Protein. *Matrix* 1992b, *12*(5), 369–80.

Ripamonti, U.; van den Heever, B.; van Wyk, J. Expression of the Osteogenic Phenotype in Porous Hydroxyapatite Implanted Extraskeletally in Baboons. *Matrix* 1993, *13*(6), 491–502.

Ripamonti, U.; Crooks, J.; Kirkbride, A.N. Sintered Porous Hydroxyapatites with Intrinsic Osteoinductive Activity: Geometric Induction of Bone Formation. *S. Afr. J. Sci.* 1999, *95*, 335–43.

Ripamonti, U.; Kirkbride, A.N. *Biomaterial and Bone Implant for Bone Repair and Replacement*. US Patent 6,302,913 B1(October 16), 2001.

Ripamonti, U. Soluble, Insoluble and Geometric Signals Sculpt the Architecture of Mineralized Tissues. *J. Cell. Mol. Med.* 2004, *8*(2), 169–80.

Ripamonti, U. Soluble Osteogenic Molecular Signals and the Induction of Bone Formation. *Biomaterials* Leading Opinion Paper 2006, *27*(6), 807–22.

Ripamonti, U.; Richter, P.W.; Thomas, M.E.; Shape, Self-Inducing Memory Geometric Cues Embedded within Smart Hydroxyapatite-Based Biomimetic Matrices. *Plast. Reconstr. Surg.* 2007, *120*, 1796–07.

Ripamonti, U. The Concavity: The "Shape of Life" and the Control of Bone Differentiation – Feature Paper – *Science in Africa* May **2012**.

Ripamonti, U.; Roden, L.C.; Renton, L.F. Osteoinductive Hydroxyapatite-Coated Titanium Implants. *Biomaterials* 2012a, *33*(15), 3813–23.

Ripamonti, U.; Roden, L.; Renton, L.; Klar, R.M.; Petit, J.-C. The Influence of Geometry on Bone: Formation by Autoinduction. *Science in Africa* 2012b. http://www.scienceinafric a.co.za/2012/Ripamonti_bone.htm.

Ripamonti, U.; Renton, L.; Petit, J.-C., Bioinspired Titanium Implants: The Concavity – The Shape of Life. In: Ramalingam M., Vallitu P., Ripamonti U., Li W.-J. (eds.) *Tissue Engineering and Regenerative Medicine. A Nano Approach.* CRC Press Taylor & Francis, Boca Raton USA, 2013, Chapter 6, 105–23.

Ripamonti, U. Biomimetic Functionalized Surfaces and the Induction of Bone Formation. *Tissue Eng.* 2017, *23*(21,22), 1197–2009.

Ripamonti, U. Functionalized Surface Geometries Induce: "Bone: Formation by Autoinduction". *Front. Physiol.* 2018, *8*, 1084, doi:10.3389/fphys.2017.01084.

Sampath, T.K.; Reddi, A.H. Dissociative Extraction and Reconstitution of Extracellular Matrix Components Involved in Local Bone Differentiation. *Proc. Natl. Acad. Sci. U.S.A.* 1981, *78*(12), 599–603.

Sampath, T.K.; Reddi, A.H. Homology of Bone-Inductive Proteins from Human, Monkey, Bovine, and Rat Extracellular Matrix. *Proc. Natl. Acad. Sci. U.S.A.* 1983, *80*(21), 6591–95.

Sampath, T.K.; Reddi, A.H. Importance of Geometry of the Extracellular Matrix in Endochondral Bone Differentiation. *J. Cell Biol.* 1984, *98*(6), 2192–97.

Sun, J., Jamilpour, N., Wang, F.-Y., Wong, P.K. Geometric Control of Capillary Architecture via Cell-Matrix Mechanical Intercations. *Biomaterials* 2014, *35*(10), 3273–280.

van Eeden, S.; Ripamonti, U. Bone Differentiation in Porous Hydroxyapatite Is Regulated by the Geometry of the Substratum: Implications for Reconstructive Craniofacial Surgery. *Plast. Reconstr. Surg.* 1994, *93*(5), 959–66.

Vlacic-Zischke, J.; Hamlet, S.M.; Friis, T., et al. The Influence of Surface Microroughness and Hydrophilicity of Titanium on the Up-Regulation of TGFβ/BMP Signalling in Osteoblasts. *Biomaterials* 2011, *32*, 7403–10.

Watari, S.; Hayashi, K.; Wood, J.A.; Russell, P.; Nealey, P.F.; Murphy, C.J.; Genetos, D.C. Modulation of Osteogenic Differentiation in hMSCs Cells by Submicron Topographically-Patterned Ridges and Grooves. *Biomaterials* 2012, *33*(1), 128–36.

Wilkinson, A.; Hewitt, R.N.; McNamara, L.E.; McCloy, D.; Meek, D.R.M.; Dalby, M.J. Biomimetic Microtopography to Enhance Osteogenesis *In Vitro*. *Acta Biomater.* 2011, *7*(7), 2919–25.

Williams, D.F. The Continuing Evolution of Biomaterials 2011, *31*(1), 1–2, doi:10.1016/j.bio-materials.2010.09.048. Epub 2010 October 8, PMID:20933267.

Zhang, J.; Luo, X.; Barbieri, D.; Barradas, A.M.C.; de Bruijn, J.; van Blitterswijk, A.; Yuan, H. The Size of Surface Microstructure as an Osteogenic Factor in Calcium Phosphate Ceramics. *Acta Biomater.* 2014, *10*(7), 3254–63.

Zhang, J.; Dalbay, M.T.; Luo, X.; Vrij, E.; Barbieri, D.; Moroni, L.; de Bruijn, J.D.; van Blitterswijk, C.A.; Chapple, J.P.; Knight, M.M.; Yuan, H. Topography of Calcium Phosphate Ceramics Regulates Primary Cilia Length and TGF Receptor Recruitment Associated with Osteogenesis. *Acta Biomater.* 2017a, *57*, 487–97.

Zhang, J.; Sun, L.; Luo, X.; Barbieri, D.; de Bruijn, J.D.; van Blitterswijk, C.A.; Moroni, L.; Yuan, Y. Cells Responding to Surface Structure of Calcium Phosphate Ceramics for Bone Regeneration. *J. Tissue Eng. Reg. Med.* 2017b, *11*(11):3273–83. doi:10.1002/term.2236.

Zhang, Y.; Chen, S.E.; Shao, J.; van den Beucken, J.J.J.P. Combinatorial Surface Roughness Effects on Osteoclastogenesis and Osteogenesis. *ACS Appl. Mater. Interfaces* 2018, *10*, 36652–663.

1 The New Frontiers in Bone Tissue Engineering
Functionalized Biomimetic Surfaces beyond Morphogens and Stem Cells

Ugo Ripamonti and Laura C. Roden

1.1 TISSUE INDUCTION AND MORPHOGENESIS

Serendipitously, one of the authors came across key papers on the induction of bone formation, the paper in the *Proceedings of the National Academy of Sciences* USA by Reddi and Huggins (1972) and the paper of Urist in *Science* (1965). Further searches led to two critical papers by Sampath and Reddi in the *Proceedings of the National Academy of Sciences* USA 1981 and 1983. Both papers described the induction of bone formation as a combinatorial molecular protocol recombining, or reconstituting, soluble and insoluble signals to trigger the ripple-like cascade of the induction of bone formation (Sampath and Reddi 1981; Sampath and Reddi 1983). The critical importance of the dissociative extraction and reconstitution of the extracellular matrix of bone is what propelled the experimental and clinical progression of the "bone induction principle" (Urist et al. 1967) from pre-clinical to clinical studies (Sampath and Reddi 1981; Sampath and Reddi 1983; Ripamonti and Reddi 1995; Reddi 1997; Reddi 2000; Ripamonti et al. 2001; Ripamonti 2006; Ripamonti et al. 2006; Ripamonti et al. 2007).

However, it was the paper by Piecuch entitled "Extraskeletal implantation of a porous hydroxyapatite ceramic" published in the *Journal of Dental Research* (1982) that inspired the author's study of the biology of the incorporation of such macroporous biomatrices not only in bony sites but also in intramuscular heterotopic sites of the Chacma baboon *Papio ursinus*. Macroporous hydroxyapatite constructs derived from corals, with an average pore size of 500 µm, were prepared by Interpore International (Irvine, CA) and implanted in calvarial defects and in *rectus abdominis* intramuscular sites (Ripamonti 1991; Ripamonti et al. 2001; Ripamonti 2009). Experiments with these constructs demonstrated the spontaneous osteoinductivity of the macroporous constructs (Ripamonti 2006; Ripamonti 2009; Ripamonti 2017).

When implanted in heterotopic intramuscular sites, the coral-derived constructs initiated the induction of bone formation within the macroporous spaces. The induced bone was evident within the pores of the specimens' harvested 90 days post-intramuscular implantation (Fig. 1.1) (Ripamonti 1990; Ripamonti 1991). Bone had formed tightly attached to the macroporous surfaces and extended with fine trabeculae across the macroporous spaces, supported by a closely associated rich vascular network (Fig. 1.1). These coral-derived substrata were later described as "macroporous bioreactors" (Ripamonti et al. 2009; Ripamonti et al. 2010; Ripamonti 2017), but it took almost 25 years to resolve the morphological data to understand and unravel the mechanism of the spontaneous induction of bone formation within the macroporous constructs (Klar et al. 2013).

In collaboration with the Council for Scientific and Industrial Research (CSIR), Pretoria, we designed and tested highly crystalline sintered hydroxyapatites that also resulted in the spontaneous induction of bone formation (Ripamonti et al. 1999). This was *par force* the scientific evolution to further understand the induction of bone formation in coral-derived macroporous bioreactors (Ripamonti 1990; Ripamonti 1991; Ripamonti 1993; Ripamonti et al. 1993; Ripamonti 1996) and highly sintered crystalline hydroxyapatites for potential clinical applications (Ripamonti 1994; Ripamonti et al. 1999; Ripamonti 2004; Ripamonti et al. 1995; Ripamonti et al. 1997; Ripamonti et al. 2000; Ripamonti and Kirkbride 2001).

Experiments described in Ripamonti (1991) and Ripamonti et al. (1993) revealed the critical role of mesenchymal collagenous condensations that developed by day 30 post-implantation (Fig. 1.2). This was followed by detectable alkaline phosphatase expression by day 60 at the hydroxyapatite interface. These results indicated morphogenetic events pre-dating the induction of bone formation (Fig. 1.2) (Ripamonti 1990; Ripamonti 1991; Ripamonti et al. 1993). Furthermore, undecalcified sections showed alkaline phosphatase staining of the invading vasculature in close association with the coral-derived substrate by day 30 (Fig. 1.3). The stained multicellular layers of the invading capillaries were identified as the "osteogenetic vessels" described by Trueta (Trueta 1963). Laminin, a prominent vascular basement membrane component significantly associated with differentiation of the osteogenic phenotype (Foidart and Reddi 1980; Wlodarski and Reddi 1986), was also localized within the invading capillaries on days 30 and 60 (Fig. 1.3f).

Inspired by Trueta's paper on the role of the vessels in angiogenesis (Trueta 1963), we hypothesized that the alkaline phosphatase expression within the endothelial and sub-endothelial compartments of the osteogenetic vessels might provide a temporally regulated flow of bone precursor cells (Ripamonti et al. 1993). These precursor cells would be capable of the expression of the osteogenic phenotype when in contact with the calcium phosphate-based matrix of the implanted bioreactors (Figs. 1.3, 1.4). The morphological and biochemical data from these implants support this hypothesis (Figs. 1.4a,b) (Ripamonti et al. 1993; Ripamonti 2009). The induction and alignment of mesenchymal tissue condensations against the calcium phosphate-based surfaces are critical for the subsequent induction of bone formation (Ripamonti 1990; Ripamonti 1991; Ripamonti et al. 1993) (Figs. 1.2, 1.4).

The condensations which form by day 30 do not have detectable alkaline phosphatase expression (Figs. 1.3a,b), but within the subsequent 30 days, osteoblast differentiation and bone formation take place (Figs. 1.4b,e). In contrast, alkaline phosphatase

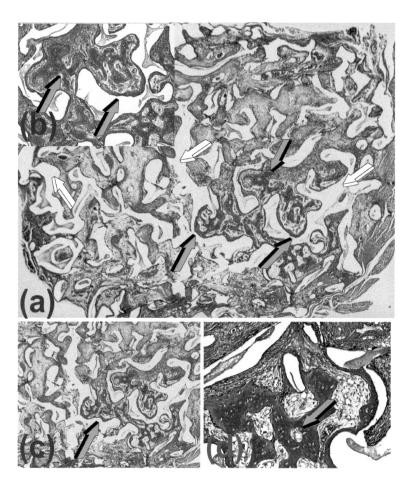

FIGURE 1.1 Bone induction and morphogenesis spontaneously initiating within the macroporous spaces of a coral-derived bioreactor implanted heterotopically in the *rectus abdominis* muscle of a Chacma baboon *Papio ursinus* on day 90 after intramuscular implantation. (a) The spontaneous induction of bone formation without the exogenous application of osteogenic proteins of the transforming growth factor-β (TGF-β) supergene family. White arrows indicate the calcium phosphate biomatrix dissolved after histological processing (Ripamonti 1991). Vascularized trabeculae of newly formed woven bone (*light blue* arrows) formed by day 90 within the macroporous spaces supported by prominent angiogenesis (inset b). (c, d) High-power views of (a) detailing the induction of hyper cellular woven bone formation (*light blue* arrows) tightly attached to the biomimetic matrix substratum. Decalcified sections cut at 6 μm from paraffin-embedded blocks stained with toluidine blue in 70% ethanol. The images are from histological section 435/88 in honour of Barbara van den Heever who prepared impeccable sections after several experiments in *Papio ursinus* and alerted the writer to the morphogenesis of bone within the macroporous spaces of that section (Ripamonti 1991).

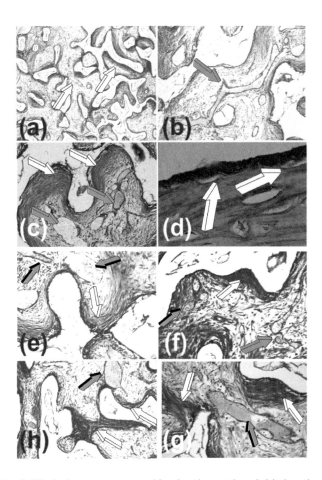

FIGURE 1.2   Self-inducing macroporous biomimetic matrices initiating tissue induction and morphogenesis within the macroporous spaces on day 30 after heterotopic implantation, leading to the spontaneous and/or intrinsic induction of bone formation (Ripamonti 1990; Ripamonti 1991; Ripamonti et al. 1993; Ripamonti 1996). (a) Low-power view of a macroporous bioreactor harvested on day 30 after *rectus abdominis* implantation with differentiation of mesenchymal collagenous condensations at the hydroxyapatite interface (*white* arrows). (b) High-power view detailing the prominent vascular invasion (*light blue* arrow) within the fibrovascular tissue invading the macroporous spaces of the coral-derived bioreactor. (c) Collagenous condensations tightly attached to the macroporous substratum (*white* arrows) supported by prominent vascular invasion and angiogenesis (*light blue* arrow). (d) Early differentiating events at the hydroxyapatite interface with cellular differentiation towards the osteogenic phenotype (*white* arrows) in tight relationship with the substratum on day 30 after heterotopic implantation. (e,f,g,h) Patterns of vascular invasion, collagenous condensations' distribution and alignment (*white* arrows) within the macroporous spaces of the coral-derived biomimetic matrices. There is always prominent angiogenesis (*light blue* arrows) continuously supporting tissue induction and morphogenesis within the macroporous spaces. Decalcified sections cut at 6 μm from paraffin-embedded blocks harvested and processed on day 30 after heterotopic implantation and stained with toluidine blue in 70% ethanol.

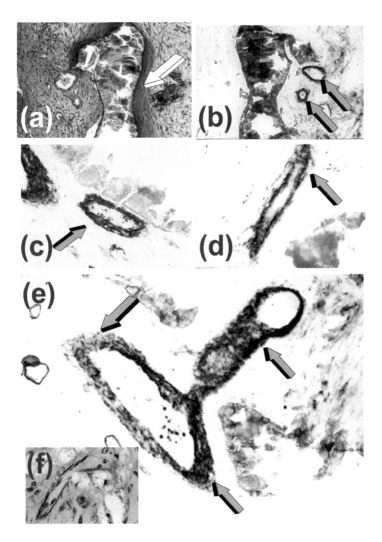

FIGURE 1.3   Early morphogenetic events on day 30 leading to the induction of bone formation by macroporous coral-derived bioreactors implanted in the *rectus abdominis* muscle of
the Chacma baboon *Papio ursinus*. (a) Mesenchymal condensations on day 30 after heterotopic implantation (*white* arrow). (b) Consecutive serial section approximately 20 μm from
(a) showing alkaline phosphatase staining (*light blue* arrows) of the invading osteogenetic
vessels of Trueta's definition (Trueta 1963). There is, however, a lack of alkaline phosphatase staining of the collagenous condensations against the hydroxyapatite interface. (c,d,e)
Significant alkaline phosphatase staining (*light blue* arrows) of the invading sprouting capillaries within the macroporous spaces in tight relationship (c) with the macroporous biomimetic matrix. Note in (e) several layers of endothelial and perivascular cells expressing
alkaline phosphatase whilst invading the macroporous spaces. Inset (f) immunolocalization
of laminin within the basement membrane of an invading capillary (Ripamonti et al. 1993).
Undecalcified sections from tissue blocks fixed in 70% ethanol, embedded, undecalcified, in
historesin (LKB Bromma, Sweden), and cut at 3–5 μm (Ripamonti et al. 1993).

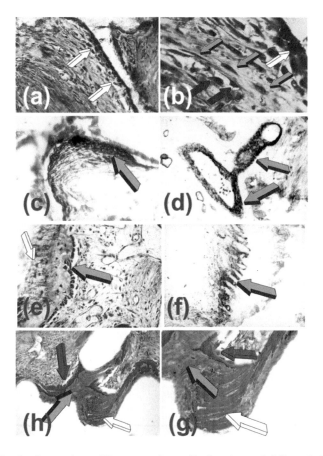

FIGURE 1.4 Angiogenesis, capillary sprouting, cell migration and differentiation from the angiogenetic microenvironment to the bone induction microenvironment initiated by cellular differentiating and de-differentiating events by the "morphogenetic" and "osteogenetic vessels" of Aristotle's (Crivellato et al. 2007) and Trueta's definitions (Trueta 1963) at the osteogenetic microenvironment (*white* arrow in b). Vascular invasion and capillary sprouting provide the molecular, cellular and morphological templates for cell migration and differentiation with the induction of bone formation against the hydroxyapatite interface. (a,b) High-power views of the angiogenetic and osteogenetic microenvironments initiated by the calcium phosphate-based surface topography of the heterotopically implanted macroporous bioreactors on day 60 after heterotopic implantation. Middle-power view (a) with osteoblastic-like cell differentiation (*white* arrows) at the hydroxyapatite interface. (b) Detail of (a) highlighting osteoblastic-like cells differentiation at the interface (*white* arrow) and cellular trafficking in the angiogenetic compartment with highly hyperchromatic endothelial nuclei (*dark blue* arrows) seemingly migrating and de-differentiating from the penetrating vessels (*light blue* arrows) into the osteogenic microenvironment, providing a continuous flow of cellular progenitors for the induction of bone formation. (c) Consecutive serial section approximately 20 μm from (a) showing pronounced alkaline phosphatase expression in the collagenous condensation as shown in (a) on day 60. (d) Time sequences of alkaline phosphatase expression events showing localization of the invading sprouting capillaries on day 30. The developmental patterns

FIGURE 1.4 (CONTINUED)

of tissue induction and morphogenesis with expression of alkaline phosphatase and laminin biochemical markers point to multiple series of sequential events regulated by the surface characteristics of the heterotopically implanted biomatrices when in contact with the intramuscular *rectus abdominis* microenvironment (Ripamonti et al. 1993). (e) Induction of bone formation at the hydroxyapatite interface with contiguous hyperchromatic osteoblasts secreting newly formed bone tightly attached to the hydroxyapatite surface, with penetration of newly formed bone within the hydroxyapatite substratum (*white* arrow). (f) Consecutive serial section approximately 20 μm apart showing alkaline phosphatase staining of the secreting osteoblasts as shown in (e). (h,g) Undecalcified sections stained free-floating with Goldner's trichrome (Ripamonti et al. 1993) illustrating mineralization within newly formed condensations at the hydroxyapatite interface. (h) Low-power view showing newly formed mineralized bone in blue (*light blue* arrow) within the macroporous spaces and attached to the implanted substratum. (g) Detail of (h) depicting mineralized newly formed bone (*light blue* arrow) developing within the tightly packed collagenous condensations (*white* arrow) with secreted osteoid seams (*orange* arrow) surfacing the newly formed mineralized bone. Undecalcified sections cut at 3 to 6 μm from K-Plast resin-embedded specimens and stained, free-floating with Goldner's trichrome (Ripamonti et al. 1993).

expression is detectable along the invading capillaries by day 30 (Figs. 1.3b,c,d,e) but not on day 90 (Ripamonti et al. 1993). On day 60, the most salient results, however, were the differentiation of osteoblast-like cells with alkaline phosphatase expression at the hydroxyapatite interface separated by a continuous layer of cemental matrix into which cells are tightly embedded along the differentiating substratum (Figs. 1.4a,b).

The angiogenic induction in mesenchymal condensations is noteworthy and correlates with the alkaline phosphatase staining. High-power views indicate the migration of mesenchymal stem cells from the angiogenic compartment of the invading mesenchymal condensations to the osteogenic compartment of the differentiating osteogenic cells at the hydroxyapatite interface (Fig. 1.4b). It was noteworthy that alkaline phosphatase expression of mesenchymal condensations was only shown together with differentiating osteoblast-like cells adjacent to the hydroxyapatite interface by day 60 (Fig. 1.4c). The distribution patterns of alkaline phosphatase expression suggested a specific temporal and developmental sequence of tissue patterning of mesenchymal cellular condensations, ultimately regulating the differentiation of osteoblast-like cells within condensations and the induction of bone formation against the hydroxyapatite interface (Figs. 1.4e,f,h,g).

Undecalcified sections cut from block embedded in K-Plast resin showed developmental transitional phases of collagenous condensations against the hydroxyapatite interface, firstly non-mineralized condensations (Figs. 1.4h,g *white* arrows) with foci of mineralization across the condensations (Figs. 1.4h,g *light blue* arrows) with osteoid synthesis surfaced by contiguous osteoblasts (Figs. 1.4h,g *orange* arrows).

To investigate the signalling aspect of the spontaneous induction of bone formation within the macro pores of the coral-derived bioreactors, doses of 125 or 150 μg hNoggin, a BMP antagonist, were adsorbed onto the constructs prior to implantation (Klar et al. 2014; Ripamonti et al. 2015). There was prominent inhibition of tissue patterning in hNoggin-treated bioreactors, consistently showing lack of fibrovascular invasion and haphazardly generated tissue invasion with lack of connective tissue alignment and patterning at the hydroxyapatite interface (Fig. 1.5). This indicated that the induction of bone formation by coral-derived macroporous bioreactors is *via* the BMP pathway, as the applied hNoggin would block the biological activity of any BMPs secreted and adsorbed onto the substratum (Fig. 1.5) (Klar et al. 2014; Ripamonti et al. 2015). The lack of tissue patterning with poorly remodelled mesenchymal condensations and lack of fibrovascular angiogenic invasion, as identified in hNoggin-pre-treated macroporous bioreactors (Fig. 1.5b), dramatically points to the critical patterning role of BMPs before the induction of bone formation. The induction of bone formation must thus be pre-dated by the induction and remodelling of mesenchymal condensations, vascular invasion and the generation of Trueta's osteogenetic vessels. The lack of capillary sprouting and vascular invasion together with poorly constructed tissue patterned collagenous condensations with lack of alignment along the hydroxyapatite surface in hNoggin-treated constructs indirectly supports the critical role of tissue patterning in angiogenesis by selected BMPs initiating the spontaneous and/or intrinsic induction of bone formation (Fig. 1.5) (Klar et al. 2013; Ripamonti et al. 2015; Ripamonti 2016).

Unexpectedly, data from different experiments provided clues about the critical role of angiogenesis, capillary sprouting, invasion and differentiation during the induction of bone formation by coral-derived macroporous bioreactors. When macroporous bioreactors, super activated by the addition of 125 µg hTGF-β_3, were implanted in calvarial defects, the induction of bone formation was minimal (Ripamonti et al. 2016).

In previous experiments using insoluble collagenous bone matrices delivering 125 µg hTGF-β_3 (Ripamonti et al. 2008), the limited induction of bone formation on day 30 was correlated to the expression of Smad-6 and Smad-7, downstream antagonists of the TGF-β signalling pathway (Ripamonti et al. 2008). On day 90, there was some restoration of the bone induction cascade but only pericranially (Ripamonti et al. 2008). Mechanistically, the partial restoration of the induction of bone formation on day 90, correlated to the limited if any expression of *Smad-6* and *-7* (Ripamonti et al. 2008). In more recent experiments, the limited induction of bone formation as seen on day 30 after calvarial implantation of 125 µg hTGF-β_3 correlated to downregulation of *OP-1, Osteocalcin, RUNX-2, Id2 and Id3* (Ripamonti et al. 2016). It was noteworthy that the morphology of calvarial repair using collagenous matrices reconstituted with doses of hTGF-β_3, with bone induction preferentially pericranially, with a lack of bone formation endocranially above the dura, suggested a radius of activity set into motion by inhibitory mechanisms originating from the dura mater and/or the highly vascularized leptomeninges below (Ripamonti et al. 2008). The morphological-molecular correlation indicated that the microenvironment of the calvarial defects implanted with hTGF-β_3-loaded bioreactors resulted in a non-osteogenic state, downregulating critical osteogenic mRNA markers. This was clearly confirmed in experiments where segregation of the *dura mater* from the osteogenic calvarial diploic microenvironment with a supradural impermeable nylon foil membrane restored the bone induction cascade with upregulation of osteogenic

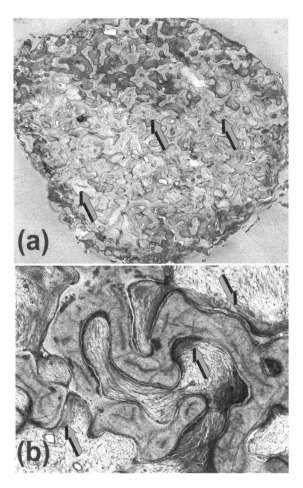

FIGURE 1.5   Which are the mechanisms that initiate the spontaneous and/or intrinsic induction of bone formation by coral-derived macroporous bioreactors when implanted in heterotopic intramuscular sites of the *rectus abdominis* muscle in *Papio ursinus*? To mechanistically resolve the intrinsic induction of bone formation, coral-derived bioreactors were preloaded with 125 or 150 µg doses of recombinant human Noggin (hNoggin), a BMP antagonist, and implanted in the *rectus abdominis* muscle of *Papio ursinus* (Klar et al. 2013; Ripamonti et al. 2015). Harvested tissues on day 90 were processed for undecalcified histology on the Exakt diamond saw cutting and polishing equipment (Klar et al. 2013; Ripamonti et al. 2015). (a) Low-power view of a hNoggin-treated bioreactor showing lack of bone differentiation throughout the macroporous spaces characterized by limited vascular invasion (*light blue* arrows). (b) Detail of (a) highlighting limited patterning of mesenchymal condensations with haphazardly constructed collagenous condensations (*light blue* arrows). Note limited vascular invasion and capillary sprouting with lack of bone differentiation. Undecalcified section prepared by the Exakt cutting and grinding diamond saw system (Donath and Breuner 1982), ground and polished to 30 µm, stained with toluidine blue (Klar et al. 2013; Ripamonti et al. 2015).

markers (Ripamonti et al. 2016). Furthermore, the experiments also showed that tissue patterning and morphogenesis as initiated by the recombinant hTGF-β_3 isoform are independent of the carrier matrix used.

Our systematic studies in calvarial defects of *Papio ursinus* using the three mammalian hTGF-β isoforms (Ripamonti et al. 1997; Ripamonti et al. 2000; Ripamonti et al. 2008; Ripamonti et al. 2016) have suggested novel regulatory pathways through *Id2* and *Id3* expression (Ripamonti et al. 2016). The molecular data have also shown that when TGF-β3 expression decreases, Id2 and Id3 expression are upregulated (Ripamonti et al. 2016). This has suggested that TGF-β signalling to regulate *Id2* and *Id3* gene expression promotes or inhibits the induction of calvarial bone formation. Interplays of *Id* genes, *TGF-βs*, *Notch* and *BMPs* control endothelial cell migration and differentiation (Itoh et al. 2004). Upregulation of *Id* gene expression promotes endothelial cell migration, with molecular interplay between Notch and BMP receptor signalling pathways. This regulates endothelial cell migration *via* upregulation of *Id1* (Itoh et al. 2004). Segregated calvarial defects showed the induction of bone formation with capillary sprouting and invasion, directly correlating to *Id2* and *Id3* upregulation (Ripamonti et al. 2016). The application of exogenous hTGF-β_3 to defects segregated from the dura mater creates a microenvironment that is permissive for bone formation (Ripamonti et al. 2016).

Interestingly, implanted coral-derived bioreactors show a series of gene expression pathways correlating to vascular invasion and capillary sprouting that correlate with the morphological traits that often characterize the harvested bioreactors (Ripamonti et al. 2009; Klar et al. 2013; Ripamonti et al. 2016). Capillary sprouting and invasion are seen by days 15 and 30 across the macroporous spaces (Ripamonti et al. 1993a; Klar et al. 2013), with further angiogenesis and capillary invasion by days 60 and 90 (Ripamonti et al. 1993; Klar et al. 2003) (Figs. 1.3, 1.4, 1.6). Angiogenesis and capillary invasion are still prominent on days 60 and 90 correlating to *Type IV Collagen* expression as shown by qRT-PCR (Klar et al. 2013). Capillary invasion often seemingly touches the calcium phosphate-based macroporous surface with newly formed bone by days 30 and 60 after heterotopic intramuscular implantation (Fig. 1.6). Between the newly formed bone and the exquisite capillary elongations, there are several multinucleated osteoclastic cells tightly attached to the hydroxyapatite macroporous surface (Fig. 1.7). Capillary invasion and angiogenesis throughout the macroporous spaces of the coral-derived macroporous bioreactors are critical key for the spontaneous and/or intrinsic induction of bone formation (Ripamonti 1991; Ripamonti et al. 1993a; Ripamonti 2017).

Osteoclastic activity re-patterning the surface micro-geometry of the implanted macroporous bioreactors is the key factor resulting in the spontaneous and/or intrinsic induction of bone formation (Ripamonti et al. 2010; Klar et al. 2013; Ripamonti 2017). Pre-treatment of coral-derived macroporous bioreactors with 240 µg bisphosphonate zoledronate Zometa®, an osteoclastic inhibitor (Ripamonti et al. 2010), blocks the induction of bone formation (Ripamonti et al. 2010; Ripamonti et al. 2015). In later chapters, the role of osteoclastic activity topographically and geometrically modifying the hydroxyapatite surface will be discussed at length.

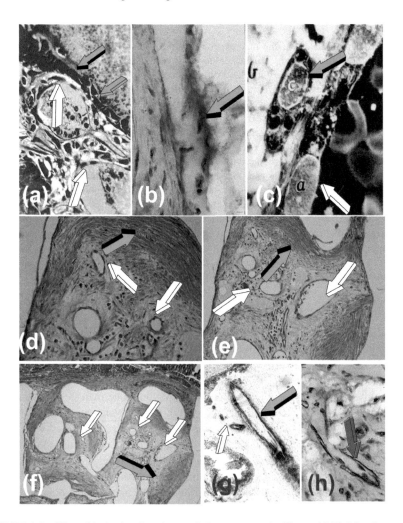

FIGURE 1.6   The critical role of angiogenesis in osteogenesis (Trueta 1963). Morphogenetic and osteogenetic vessels of Aristotelian and Trueta's definitions set into motion osteogenesis in angiogenesis as previously defined (Ripamonti 2006; Ripamonti et al. 2006; Ripamonti et al. 2007). (a) Intimate and exquisite relationships between invading capillaries and hyperchromatic nuclei of invading endothelial cells (*white* arrows) seemingly migrating from the invading vascular microenvironment to the differentiated osteoblastic-like cells (*light blue* arrows) onto the insoluble collagenous bone matrix. Collagenous bone matrix was used as a carrier for doses of osteogenin purified to apparent homogeneity from baboon crude extracts after chaotropic extraction in 4 M guanidinium hydrochloride and purified more than 60,000-fold (Ripamonti et al. 2001; Ripamonti 2006). The image reveals the complex pleiotropic function of the mechanisms responsible for osteogenesis in angiogenesis, not least the yet unknown molecular direct cell-to-cell interactions between osteoblasts and endothelial cells within the three-dimensional architecture of the newly induced ossicles in angiogenesis. (b) Further intimate and direct relationships between invading capillary resting along the micro patterned hydroxyapatite interface (*light blue* arrow) expressing laminin immunostaining on

FIGURE 1.6 (CONTINUED)

day 60 after implantation of the macroporous biomimetic matrix (Ripamonti et al. 1993). (c) The osteogenetic vessels of Trueta's definition (from Trueta 1963). Capillaries and endothelial cells (*white* arrow) in tight relationships with osteoblastic cells differentiation (*light blue* arrow). Osteoblasts and osteoprogenitors are in contact with the endothelium basement membrane (Trueta 1963) within newly forming bone (c). (d,e,f) The osteogenetic vessels of Trueta's definition (*white* arrows) (Trueta 1963) invading the macroporous spaces with osteoprogenitor cells supporting the induction of mesenchymal condensations at the hydroxyapatite interface (*light blue* arrows). There is a tight relationship between the endothelial cells and active osteoblasts (a,c). Cells are only separated by basement membrane components of the invading capillaries (c). The work of Reddi and colleagues (1994) suggested that, during bone regeneration osteocytes, a developmental cell of the osteoprogenitor-osteoblast lineage might retain a "memory" of the initial contact of osteoblasts with specific laminin domains (Vukicevic et al. 1900). Invading capillaries supporting the induction of bone formation (g and h), with alkaline phosphatase (g) and laminin (h) expression, present basement membrane components with a specific set of amino-acid domains which may be identifiable by differentiating osteoblastic cells. This may set into motion the ripple-like cascade of cellular differentiation and the induction of bone formation (Vukicevic et al. 1990).

1.2   THE CONCAVITY: THE SHAPE OF LIFE

In parallel with experiments, we implanted coral-derived constructs in two geometric configurations: solid blocks three or four millimetres in diameter were tested vs. particulate or granular hydroxyapatites with identical surface and biomaterial characteristics (Ripamonti et al. 1992). Blocks in both configurations were reconstituted with highly purified osteogenic fractions from bovine bone matrices (Luyten et al. 1989) and implanted bilaterally under the skin of Long–Evans rats (Ripamonti et al. 1992a). Coral-derived blocks in disc configuration, when reconstituted with highly purified osteogenic fractions, resulted in the induction of bone formation by day 7 after heterotopic implantation in the rodent bioassay (Figs. 1.8, 1.9). In marked contrast, coral-derived constructs in granular/particulate configuration recombined with identical osteogenic proteins failed to initiate the induction of bone formation (Ripamonti et al. 1992a; Ripamonti et al. 2012a).

Similar experiments in the Chacma baboon *Papio ursinus* using cylinders vs. granular particulate coral-derived constructs with identical surface characteristics but without the exogenous application of osteogenic soluble molecular signals also showed the lack of bone formation in particulate/granular geometric constructs (Fig. 1.10) (van Eeden and Ripamonti 1994). There was, however, a specimen of particulate/granular hydroxyapatite that showed, albeit minimally, the intrinsic and spontaneous induction of bone formation within a concavity of the substratum (Figs. 1.10c,d). The represented image was instrumental to define a geometric configuration responsible for the induction of bone formation, that is, that the concavity is an inductive geometric signal that set into motion the induction of bone formation within the macroporous spaces of the coral-derived bioreactors (Figs. 1.10c,d) (van Eeden and Ripamonti 1994).

The discovery that calcium phosphate-based macroporous constructs were endowed with the capacity to spontaneously initiate the induction of bone formation

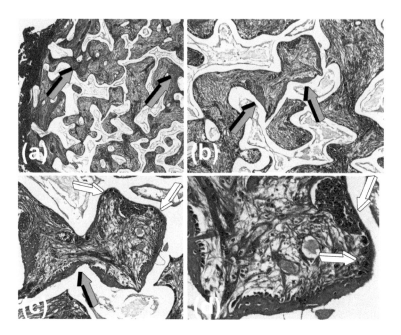

FIGURE 1.7 Vascular invasion, capillary sprouting, osteoclastic cell differentiation and bioactivity of the calcium phosphate-based macroporous bioreactor initiating the "spontaneous and/or intrinsic" induction of bone formation by topographically geometrically modified osteoclastic-resorbed substrata. Images on day 90 represent maturational gradients of tissue induction and morphogenesis controlling the development of tissue formation from day 30 to day 90 after heterotopic implantation, including but not limited to fibrovascular invasion, angiogenesis, osteoclastogenesis, induction of collagenous condensations, culminating in the induction of bone formation on topographically modified substrata by osteoclastogenesis. (a) Low-power digital image of a coral-derived bioreactor showing the induction of mesenchymal collagenous condensations at the hydroxyapatite interface (*light blue* arrows). (b) There is pronounced angiogenesis and capillary sprouting within the invading fibrovascular tissue with the induction of bone formation (*light blue* arrows) along the coral-derived bioreactor. (c) Detail of a macroporous space with invaded fibrovascular tissue showing osteoclastogenesis (*white* arrows) in concavities of the substratum in close relationship with newly forming bone and vascular invasion. (d) Detail of (c) depicting the morphological and thus molecular relationships between vascular invasion, osteoclastogenesis and osteogenesis within the coral-derived concave-shaped macroporous space (*white* arrows). Wax-embedded decalcified tissue blocks, paraffin-embedded sections cut at 3–6 μm and stained with Goldner's trichrome.

when implanted in heterotopic *rectus abdominis* sites of the Chacma baboon *Papio ursinus* was duly patented to possibly develop licensing for the biotechnology industry (Ripamonti 1994). The paper published in *Plastic and Reconstructive Surgery* (van Eeden and Ripamonti 1994) raised once again the critical role of the geometric configuration in the induction of bone formation. However, the biotechnology industries still market granular and/or particulate calcium phosphate-based biomaterials to the oral, maxillofacial and orthopaedic communities as bone-filler material,

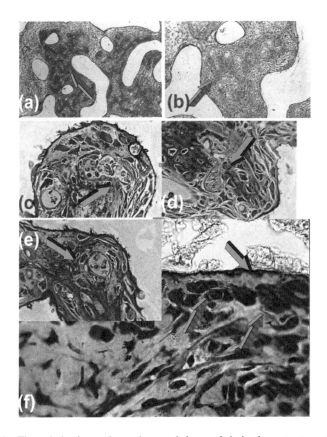

FIGURE 1.8 Tissue induction and morphogenesis by coral-derived constructs preloaded with doses of naturally derived highly purified osteogenic protein fractions after chromatography on hydroxyapatite Ultrogel, heparin-Sepharose adsorption and affinity columns (Ripamonti et al. 2001; Ripamonti 2006). After affinity chromatography on Pharmacia heparin-Sepharose, 500 mM eluted bioactive fractions were exchanged with 4 mM guanidinium-HCl and up-loaded onto tandem Sephacryl S-200 gel filtration chromatography columns eluted and washed in 4 mM guanidinium-HCl. Coral-derived bioreactors in discs and granular particulate configurations were preloaded with 50 μg highly purified naturally derived bovine osteogenic fractions (Luyten et al. 1989) in 50 μl hydrochloric acid and implanted in the subcutaneous tissue of Long–Evans rats (Ripamonti et al. 1992). (a,b) Low-power views highlighting early morphogenetic events within the macroporous spaces of a coral-derived disc (500 μm pore size) on day 7 after subcutaneous implantation. Arrows indicate the genesis of trabeculations surrounding invading capillaries within the macroporous spaces. (b) Low-power view identifying the spatial relationship of contiguous osteoblasts and the central blood vessels enveloped by the newly formed trabeculae of bone. (c,d,e) Osteoblasts differentiation, matrix synthesis, capillary sprouting and invasion (*light blue* arrows) and induction of newly formed bone matrix tightly embedded against the coral-derived substratum. Note the proximity of the endothelium (*light blue* arrows) with the secreting osteoblasts. (f) Newly differentiating osteoblast-like cells on day 11 in direct apposition to the hydroxyapatite substratum. Large plumped osteoblasts (*blue* and *magenta* arrows) embedded in the newly deposited matrix at the metachromatic cemental line (*light blue* arrow in f) of the interface of the coral-derived substratum. Undecalcified specimen blocks embedded in JB4 resin. Undecalcified sections, cut at 2 μm, stained with toluidine blue (Ripamonti et al. 1992).

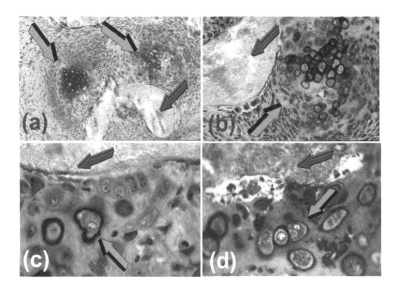

FIGURE 1.9 Induction of chondrogenesis by highly purified naturally derived osteogenic proteins purified by liquid chromatography and bioassayed in the subcutaneous space of Long–Evans rats (Ripamonti et al. 1992). Cartilage with hypertrophic chondrocytes develops in close proximity to the coral-derived hydroxyapatite substratum. On days 7–11, chondrogenesis was observed as a distinct developmental phase from the predominantly intramembranous pattern of bone differentiation (see Figs. 1.8a,b). (a,b) Islands of cartilage induction (*light blue* arrows) after preloading macroporous coral-derived discs (*magenta* arrows) with 50 μl osteogenic fractions in 50 μl hydrochloric acid (HCl). (c,d) High-power views highlighting hypertrophic chondrocytes (*light blue* arrows) with tight secretion of metachromatic matrix at the hydroxyapatite interface (*magenta* arrows). Undecalcified specimen blocks embedded in JB4 resin, cut at 2 μm and stained with toluidine blue (Ripamonti et al. 1992).

disregarding the biological knowledge of the concavity as a spontaneous inducer of bone formation (Ripamonti et al. 1999; Ripamonti 2004).

The critical role of the concavity spontaneously initiating the induction of bone formation was confirmed by intramuscular *rectus abdominis* implantation of sintered crystalline hydroxyapatites with concavities of specific dimensions on both planar surfaces of the sintered constructs (Ripamonti et al. 1999; Ripamonti 2004; Ripamonti 2012; Ripamonti et al. 2012a; Ripamonti et al. 2012b; Ripamonti et al. 2013). These findings are described in detail in Chapter 5. The concept of geometric inductive microenvironments was further tested by culturing *in vitro* macroporous fragments of coral-derived constructs combined with mouse-derived fibroblasts (NIH3T3) and pre-osteoblasts (MC 3T3-E1) cells (Ripamonti et al. 2012a; Ripamonti et al. 2012b). We found that concavities of the substratum oriented and polarized MC 3T3-E1 cells when cultured along concavities of the nanotopographically designed crystalline substratum (Fig. 1.11a inset) (Ripamonti et al. 2012a; Ripamonti et al. 2012b). The *in vitro* study was set to provide an *ex-vivo* bioreactor for later transplantation in rodents and in primates (Ripamonti et al. 2012b; Heliotis et al. 2006).

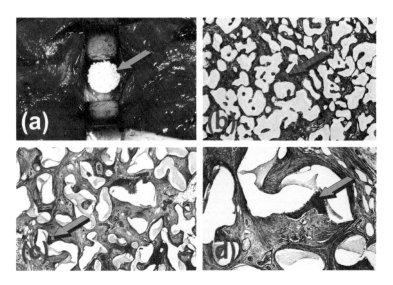

FIGURE 1.10 The critical role of the geometry of the inductor on the expression of the osteogenic phenotype and the induction of bone formation. Coral-derived bioreactors with two different geometric configurations were implanted in the *rectus abdominis* heterotopic sites of *Papio ursinus* (van Eeden and Ripamonti 1994). Substrata consisted of blocks in cylinder configuration (20 mm in length and 7 mm in diameter) and particulate granular configuration (460 to 620 µm in diameter) of coral-derived macroporous bioreactors (van Eeden and Ripamonti 1994; Ripamonti 2006; Ripamonti 2017). To avoid premature migration of individual granules during heterotopic intramuscular implantation (a), particulate granular hydroxyapatites were pre-formed by adding 1 mg of chondroitin-6-sulfate (Sigma Co., St. Louis, Mo) and 2 mg of baboon type I collagen to 400 mg of granular hydroxyapatite per implant, dispensed in sterile propylene tubes (van Eeden and Ripamonti 1994). After absolute ethanol precipitation and centrifugation, the implants were washed three times with chilled 85% ethanol, dried in a SpeedVac SC100 concentrator (Savant Instruments, Farmington, NY), and stored lyophilized at 4°C until implantation (van Eeden and Ripamonti 1994). Implants were harvested on days 60 and 90 after heterotopic intramuscular implantation. Specimens were fixed in 10% neutral buffered formaldehyde, were decalcified in formic-hydrochloric acid mixture and double embedded in celloidin and paraffin wax. Serial sections cut at 5 µm were cut longitudinally along the flat surface of the specimens of particulate granular hydroxyapatite (van Eeden and Ripamonti 1994) and stained with 0.1% toluidine blue in 30% ethanol or with Goldner's trichrome. (b) Low-power view of a granular particulate tissue block on day 90 after *rectus abdominis* implantation. Fibrovascular tissue invasion (*magenta* arrow) but lack of bone differentiation. (c) Low-power view of another granular particulate coral-derived hydroxyapatite specimen showing fibrovascular tissue invasion with pronounced capillary sprouting with the induction of bone formation within the concavity of a particulate hydroxyapatite granule (*light blue* arrow). (d) High-power view highlighting the induction of bone formation (*light blue* arrow) within a concavity of the coral-derived bioreactor. Decalcified sections, Goldner's trichrome stain.

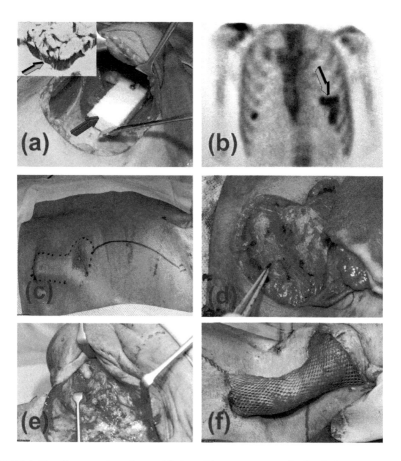

FIGURE 1.11 Construction of a prefabricated human osteogenic flap in heterotopic intramuscular sites for later transplantation into a mandibular recipient bed for reconstruction after squamous cell carcinoma debridement and surgical ablation of a large mandibular segment (Heliotis et al. 2006). The favoured intramuscular sites are highly vascularized with niches of responding perivascular and vascular stem cells (reviewed in Ripamonti et al. 2007), ideal for manufacturing prefabricated heterotopic bone grafts for autologous transplantation (reviewed in Ripamonti et al. 2006; Ripamonti et al. 2007). Inset in (a) shows optimal polarization and alignment of pre-osteoblast (MC 3T3-E1) cells *in vitro* self-assembling within a concavity of the coral-derived substratum (*light blue* arrow). (a) L-shaped coral-derived macroporous construct (*dark blue* arrow) is implanted in the muscular tissue of the chest of a human patient after reconstitution with 2.5 mg doses of osteogenic protein-1 (hOP-1; Stryker Biotech, USA). (b) Scintigraphy demonstrates the induction of bone formation (*light blue* arrow) in the L-shaped prefabricated flap in the *pectoralis* muscle. (c,d,e) Surgical debridement and harvest of the newly generated L-shaped sagomated heterotopic prefabricated graft and autologous transplantation into the surgically prepared recipient bed (c,d,e) after the preparation of a pedicled flap (c,f) into the mandibular recipient bed (Heliotis et al. 2006).

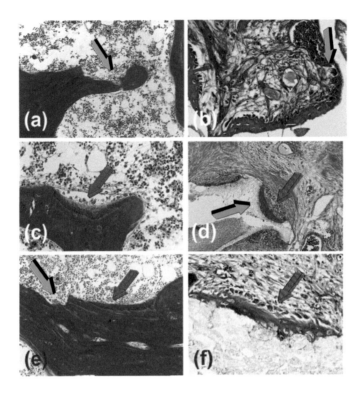

FIGURE 1.12   The induction of bone formation mimics the remodelling cycle of the cortico-cancellous bone. Lacunae, pits and concavities as cut by osteoclastogenesis (*light blue* arrows in a,e) initiate the remodelling of the cortico-cancellous bone and provide resorption lacunae in the form of concavities that initiate the formative phase with osteoid deposition within concavities cut by osteoclastogenesis (*magenta* arrows c,e). Similarly, biomimetic matrices replicate the geometries of cortico-cancellous bone remodelling whereby the calcium phosphate-based matrices sustain osteoclastogenesis (*light blue* arrow in b), Ca^{2+} release, angiogenesis, somatic cell de-differentiation and the induction of bone formation as a secondary response (*magenta* arrows d,f) (Ripamonti et al. 2010; Ripamonti et al. 2012a; Klar et al. 2013; Ripamonti et al. 2015; Ripamonti 2017; Ripamonti 2018). Undecalcified section cut at 3 to 6 μm stained with Goldner's trichrome (a,b,c,e) or toluidine blue in 70% ethanol.

A prefabricated human bioreactor was constructed with an L-shaped coral-derived macroporous bioreactor mimicking a reconstructed mandible, later transplanted to a mandibular defect using a vascularized hydroxyapatite/osteogenic protein-1 composite graft (Fig. 1.11) (Heliotis et al. 2006).

The central question in developmental biology and thus tissue engineering and regenerative medicine is the molecular basis of pattern formation (Reddi 1984; Lander 2007; De Robertis 2008). It became apparent to one of the authors that the correspondence of the concavity within the macroporous calcium phosphate-based constructs with the concavities generated by osteoclastogenesis during the remodelling cycle of the cortico-cancellous osteonic bone was key to understanding the bone-inductive capacity (Fig. 1.12). There is a profound biomimetism of the "geometric induction of bone formation" with the remodelling cycle of the cortico-cancellous osteonic bone (Figs. 1.12). It is well known that the properties of a substrate affect not only cell attachment and growth, but also stem cell differentiation. The molecular basis of this has been explored *in vitro* in the context of osteogenic and adipogenic differentiation from mesenchymal stem cells (McBeath et al. 2004; Kilian et al. 2010; Kress et al. 2012; Werner et al. 2016). When the geometry of a cell substrate increases actomyosin contractility, this alters the compression of the cell's nucleus and osteogenesis is promoted (Kilian et al. 2010; Werner et al. 2017). Mechanical cues from substrate topography that alter cell shape have been shown to influence mesenchymal stem cell fate through a RhoA signalling-dependent mechanism (McBeath et al., 2004) stimulating MAPK cascades and Wnt signalling (Kilian et al. 2010). This conversion of a mechanical signal into a biochemical response, or cellular mechanotransduction, drives tissue patterning (Ghosh et al. 2013). The geometric properties of the substrate that promote osteogenic differentiation simulate those of the microenvironment of the differentiated cells within bone (Ripamonti et al. 1999; Kilian et al. 2010; Klar et al. 2013; Ripamonti et al. 2015; Ripamonti 2017)

The Bone Research Laboratory at the University of the Witwatersrand has demonstrated that it is not necessary to combine soluble molecular signals with insoluble signals or substrata for the induction of bone formation. The induction of bone formation is also achieved by constructing macroporous biomimetic matrices that *per se* initiate the ripple-like cascade of bone differentiation by induction (Ripamonti 1990; Ripamonti 1991; Ripamonti et al. 1993; Ripamonti 1996; Ripamonti et al. 1999; Ripamonti 2004; Ripamonti 2006; Ripamonti 2009; Ripamonti et al. 2009; Ripamonti et al. 2010; Ripamonti 2012; Ripamonti et al. 2012a; Ripamonti et al. 2012b; Ripamonti 2017; Ripamonti 2018).

ACKNOWLEDGEMENTS

The discovery by serendipity of the so-defined spontaneous and/or intrinsic induction of bone formation by coral-derived macroporous bioreactors was the beginning of a fascinating and intriguing scientific ride across cell differentiation, tissue induction and morphogenesis, morphogens, stem cells and de-differentiation, angiogenesis and organogenesis that has highlighted to no end the scientific and intellectual journey of the writer of this chapter and of this volume. Many thanks to many students, scientists, technologists and Interpore International Clayton Shors and Zimmer Biomet Mike Ponticiello for the continuous supply of coral-derived bioreactors for implantation in *Papio ursinus*. I would like to thank two senior technologists that shaped the Bone Research Laboratory into a fully fledged Unit of the SA Medical Research Council and the hosting University: Barbara van den Heever, who started the precision cutting of undecalcified sections using Leica sledge microtomes with carbide-tungsten knifes, and Ruqayya Parak for adding the third dimension of undecalcified histology by cutting impeccable sections on the Exakt precision cutting and grinding diamond saw. To both, special recognition is offered for also being capable of handling the time constraints of continuously preparing impeccable undecalcified whole mounted and stained sections. To Hari Reddi special recognition is offered, for walking with the writer into the critical role of geometry initiating the induction of bone formation. Special recognition is given to the Laboratories of Molecular and Cellular Biology of the University, School of Clinical Medicine, for finally assigning the molecular mechanisms of the induction of bone formation by coral-derived macroporous constructs. Special thanks to Raquel Duarte, Caroline Dickens, Therese Dix-Peek and Rolando Klar, who molecularly assigned the spontaneous induction of bone formation by coral-derived bioreactors. Finally, UR would like to recognize and thank the University of the Witwatersrand for sustaining the often-perilous development of the Bone Research Unit across the century and his daughter Daniella Bella for the creativity to write.

REFERENCES

Crivellato, E.; Nico, B.; Ribatti, D. Contribution of Endothelial Cells to Organogenesis: A Modern Reappraisal of an Old Aristotelian Concept. *J. Anat.* 2007, *211*(4), 415–27.

De Robertis, E. Evo-Devo: Variations on Ancestral Themes. *Cell* 2008, *132*(2), 185–95.

Donath, K.; Breuner, G. A Method for the Study of Undecalcified Bone and Teeth with Attached Soft Tissue – The "Säge Schliff" (Sawing and Grinding) Technique. *J. Oral Path.* 1982, *11*, 318–26.

Foidart, J.M.; Reddi, A.H. Immunofluorescent Localization of Type IV Collagen and Laminin During Endochondral Bone Differentiation and Regulation by Pituitary Growth Hormone. *Dev. Biol.* 1980, *75*(1), 130–36.

Ghosh, K.; Thodeti, C.K.; Ingber, D.E. Micromechanical Design Criteria for Tissue Engineering Biomaterials. In: B.D. Ratner, A.S. Hoffman, F.J. Schoen, J.E. Lemons (eds.) *Biomaterials Science: An Introduction to Materials in Medicine* (3rd Edition), Academic Press, Elsevier, Oxford, UK, 2013, 1165–78.

Heliotis, M.; Laveru, K.M.; Ripamonti, U.; Tsiridis, E.; di Silvio, L. Transformation of a Prefabricated Hydroxyapatite/Osteogenic protein-1 Implant into a Vascularized Pedicled Bone Flap in the Human Chest. *Int. J. Oral Maxillodac Surg.* 2006, *35*(3), 265–59.

Itoh, F.; Itoh, S.; Goumans, J.; Valdimarsdottir, G.; Iso, T.; Dotto, G.P.; Hamamoru, Y.; Kedes, L.; Kato, M.; ten Dijke, P. Synergy and Antagonism Between Notch and BMP. *EMBO J.* 2004, *3*, 541–51. www.embojournal.org.

Kilian, K.A.; Bugarija, B.; Lahn, B.T.; Mrksich, M. Geometric Cues for Directing the Differentiation of Mesenchymal Stem Cells. *Proc. Natl. Acad. Sci. U.S.A.* 2010, *107*(11), 4872–877.

Klar, R.M.; Duarte, R.; Dix-Peek, T.; Dickens, C.; Ferretti, C.; Ripamonti, U. Calcium Ions and Osteoclastogenesis Initiate the Induction of Bone Formation by Coral-Derived Macroporous Constructs. *J. Cell. Mol. Med.* 2013, *17*(11), 1444–57.

Klar, R.M.; Duarte, R.; Dix-Peek, T.; Ripamonti, U. The Induction of Bone Formation by the Recombinant Human Transforming Growth Factor-β3. *Biomaterials* 2014, *17*(9): 2773–88.

Kress, S.; Neumann, A.; Weyand, B.; Kasper, C. Stem Cell Differentiation Depending on Different Surfaces. *Adv. Biochem. Eng. Biotechnol.* 2012, *126*, 263–83.

Lander, A.D. Morpheus Unbound: Reimagining the Morphogen Gradient. *Cell* 2007, *128*(2), 245–56.

Luyten, F.P.; Cunningham, N.S.; Ma, S.; Muthukumaran, N.; Hammonds, R.G.; Nevins, W.B.; Woods, W.I.; Reddi, A.H. Purification and Partial Amino Acid Sequence of Osteogenin, a Protein Initiating Bone Differentiation. *J. Biol. Chem.* 1989, *264*(23), 13377–80.

McBeath, R.; Pirone, D.M.; Nelson, C.M.; Bhadriraju, K.; Chen, C.S. Cell Shape, Cytoskeletal Tension, and RhoA Regulate Stem Cell Lineage Commitment. *Dev. Cell* 2004, *6*(4), 483–95.

Piecuch, J.F. Extraskeletal Implantation of a Porous Hydroxyapatite Ceramic. *J. Dent. Res.* 1982, *61*(12), 1458–60.

Reddi, A.H.; Huggins, C. Biochemical Sequences in the Transformation of Normal Fibroblasts in Adolescent Rats. *Proc. Natl. Acad. Sci. U.S.A.* 1972, *69*(6), 1601–5.

Reddi, A.H. Extracellular Matrix and Development. In: K.A. Piez, Reddi, A.H. (eds.) *Extracellular Matrix Biochemistry*, Elsevier, New York, 1984, 247–291.

Reddi, A.H. Symbiosis of Biotechnology and Biomaterials: Applications in Tissue Engineering of Bone and Cartilage. *J. Cell. Biochem.* 1994, *56*(2), 192–5.

Reddi, A.H. Bone Morphogenesis and Modeling: Soluble Signals Sculpt Osteosomes in the Solid State. *Cell* 1997, *89*(2), 159–61.

Reddi, A.H. Morphogenesis and Tissue Engineering of Bone and Cartilage: Inductive Signals, Stem Cells, and Biomimetic Biomaterials. *Biomater. Tissue Eng.* 2000, *6*(4), 351–9.

Ripamonti, U. Inductive Bone Matrix and Porous Hydroxyapatites Composites in Rodents and Non-Human Primates. In: T. Yamamuro, Wilson-Hench, J., Hench, L.L. (eds.) *Handbook of Bioactive Ceramics: Calcium Phosphate and Hydroxyapatite Ceramics* (Vol. II). CRC Press, Boca Raton, 1990, 245–53.

Ripamonti, U. The Morphogenesis of Bone in Replicas of Porous Hydroxyapatite Obtained from Conversion of Calcium Carbonate Exoskeleton of Coral. *J. Bone Joint Surg. Am* 1991, *73*(5), 692–703.

Ripamonti, U.; Ma, S.; van den Heever, B.; Reddi, A.H. The Critical Role of Geometry of Porous Hydroxyapatite Delivery System in Induction of Bone by Osteogenin, a Bone Morphogenetic Protein. *Matrix* 1992, *12*(3), 202–12.

Ripamonti, U. *The Generation of Bone in Nonhuman Primates. Experimental Studies on the Baboon* (Papio ursinus). PhD (Med.), Faculty of Medicine, University of the Witwatersrand, Johannesburg, 1993.

Ripamonti, U.; van den Heever, B.; van Wyk, J. Expression of the Osteogenic Phenotype in Porous Hydroxyapatite Implanted Extraskeletally in Baboons. *Matrix* 1993, *13*(6), 491–502.

Ripamonti, U. *A Method for Screening a Selected Material for Its Osteoconductive and Osteoinductive Potential.* South African Patent 92/3982, May 25 1994; US patent 5,355,898, October 18, 1994.

Ripamonti, U.; Reddi, A.H. Bone Morphogenetic Proteins: Applications in Plastic and Reconstructive Surgery. Adv. *Plast. Reconstr. Surg.* 1995, *11*, 47–65.

Ripamonti, U.; Kirkbride, A.N. *Biomaterial and Bone Implant for Bone Repair and Replacement.* PCT/NL95/00181, WO9532008A1, November 30 1995.

Ripamonti, U. Osteoinduction in Porous Hydroxyapatite Implanted in Heterotopic Sites of Different Animal Models. *Biomaterials* 1996, *17*(1), 31–5.

Ripamonti, U.; Kirkbride, A.N. *Biomaterial and Bone Implant for Bone Repair and Replacement.* EP760687A1, March 12 1997.

Ripamonti, U.; Duneas, N.; van den Heever, B.; Bosch, C.; Crooks, J. Recombinant Transforming Growth Factor-β_1 Induces Endochondral Bone in the Baboon and Synergizes with Recombinant Osteogenic protein-1 (Bone Morphogenetic protein-7) to Initiate Rapid Bone Formation. *J. Bone Miner. Res.* 1997, *12*(10), 1584–95.

Ripamonti, U.; Crooks, J.; Kirkbride, A.N. Sintered Porous Hydroxyapatites with Intrinsic Osteoinductive Activity: Geometric Induction of Bone Formation. *S. Afr. J. Sci.* 1999, *95*, 335–43.

Ripamonti, U.; Kirkbride, A.N. *Biomaterial and Bone Implant for Bone Repair and Replacement.* US Patent 6,117,172, September 12 2000.

Ripamonti, U.; Crooks, J.; Matsaba, T.; Tasker, J. Induction of Endochondral Bone Formation by Recombinant Human Transforming Growth Factor-β_2 in the Baboon (*Papio ursinus*). *Growth Factors* 2000, *17*(4), 269–85.

Ripamonti, U.; Kirkbride, A.N. *Biomaterial and Bone Implant for Bone Repair and Replacement*. US Patent 6,302,913 B1, October 16 2001.

Ripamonti, U.; Ramoshebi, L.N.; Matsaba, T.; Tasker, J.; Crooks, J.; Teare, J. Bone Induction by BMPs/OPs and Related Family Members in Primates. *J. Bone Joint Surg. Am* 2001, 83-A(Suppl. 1 (Pt 2)), S116–27.

Ripamonti, U. Soluble, Insoluble and Geometric Signals Sculpt the Architecture of Mineralized Tissues. *J. Cell. Mol. Med.* 2004, *8*(2), 169–80.

Ripamonti, U. Soluble Osteogenic Molecular Signals and the Induction of Bone Formation. *Biomaterials* Leading Opinion Paper 2006, *27*(6), 807–22.

Ripamonti, U.; Ferretti, C.; Heliotis, M. Soluble and Insoluble Signals and the Induction of Bone Formation: Molecular Therapeutics Recapitulating Development. *J. Anat.* 2006, *209*(4), 447–68.

Ripamonti, U.; Heliotis, M.; Ferretti, C. Bone Morphogenetic Proteins and the Induction of Bone Formation: From Laboratory to Patients. *Oral Maxillofac. Surg. Clin. North Am.* 2007, *19*(4), 575–89.

Ripamonti, U.; Ramoshebi, L.N.; Teare, J.; Renton, L.; Ferretti, C. The Induction of Endochondral Bone Formation by Transforming Growth Factor-β_3: Experimental Studies in the Non-Human Primate *Papio ursinus*. *J. Cell. Mol. Med.* 2008, *12*(3), 1029–48.

Ripamonti, U. Biomimetism, Biomimetic Matrices and the Induction of Bone Formation. *J. Cell. Mol. Med.* 2009, *13*(9B), 2953–72.

Ripamonti, U.; Crooks, J.; Khoali, L.; Roden, L. The Induction of Bone Formation by Coral Derived Calcium Carbonate/Hydroxyapatite Constructs. *Biomaterials* 2009, *30*(7), 1428–39.

Ripamonti, U.; Klar, R.M.; Renton, L.F.; Ferretti, C. Synergistic Induction of Bone Formation by hOP-1 and TGF-β_3 and Inhibition by Zoledronate in Macroporous Coral- Derived Hydroxyapatite Constructs. *Biomaterials* 2010, *31*(25), 6400–10.

Ripamonti, U. The Concavity: The "Shape of Life" and the Control of Bone dDfferentiation – Feature Paper – *Science in Africa*, 2012. http://www.scienceinafrica.co.za/2012/Ripamonti_bone.htm.

Ripamonti, U.; Roden, L.C.; Renton, L.F. Osteoinductive Hydroxyapatite-Coated Titanium Implants. *Biomaterials* 2012a, *33*(15), 3813–23.

Ripamonti, U.; Roden, L.; Renton, L.; Klar, R.M.; Petit, J.-C. The Influence of Geometry on Bone: Formation by Autoinduction. *Science in Africa* 2012b. http://www.scienceinafrica.co.za/2012/Ripamonti_bone.htm

Ripamonti, U.; Renton, L.; Petit, J.-C. Bioinspired Titanium Implants: The Concavity - The Shape of Life. In: M. Ramalingam, P. Vallitu, U. Ripamonti, W.-J. Li (eds.) *CRC Press Taylor & Francis, Boca Raton USA;* Tissue Engineering and Regenerative Medicine. A Nano Approach, CRC Press, Boca Raton 2013, Chapter 6, 105–23.

Ripamonti, U.; Dix-Peek, T.; Parak, R.; Milner, B.; Duarte, R. Profiling Bone Morphogenetic Proteins and Transforming Growth Factor-Bs by hTGF-β3 Pre-Treated Coral-Derived Macroporous Bioreactors: The Power of One Morphogenetic Proteins and Transforming Growth Factor-βs by hTGF-β3 Pre-Treated Coral-Derived Macroporous Constructs: The Power of One. *Biomaterials* 2015, *49*, 90–02. http://dx.doi.org./10.1016/j.biomaterials.2015.01.056.

Ripamonti, U. *Induction of Bone Formation in Primates. The Transforming Growth Factor-beta3*. U. Ripamonti (ed.), CRC Press, Boca Raton, 2016.

Ripamonti, U.; Klar, M.R.; Parak, R.; Dickens, C.; Dix-Peek, T.; Duarte, R. Tissue Segregation Restores the Induction of Bone Formation by the Mammalian Transforming growth factor-β3 in Calvarial Defects of the Non-Human Primate Papio ursinus *Biomaterials* 2016, 86, 21–32. doi: 10.1016/j.biomaterials2016.01.071.

Ripamonti, U. Biomimetic Functionalized Surfaces and the Induction of Bone Formation. *Tissue Eng.* 2017, *23*(21,22), 1197–2009.

Ripamonti, U. Functionalized Surface Geometries Induce: *"Bone: Formation by Autoinduction"*. *Front. Physiol.* 2018, *8*, 1084, doi:10.3389/fphys.2017.01084.

Sampath, T.K.; Reddi, A.H. Dissociative Extraction and Reconstitution of Extracellular Matrix Components Involved in Local Bone Differentiation. *Proc. Natl. Acad. Sci. U.S.A.* 1981, *78*(12), 599–603.

Sampath, T.K.; Reddi, A.H. Homology of Bone-Inductive Proteins from Human, Monkey, Bovine, and Rat Extracellular Matrix. *Proc. Natl. Acad. Sci. U.S.A.* 1983, *80*(21), 6591–95.

Trueta, J. The Role of the Vessels in Osteogenesis. *J. Bone Joint. Surg.* 1963, *45B*, 402–18.

Urist, M.R. Bone: Formation by Autoinduction. *Science* 1965, *150*(698), 893–9.

Urist, M.R.; Silverman, B.F.; Buring, K.; Dubuc, F.L.; Rosenberg, J.M. The Bone Induction Principle. *Clin. Orthop. Relat. Res.* 1967, *53*, 243–83.

van Eeden, S.; Ripamonti, U. Bone Differentiation in Porous Hydroxyapatite Is Regulated by the Geometry of the Substratum: Implications for Reconstructive Craniofacial Surgery. *Plast. Reconstr. Surg.* 1994, *93*(5), 959–66.

Wlodarski, I.; Reddi, A.H. Alkaline Phosphatase as a Marker of Osteoinductive Cells. *Calcif. Tissue Int.* 1986, *39*(6), 259–62.

Vukicevic, S.; Luyten, F.P.; Kleinman, H.K.; Reddi, A.H. Differentiation of Canalicular Cell Processes in Bone Cells by Basement Membrane Matrix Components: Regulation by Discrete Domains of Laminin. *Cell* 1990, *63*(2), 437–45.

Werner, M.; Blanquer, S.B.; Haimi, S.P.; Korus, G.; Dunlop, J.W.; Duda, G.N.; Grijpma, D.W.; Petersen, A. Surface Curvature Differentially Regulates Stem Cell Migration and Differentiation via Altered Attachment Morphology and Nuclear Deformation. *Adv. Sci.* (Weinh) 2016, *4*(2), 1600347.

2 The Induction of Bone Formation

When and Why Bone Forms and Sometimes Repairs and Regenerates: The Enigmatic Myth of Bone Tissue Engineering and the Dream of Regenerative Medicine

Ugo Ripamonti

2.1 THE INDUCTION OF BONE FORMATION

The understanding of the molecular and cellular mechanisms of cell differentiation and morphogenesis is central to the biology of tissue regeneration (King and Newmark 2012). The ability of bone to regenerate and heal without scarring has been known since antiquity and Hippocratic times (reported by Reddi 2000).

Still to our days, and in spite of extensive research into the supramolecular assembly of the bone extracellular matrix, little is known about the capacity to regenerate large cranio-mandibulo-facial defects in human patients, together with its regenerative potential upon implantation of the cloned osteogenic molecular signals of the transforming growth factor-β (TGF-β) supergene family (Ripamonti 2003).

Much less is known about the apparent lack of translational research on the often outstanding results in pre-clinical animal models including non-human primates. On the other hand, translational research from non-human to human primates *Homo sapiens* is often limited and inadequate when compared to pre-clinical results in a variety of animal models (Williams 2006; Ripamonti et al. 2006; Ripamonti et al. 2007; Ferretti et al. 2010; Manologas and Kroneberg 2014; Ripamonti et al. 2014). Bone and large critical size defects lack a proper template for an orchestrated tissue regeneration cascade and require autogenous bone grafting with harvested-related morbidity (Habal 1994; Burchardt 1987).

Reconstruction of large craniofacial and appendicular skeletal defects in humans still requires the harvesting of autogenous bone from a distant donor site, most often the iliac crest (Habal 1994; Burchardt 1987; Ferretti et al. 2016). We recently reported the successful use of autogenous bone grafting to reconstruct large mandibular defects in human patients with autologous compressed particulate cortico-cancellous bone grafts, with *restitutio ad integrum* of the often large if not massive mandibular defects (Ferretti et al. 2016).

Supra physiological doses of a single recombinant human bon morphogenetic protein (hBMP) are needed to often induce unacceptable tissue regeneration whilst incurring significant costs without achieving equivalence to autogenous bone grafts (Ferretti et al. 2010; Ripamonti et al. 2006; Ripamonti et al. 2007; Ripamonti et al. 2014). Large bony defects, in particular severe mandibular discontinuities, can be however re-engineered with autogenous bone grafts, highlighting the concept of clinically significant osteoinduction whereby the regenerated bone is readily identifiable on radiographic examination by virtue of its opacity and trabecular architecture (Fig. 2.1) (Ripamonti et al. 2014).

The quality and extent of regeneration of treated large, often massive, mandibular defects (Fig. 2.1) are not comparable to mandibular regenerates after implantation of high doses of a single recombinant hBMP (Ferretti et al. 2010; Ripamonti et al. 2014; Ferretti et al. 2016; Ferretti and Ripamonti 2020).

Is *restitutio ad integrum* possible after implantation of osteogenic soluble molecular signals reconstituted with an appropriate delivery system in large skeletal defects of the cranio-maxillo-facial and of the axial skeletons? In previous contributions we have often tried to objectively examine the often far too optimistic statements about tissue engineering advances, with several statements about successful regeneration in human patients canvassing "molecular biologist and surgeons alike that the era of tissue reconstruction of spare parts of the human body is finally close at hand" (Ripamonti et al. 2014).

In previous contributions we have often tried to objectively analyze the hype of regenerative medicine and tissue engineering, from lay press reports to formal scientific journals in the field (Ferretti et al. 2010; Ripamonti et al. 2006; Ripamonti et al. 2007; Ripamonti et al. 2012a; Ripamonti et al. 2014). Several papers are routinely published highlighting how tissue engineering is helping patents affected by a variety of functional and morphological organ and tissue damage, indicating the capacity of tissue engineering to correct and improve organ and tissue function (Langer 2019).

We further reported that the ultimate challenge of regenerative medicine and tissue engineering alike is to hypothesize the functional restoration of organs and tissues by exploring the development of human tissue factories (Ripamonti 2018a). In spite of the major advances in understanding cellular biology mechanistically, resolving several gene pathways in tissue induction and morphogenesis, successful translation in clinical contexts of the novel outstanding results in pre-clinical animal models is still not feasible (Ripamonti et al. 2007; Ferretti et al. 2010; Ripamonti et al. 2014; Ripamonti 2017). In our opinion it is worth repeating again what it is that we have stated previously (Ripamonti et al. 2014), i.e. that published perspectives in regenerative medicine have been published "even in the awareness that the need of such functionalities is largely not substantiated by experimental data" (Martin 2014).

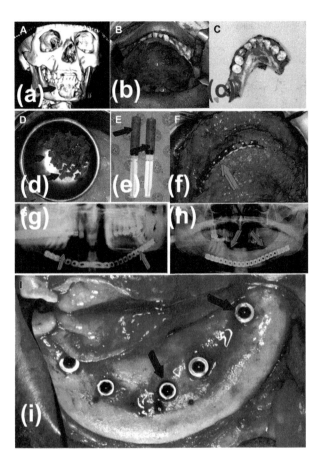

FIGURE 2.1   Is the regeneration of large cranio-mandibulo-facial defects in human patients possible after surgical debridement of large if not massive mandibular lesions? The composite image shows that substantial mandibular defects in human patients can be regenerated by the implantation of morcellated cortico-cancellous bone grafts. The iconographic plate also illustrates the concept of "clinically significant osteoinduction" whereby the regenerated bone is readily identifiable on radiographic examination by virtue of its opacity and trabecular architecture (Ferretti et al. 2010; Ripamonti et al. 2014; Ferretti et al. 2016; Ferretti and Ripamonti 2020). (a,b) A large glandular osteogenic cyst (Ripamonti et al. 2014) requires the avulsion of the affected mandible (c). (d,e,f) Fragments of autogenous bone grafts harvested from the posterior iliac crest are morcellated and implanted as compressed particulate cortico-cancellous bone grafts into 20 ml syringes (e) ejected across the vast mandibular defect after insertion of a titanium plate (*light blue* arrow in f). (g,h) Radiographic analyses after surgical debridement (g) and 6 months after implantation of compressed particulate cortico-cancellous autogenous bone harvested from the iliac crest. (h) *Restitutio ad integrum* of the large mandibular defect (*light blue* arrows) showing the quality and quantity of the regenerated bone. (i) The substantial newly formed mandibular bone after autogenous transplantation of the morcellated compressed particulate cortico-cancellous bone graft is functionally restored after the insertion of several titanium fixtures for prosthetic rehabilitation (*dark blue* arrows).

Reviews and perspectives continue to appear in the tissue engineering arena, suggesting the promise of regeneration of several organs and tissues based on *in vitro* studies or some experimentation in rodent models. The hype in regenerative medicine is a series of statements based on a promise of organ regeneration, which are simply based on hypotheticals, not proven, but only biologically and molecularly generated by several results from the bench top with minimal if any experimentation in pre-clinical contexts (Baptista and Atala 2014; Langer and Vacanti 1993; Langer 2019).

Indeed, the promise of regeneration of cranio-mandibulo-facial or axial skeletal defects in clinical contexts has so far failed, and the regeneration of bone is still unresolved (Ripamonti et al. 2014). Even our translation in clinical contexts of the rapid and substantial induction of bone formation in the chacma baboon *Papio ursinus* has shown that human patients do not respond to the hTGF-β_3 osteogenic device as compared to pre-clinical results in *Papio ursinus* (Ferretti and Ripamonti 2020).

Comparative regenerative mechanisms across different mammalian tissues (Reichman 1984; Bely 2010; Bely and Nyberg 2010; Tanaka and Reddien 2011; Lismaa et al. 2018) show that major differences do exist in the extent of tissue regeneration between animal phyla, including also differences in regenerative potential between primate species, i.e. *Papio ursinus* vs. *Homo sapiens* (Klar et al. 2014; Ferretti and Ripamonti 2020).

A more objective overview of regenerative medicine and tissue engineering at large is that the impetus of regenerative medicine has highlighted and resolved several issues still molecularly pending in tissue biology. This has however morphed into a hyperbole of promised novel regenerative treatments based only on the potential of cells' activities and function *in vitro* or experiments in rodents without proper experimental evidence of translational research in clinical contexts. Indeed, none of the unexpected brilliant results in pre-clinical models, including non-human primate species, are actually routinely used in clinical contexts (Williams 2006; Badylak and Rosenthal 2017).

Further considerable if not extraordinary hype around potential therapeutic translation in clinical contexts came with the Nobel prize-winning discoveries on the induction of pluripotent stem cells from differentiated somatic cells (Takahashi and Yamanaka 2006; Takahashi et al. 2007; Takahashi and Yamanaka 2013).

Perhaps it is worth quoting again Badylak and Rosenthal's statement in *Nature Regenerative Medicine* that "for all the progress we have made in the fast moving field of regenerative medicine, the capacity to routinely restore functional tissue following traumatic injury or degenerative disease is still beyond reach" (Badylak and Rosenthal 2017).

When discussing the induction of bone or "Bone: formation by autoinduction", as per the classic studies of M.R. Urist in *Science* (Urist 1965), it is important to properly define the terminology related to the induction of bone formation (Urist 1965; Reddi and Huggins 1972; Reddi 1981).

The acid test of the induction of bone formation is the *de novo* generation of heterotopic bone after extraskeletal implantation of an osteogenic soluble molecular signal of the transforming growth factor-β (TGF-β) supergene family (Ripamonti

2003; Ripamonti 2006; Ripamonti et al. 2008) or any other putative inductive protein. Proteins and/or matrices labelled as osteoinductive must thus be endowed with the striking prerogative of initiating the induction of bone formation in heterotopic extraskeletal sites of animal models (Urist 1965; Reddi and Huggins 1972; Ripamonti 2003; Ripamonti et al. 2004 Ripamonti 2006; Reddi 2000).

Beside biomimetic *smart* self-induce biomatrices and the osteogenic proteins of the TGF-β supergene family, it is noteworthy that the uroepithelium or the transitional epithelium of the urinary bladder also initiates the induction of bone formation intramuscularly in lagomorph, canine and non-human primate models. The induction of bone formation is defined as "uroepithelial osteogenesis" (Fig. 2.2) (Sacerdotti and Frattin 1901; Huggins 1931; Friedenstein 1962; Friedenstein 1968).

In by-now classic experiments after ligation of the renal vascular pedicle of rabbits, Sacerdotti and Frattin (1901) showed that the renal parenchyma was transformed into bone, with the induction of trabeculae of bone and associated induction of hematopoietic bone marrow. True bone was observed 90 days after ligation of the rabbit renal arteries (Sacerdotti and Frattin 1901).

Uroepithelial osteogenesis differs significantly between experimental animal models. The osteogenic activity of transitional epithelium is highest in the guinea pig, feline and canine models, lower in rodents and lowest in lagomorphs (Friedenstein 1968).

Experiments by Friedenstein (Friedenstein 1962; Friedenstein 1968) ultimately asked the compelling question that *par force* defines the bone induction principle (Urist et al. 1967) or "the osteogenic activity of several transplanted tissues, including bone and dentine matrices, and uroepithelium" (Ripamonti 2006). How is the inductive influence of the transitional epithelium transferred to competent responding cells?

Friedenstein elegantly hypothesized and demonstrated the humoral nature of the osteogenic activity of transitional epithelium, that is, the presence of a soluble molecular signal or "inductor" (Friedenstein 1962). The humoral nature of the osteogenic activity of transitional epithelium was demonstrated by transfilter bone induction (Friedenstein 1962; Friedenstein 1968). Importantly, Huggins (Huggins 1931) concluded that "the proliferating newly formed epithelium is the essential factor in this osteogenesis", that is, uroepithelial osteogenesis. His conclusions were later supported by Friedenstein who made the key observation that only the epithelium lining the basement membrane possesses osteogenetic properties (Friedenstein 1968). Classically, this was demonstrated by detaching the uroepithelium from the *tunica propria* by trypsinization and showing that the epithelium induces osteogenesis whilst the *tunica propria* lacks inductive properties (Friedenstein 1968).

Of note, Friedenstein reported that proliferating osteoblasts are seen in close vicinity to cords of transitional epithelial cells in heterotopic sites of pigs (Friedenstein 1968). Uroepithelial osteogenesis in the chacma baboon *Papio ursinus* also shows the induction of bone formation across the dome of the bladder transplanted with the *rectus abdominis* fascia in close relationship with proliferating transitional epithelial cells (Fig. 2.1b *white* arrows).

The heterotopic implantation site avoids the ambiguities of the orthotopic site where some degree of bone formation by conduction may occur from the viable interfaces (Urist 1965; Reddi 2000; Ripamonti 2006). This is particularly true when

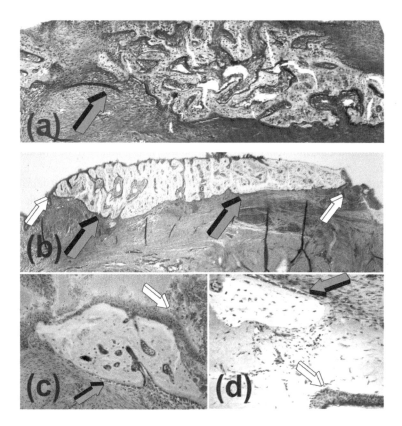

FIGURE 2.2   Uroepithelial osteogenesis in the dome of the urinary bladder of the chacma baboon *Papio ursinus*. Induction of bone formation after transplantation of the *rectus abdominis* fascia to close a full thickness defect on the dome of the bladder. The transplanted autogenous fascia sets into motion the induction of uroepithelial osteogenesis as discussed in the text, causing proliferation of the uroepithelium (*light blue* arrow in a) in a close relationship with the newly formed bone (*white* arrow in b). (b) Newly formed bone by uroepithelial osteogenesis (*light blue* arrows in b) seals the urinary bladder defect with trabeculations of the newly formed and mineralized bone. (c,d) Details of the induction of uroepithelial osteogenesis with layers of osteoblastic-like cells (*light blue* arrows) in close relationship with proliferating uroepithelium (*white* arrows). Undecalcified sections cut at 4 μm after embedding into JB4 resin.

using ostephilic macroporous substrata or *smart* biomimetic matrices (Ripamonti et al. 1999; Ripamonti 2004; Ripamonti 2006) that *per se* initiate the spontaneous and/or intrinsic induction of bone formation.

    Reporting results after implantation of demineralized bone matrices in both heterotopic and orthotopic sites in different animal models including humans, Urist states that "the orthotopic system does not offer convincing evidence of induction". The demonstration of "Bone: formation by autoinduction" must be obtained after the heterotopic extraskeletal bioassay (Urist 1965; Reddi and Huggins 1972).

The fact that the orthotopic site is fallacious when reporting osteoinductive signals and biomaterial matrices, or a combination thereof, is still poorly understood today. Yet still a plethora of research manuscripts are published, hinting at the osteoinductive capacity of newly discovered proteins and/or biomimetic matrices (for reviews see Reddi 2000; Ripamonti et al. 2001; Ripamonti et al. 2009; Ripamonti 2017).

Importantly and conclusively, the three mammalian TGF-β isoforms do not initiate the induction of bone formation when implanted in extraskeletal sites of rodents (Roberts et al. 1986; Shinozaki et al. 1997; Reddi 2000). Similarly, heterotopic implantation of calcium phosphate-based biomimetic matrices in rodents does not result in the induction of bone formation within the macroporous spaces (for reviews see Ripamonti 2004; Ripamonti 2006; Ripamonti 2009; Ripamonti et al. 2009; Ripamonti 2017).

Contrary to rodents, lagomorphs and canines, the three mammalian TGF-β isoforms initiate the rapid and substantial induction of bone formation but in primates only (Ripamonti et al. 1997; Ripamonti et al. 2000; Ripamonti 2003; Ripamonti et al. 2008; Ripamonti and Roden 2010; Klar et al. 2014; Ripamonti et al. 2015).

The molecular cloning of the osteogenic proteins of the TGF-β supergene family (Ripamonti 2003) and the results in numerous pre-clinical studies in mammalian species including non-human primates have prematurely convinced molecular biologists, tissue engineers and skeletal reconstructionists alike that a single recombinant human bone morphogenetic protein (BMP) would result in clinically acceptable tissue induction and morphogenesis in human patients (Wozney et al. 1988; Celeste et al. 1990; Özkaynak et al. 1990; Sampath et al. 1990).

Purification to homogeneity of naturally derived BMPs from bovine and baboon bone matrices (Wang et al. 1988; Luyten et al. 1999; Ripamonti et al. 1992) allegedly resolved "The reality of a nebulous enigmatic myth" (Urist 1968). Molecular and pre-clinical research data have dispelled the myth and, as Lacroix hypothesized in the middle of last century, "the possibility of promoting osteogenesis at will is really within easy reach" in the current century (Lacroix 1945).

Expression cloning of the bone morphogenetic protein members of the TGF-β supergene family (Wozney et al. 1988; Celeste et al. 1990; Özkaynak et al. 1990) has failed, however, the translational research in clinical contexts (Ferretti et al. 2010; Ripamonti et al. 2012a; Ripamonti et al. 2014; Ripamonti 2017; Ferretti and Ripamonti 2020). Selected clinical trials and a series of case reports have shown limited and substandard induction of bone formation in clinical contexts in spite of the often highly successful pre-clinical animal studies in several animal models including non-human primates (Ferretti et al. 2010; Ripamonti 2006; Ripamonti et al. 2006; Ripamonti et al. 2007; Ripamonti et al. 2012a; Ripamonti et al. 2014).

This theoretical potential has not been translated into acceptable results in human patients. Clinical trials in craniofacial and orthopaedic applications such as mandibular reconstructions and sinus-lift operations have indicated that supra physiological doses of a single recombinant human protein are needed to induce often unacceptable tissue regeneration whilst incurring significant costs without achieving equivalence to autogenous bone grafts (Ripamonti et al. 2007; Ferretti et al. 2010; Ripamonti et al. 2012a; Ripamonti et al. 2014).

The need for alternatives to recombinant human bone morphogenetic proteins (hBMPs) is now felt more acutely following the reported complication and performance failure associated with the clinical use of hBMP-2 and hOP-1, also known as hBMP-7 (Ripamonti et al. 2012a; Ripamonti 2017; Seeherman et al. 2019). Reconstruction of large craniofacial and appendicular skeletal defects in humans still requires the harvesting of autogenous bone from a distant site, most often the iliac crest (Habal 1994). A further limitation is the finite volume of bone available from any donor site (Habal 1994; Burchardt 1997).

Supra physiological doses of a single recombinant hBMP are needed to often induce unacceptable tissue regeneration without achieving equivalence to autogenous bone grafts. We have shown that large bony defects, in particular severe mandibular discontinuities, can be re-engineered with autogenous bone grafts, highlighting the concept of clinically significant osteoinduction whereby the regenerated bone is readily identifiable on radiographic examination by virtue of its opacity and trabecular architecture (Fig. 2.1) (Ferretti et al. 2010; Ripamonti et al. 2014; Ferretti et al. 2016).

In an attempt to obviate the use of supra physiological doses of hBMPs in clinical contexts associated with local and systemic adverse effects (Seeherman et al. 2019), a BMP/activin A chimera has been recently developed particularly to reduce the amount of doses implanted in pre-clinical and clinical contexts (Seeherman et al. 2019). BV-265, a BMP-2/BMP-6/activin A chimera, delivered by calcium-deficient hydroxyapatite granules suspended in a macroporous fenestrated polymer mesh-reinforced recombinant type I collagen matrix, showed repair of critically sized fibular defects in non-human primate models at concentrations ranging from 1/10 to 1/30 of the hBMP-2/absorbable collagen sponge previously used in clinical trials (Seeherman et al. 2019).

2.2 MOLECULAR REDUNDANCY AND THE INDUCTION OF BONE FORMATION BY THE HUMAN TRANSFORMING GROWTH FACTOR-β_3

Are bone morphogenetic proteins, either naturally derived from mammalian bone matrices or recombinantly produced by DNA technology, the only proteins endowed with the striking prerogative of inducing bone formation when implanted heterotopically in animal models to set into motion the induction of bone formation where there is no bone? That is, in intramuscular or subcutaneous sites?

Heterotopic implantation of hTGF-β_3 in the non-human primate *Papio ursinus* combined with insoluble collagenous bone matrix as carrier results in the rapid and substantial induction of bone formation (Fig. 2.3) (Ripamonti et al. 2008).

What is the significance of this apparent molecular redundancy of several different yet molecularly homologous proteins initiating the induction of bone formation in primates by members of the TGF-β supergene family (Ripamonti et al. 1997; Ripamonti et al. 2000; Ripamonti 2003; Ripamonti et al. 2008; Klar et al. 2014; Ripamonti et al. 2015)?

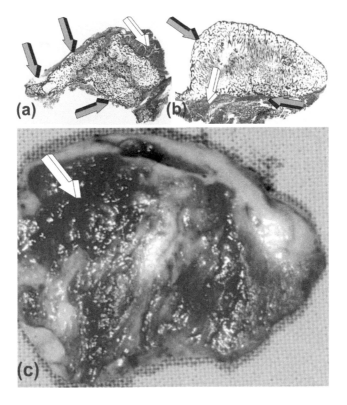

FIGURE 2.3   The remarkable and substantial induction of bone formation by the recombinant human transforming growth factor-β_3 (hTGF-β_3) on day 30 after heterotopic *rectus abdominis* intramuscular implantation in the chacma baboon *Papio ursinus* (Ripamonti et al. 2008). (a,b) Undecalcified sections of large ossicles generated by reconstituting 125 μg hTGF-β_3 with allogeneic baboon insoluble collagenous matrix (ICBM), the chaotropically extracted and inactive collagenous bone matrix. The newly generated intramuscular ossicle shows pronounced corticalization of the newly formed bone (*light blue* arrows) embedded within the *rectus abdominis* muscle (*white* arrows). The newly formed corticalized structures surround scattered remnants of the collagenous matrix as carrier facing trabeculae of newly formed mineralized matrix covered by osteoid seams. (c) Tissue induction and morphogenesis of large heterotopic ossicles within the *rectus abdominis* muscle (*white* arrow). (a,b) Undecalcified sections cut at 3 to 4 μm (Leica SM2500 Polycut-S, Reichert, Heidelberg, Germany) from K-Plast embedded tissue blocks stained free-floating with modified Goldner's trichrome (Ripamonti et al. 2008).

Recombinant hTGF-β_3 induces substantial bone formation when pre-combined with biphasic hydroxyapatite/β-tricalcium phosphate (Fig. 2.4b) or coral-derived macroporous bioreactors (Figs. 2.4c,d,e) and implanted in the *rectus abdominis* muscle of the chacma baboon *Papio ursinus* (Ripamonti et al. 2012a; Klar et al. 2014; Ripamonti et al. 2014; Ripamonti et al. 2015). Of significance, the generated heterotopic constructs, though supported by significantly different carrier substrata, i.e.

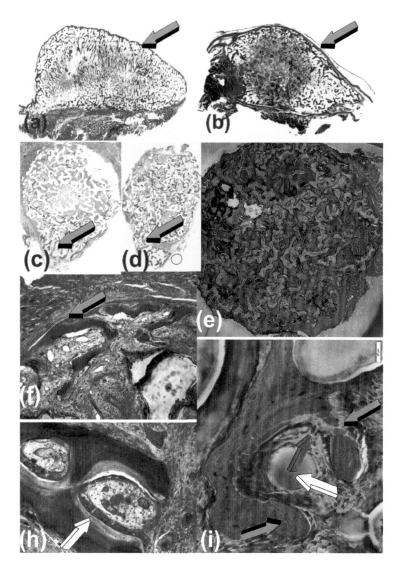

FIGURE 2.4 Iconographic landscape of the induction of bone of bone formation by the recombinant human transforming growth factor-β_3 (hTGF-β_3) after reconstitution with a variety of carriers and delivery systems for expression of the biological activity of the recombinant morphogen. (a,b) Iconographic comparison of 125 μg hTGF-β_3 reconstituted with (a) allogeneic insoluble bone matrix (ICBM) and (b) a macroporous biphasic hydroxyapatite/β-tricalcium phosphate (HA/β-TCP) (Ripamonti et al. 2012a; Ripamonti et al. 2014). Note the similar morphological appearance of the generated ossicles in spite of the significant differences in the carrier matrix used as delivery system with corticalization of both ossicles (*light blue* arrows). Note in (b) the explosive pattern of bone growth far beyond the boundaries of the implanted biphasic hydroxyapatite/β-tricalcium phosphate (HA/β-TCP) constructs. (c,d) Substantial induction of bone formation by 125 μg hTGF-β_3 reconstituted

FIGURE 2.4 (CONTINUED)

with coral-derived macroporous constructs harvested on days 60 (c) and 90 (d) after *rectus abdominis* implantation (Klar et al. 2014; Ripamonti et al. 2014). Arrows indicate the exuberant induction of bone formation outside the periphery of the heterotopically implanted macroporous constructs (*light blue* arrows c,d). (e) Low-power view of a coral-derived macroporous construct recombined with 125 μg hTGF-β₃ and harvested on day 90 after *rectus abdominis* implantation (Ripamonti et al. 2015). (f,h) Coral-derived constructs reconstituted with 250 μg hTGF-β₃ showing the rapid and substantial induction of bone formation at the periphery of the implanted macroporous constructs into the *rectus abdominis* muscle (*light blue* arrow f). (h) Detail of newly formed mineralized bone with osteoid seams populated by contiguous osteoblasts (*white* arrow). (i) Detail of a coral-derived construct super activated by 125 μg hTGF-β₃ harvested on day 90 displaying the critical role of angiogenesis in osteogenesis (Trueta 1963). Plasticity of the newly formed bone (*light blue* arrows) with a prehensile trabecula of bone with contiguous osteoblasts (*orange* arrow) surrounding the vascular central canal (*white* arrow). The digital image highlights the prehensile strength of the newly formed bone matrix around the central vascular canal highlighting the continuous molecular cross talk between the angiogenic and osteogenic compartments of the newly formed tissues by the hTGF-β₃ isoform (Ripamonti et al. 2015). (a,b) Undecalcified sections. (a) K-Plast sections. (c,d) Decalcified sections cut at 4 to 5 μm and stained with toluidine blue in 30% ethanol. (b,e,f,h,i) Undecalcified sections prepared by the Exakt cutting and grinding diamond saw system (Ripamonti et al. 2015). Sections were ground and polished to 30 μm stained with methylene blue basic fuchsin.

insoluble collagenous bone matrix (Figs. 2.3a,b, 4a) vs. biphasic hydroxyapatite/β-tricalcium phosphate (Fig. 2.4b), show remarkably similar patterns of growth and expansion within the *rectus abdominis* muscle, possibly reflecting a memory of developmental events recapitulated in post-natal osteogenesis.

The induction of bone formation by the hTGF-β₃ in the chacma baboon *Papio ursinus* has been molecularly resolved by correlating tissue induction and morphogenesis as evaluated on undecalcified sections to a time course molecular analysis. Profiled gene expression pathways by qRT-PCR showed that the induction of bone formation as initiated by the hTGF-β₃ isoform is *via* several profiled *BMPs'* genes and gene products expressed upon the heterotopic implantation of the hTGF-β₃ isoform (Klar et al. 2014; Ripamonti et al. 2014; Ripamonti et al. 2015; Ripamonti 2016; Ripamonti et al. 2016a).

Our continuous experimentation has thus shown that the rapid induction of bone formation by the hTGFβ₃ isoform in heterotopic intramuscular sites of *Papio ursinus* is *via* the expression of the *BMPs'* pathway, with hTGF-β₃ controlling the induction of bone formation by regulating the expression of *bone morphogenetic proteins* genes *via* Noggin expression (Klar et al. 2014; Ripamonti et al. 2014; Ripamonti et al. 2015; Ripamonti 2016). hTGF-β₃ elicits the induction of bone formation by upregulating endogenous bone morphogenetic proteins, and it is blocked by hNoggin, providing insights into performance failure of hBMPs in clinical contexts (Ripamonti et al. 2014; Ripamonti 2016).

Physiological expression of endogenous *BMPs'* genes and gene products upon the direct implantation of hTGF-β₃ may escape the antagonistic expression of Noggin and other inhibitors. In contrast, the direct application of high doses in mg amounts of hBMPs, representing a later by-product molecular step of the bone induction cascade as set by the hTGF-β_3 master gene in primates, set into motion Noggin antagonist action and other inhibitory genes and gene products as shown by the limited effectiveness of hBMPs in clinical contexts (Ferretti et al. 2010; Ripamonti et al. 2014; Ripamonti et al. 2015; Ferretti and Ripamonti 2020).

The induction of bone formation by the hTGF-β_3 isoform, upregulating the endogenous expression of several genes and gene products of the BMP family (Ripamonti et al. 2015), has only been shown in primates. In marked contrast, the mammalian TGF-β isoforms are not endowed with the striking prerogative of inducing bone formation in lower animal species, including rodents, lagomorphs and canines (Ripamonti et al. 1997; Ripamonti et al. 2008; Ripamonti et al. 2015).

The physiological expression of endogenous *BMPs'* genes and gene products upon the direct implantation of doses of hTGF-β_3 isoform is challenging the *status quo* of "Bone: formation by autoinduction" (Urist 1965), providing unique human biology data (Cell Editorial 2014) and necessitating to re-define the induction of bone formation in primate species, including *Homo sapiens* (Ripamonti et al. 2014; Ripamonti 2016).

2.3 TRANSLATIONAL RESEARCH ON THE RECOMBINANT TRANSFORMING GROWTH FACTOR-β_3 IN HUMAN PATIENTS

The induction of bone formation by 125 μg hTGF-β_3 in full thickness segmental mandibular defects of *Papio ursinus* with the induction of corticalized buccal and lingual newly formed plates (Ripamonti and Ferretti 2016) has called for translational research in human patients affected by massive mandibulo-facial pathology requiring extensive surgical debridement and tissue regeneration (Ferretti and Ripamonti 2020).

Of interest, however, the implantation of equal doses of hTGF-β_3 in non-healing calvarial defects in the same animals also implanted with heterotopic doses of the recombinant morphogen failed to induce calvarial regeneration (Ripamonti et al. 2016b). Our morphological and molecular studies showed complex gene expression pathways responsible for the limited induction of bone regeneration in calvarial defects, with however partial restoration of the bone induction cascade following compartmentalization and segregation of the calvarial defect unit excluding or segregating the *dura mater* during healing with a supra-dural membrane (Ripamonti et al. 2016b).

hTGF-β_3-treated and segregated bioreactors showed the induction of bone formation across the calvarial defects together with upregulation of *OP-1, BMP-2, Osteocalcin, RUNX-2, ID2* and *ID3* (inhibitor of DNA binding-2 and -3) (Ripamonti et al. 2016b). Taken together, our morphological and molecular data after undecalcified bone sectioning and qRT-PCR analyses of segregated vs. non-segregated and hTGF-β_3-treated vs. hTGF-β_3/untreated coral-derived bioreactors showed the critical role of the *dura mater*. Indeed, the *dura mater* is a common regulator, or tissue-controlling centre, of the calvarium homeostasis (Gosain et al. 2003; Gosain et al. 2011; Gosain et al. 2000; Kwan et al. 2008; Cooper et al. 2011; Takagi and Urist 1982).

Tissue segregation by the insertion of a supra-dural membrane disrupts a number of molecular and cellular signals from the underlying dura mater, and possibly from the highly vascularized leptomeninges below, effectively blocking the bone induction cascade as seen molecularly and morphologically (Ripamonti et al. 2016b). Indeed previous experiments in *Papio ursinus* with identical doses of the recombinant morphogen delivered by insoluble collagenous bone matrices showed that hTGF-β_3-treated non-segregated specimens upregulated the *Smad-6* and *Smad-7* pathways, correlating with minimal induction of bone formation (Ripamonti et al. 2008).

The implantation of coral-derived bioreactors in non-segregated defects pre-combined with relatively high doses of the recombinant morphogen, i.e. 125 µg hTGF-β_3, disrupted the homeostasis of the calvarial wound, altering the molecular and cellular expression patterns of the dura, setting into motion downregulation of the profiled osteogenic genes, blocking the induction of bone formation (Ripamonti et al. 2016b).

The substantial biological activity of the hTGF-β_3 isoform in heterotopic *rectus abdominis* intramuscular sites of *Papio ursinus* together with the failure of predictable tissue induction by hBMPs in mandibular defects of human patients, has prompted us to use a cohort of paediatric patients to reconstruct mandibular defects of such a size to preclude reconstruction with autologous bone (Ferretti and Ripamonti 2020). Large mandibular defects were implanted with 250 µg of recombinant hTGF-β_3 osteogenic device (Ferretti and Ripamonti 2020). Patients were followed up for 3 to 6 years. Three patients achieved clinically significant osteoinduction, one patient with hTGF-β_3 and two by combining hTGF-β_3 with a small supplement of autogenous bone (Ferretti and Ripamonti 2020). We concluded by stating that the hTGF-β_3 osteogenic device delivered by human demineralized bone matrix initiates the induction of bone formation in mandibular defects (Fig. 2.5) and potentiates autologous bone graft activity in large mandibular defects in paediatric patients (Ferretti and Ripamonti 2020).

Yet the biological activity of the hTGF-β_3 osteogenic device is not predictable as in *Papio ursinus* when implanted in human patients (Ferretti and Ripamonti 2020). The induction of bone formation by the hTGF-β_3 implanted in the *rectus abdominis* of *Papio ursinus* is truly substantial (Ripamonti et al. 2008; Klar et al. 2014; Ripamonti et al. 2015; Ripamonti 2016). This prominent induction of bone formation by hTGF-β_3 applied singly resides in the upregulation of selected genes involved in tissue induction and morphogenesis, and in particular *Osteocalcin, RUNX-2, OP-1, BMP-3, BMP-2, TGF-β_1, TGF-β_3*, with, however, of note, a lack of *TGF-β_2* upregulation (Ripamonti et al. 2015; Ripamonti et al. 2016a). Remarkably, we have shown that the induction of bone formation when applying doses of hTGF-β_3 singly recapitulates the synergistic induction of bone formation as shown by binary applications of hOP-1 with hTGF-β_3 at the ratio by weight of 20:1 hOP-1:hTGF-β_3 (Ripamonti et al. 2010; Ripamonti et al. 2016a) or 20:1 hOP-1: hTGF-β_1 (Ripamonti et al. 1997).

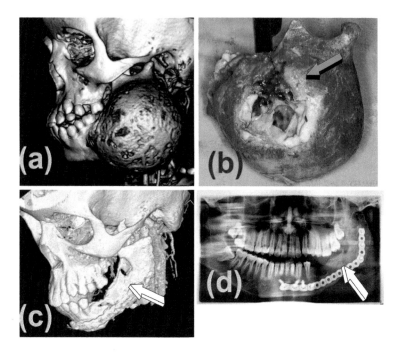

FIGURE 2.5   From the bench-top and pre-clinical studies in the chacma baboon *Papio ursinus* (Ripamonti et al. 2008; Klar et al. 2014; Ripamonti et al. 2015; Ripamonti et al. 2016) highlighting the mechanistic insights of the as yet unreported induction of bone formation by the hTGF-β_3 isoform in primates, to translational research into human patients affected by severe mandibular discontinuities implanted with 250 μg of hTGF-β_3 osteogenic device (Ferretti and Ripamonti 2020). (a,b) Tri-dimensional CT scan revealing a massive lesion on the left mandibular ramus and body. (c) μCT tomogram of the reconstituted mandible (*white* arrow) 6 months after implantation of 250 μg of the hTGF-β_3 osteogenic device. (d) Panoramic radiograph of the engineered mandible (*white* arrow) 6 years after reconstruction with the hTGF-β_3 osteogenic device.

2.4 MOLECULAR MICROENVIRONMENTS SUPER ACTIVATED BY MORPHOGENETIC SIGNALS

Experiments in the *rectus abdominis* muscle of the chacma baboon *Papio ursinus* using either insoluble collagenous bone matrices or coral-derived bioreactors super activated by single doses of the hTGF-β_3 isoform have often shown limited or absent bone differentiation in the central or internal regions of the carriers used to generate heterotopic ossicles (Figs. 2.3, 2.4) (Ripamonti et al. 1997; Ripamonti et al. 2010; Klar et al. 2014; Ripamonti et al. 2015; Ripamonti et al. 2016a). There is, however, substantial induction at the periphery of the implanted bioreactors, replicating the induction of bone formation and the maturational gradient of the synergistic induction of bone formation (Fig. 2.4) (Ripamonti et al. 1997; Ripamonti et al. 2010; Ripamonti et al. 2016a; Ripamonti 2016).

Importantly, we have shown that hTGF-β_3 combined with either insoluble collagenous bone matrices, coral-derived bioreactors or biphasic sintered calcium phosphate-based constructs (Ripamonti et al. 2008; Ripamonti et al. 2010; Ripamonti et al. 2012a; Ripamonti et al. 2016a) molecularly and morphologically equates to the synergistic induction of bone formation whereby a single human recombinant osteogenic protein is significantly super activated three- to four-fold by the binary application of relatively low doses of a hTGF-β isoform (Ripamonti et al. 1997; Ripamonti et al. 2010; Ripamonti 2016; Ripamonti et al. 2016a).

Our studies on the "bone induction principle" (Urist et al. 1967) in the past several years have also focused on the spontaneous and/or intrinsic induction of bone formation regulated by the geometry of *smart* biomimetic calcium phosphate-based bioreactors (Ripamonti et al. 1999; Ripamonti 2017).

Osteoclastogenesis results in nanotopographic calcium phosphate-based geometric configurations with Ca^{++} release within the protected microenvironment of the concavities formed by sintering the macroporous spaces (Ripamonti et al. 2010; Klar et al. 2013; Ripamonti 2017). The induction of angiogenesis together with available Ca^{++} release ultimately initiates stem cell differentiation and the induction of osteoblastic-like cells (Klar et al. 2013; Ripamonti 2017). Differentiated osteoblasts express and secrete osteogenic proteins of the TGF-β supergene family to be embedded onto the matrix, initiating bone formation as a secondary response (Ripamonti et al. 1993; Ripamonti et al. 1999; Klar et al. 2013; Ripamonti 2017).

The "geometric induction of bone formation" (Ripamonti et al. 1999) presents unique molecular and morphological characteristics when compared to the classic experiments describing "Bone: formation by autoinduction" (Levander 1945; Levander and Willestaedt 1946; Urist 1965; Reddi and Huggins 1972; Reddi 2000). The above experiments report the induction of bone formation by a diffusible morphogen, a BMP complex, from the demineralized matrix of bone in contact with infiltrating responding cells between the demineralized particles. Alternatively (reviewed by Ripamonti and Reddi 1995, Reddi 2000 and Ripamonti et al. 2004), the osteogenic soluble molecular signals are reconstituted with an insoluble signal or substratum to initiate the ripple-like cascade of the induction of bone formation (Ripamonti and Reddi 1995; Reddi 2000; Ripamonti et al. 2004).

The above reconstitution has been and still is the molecular paradigm of regenerative medicine and tissue engineering at large with prominent results in pre-clinical contexts (Figs. 2.3, 2.4) (Khouri et al. 1991; Ripamonti et al. 2000; Ripamonti et al. 2004; Ripamonti et al. 2016a). This CRC Press contribution to the "geometric induction of bone formation" (Ripamonti et al. 1999) wishes to present and review a modified regenerative paradigm in which the very insoluble signal resorbs *via* a downstream of molecular and cellular cascades that sculpt resorption pits and lacunae in the geometric form of concavities within the implanted tricalcium phosphate/hydroxyapatite biomatrices (Ripamonti et al. 2008b). The concavities, as sculpted within the biomimetic matrices, initiate the induction of bone differentiation as a secondary response (Ripamonti et al. 1999; Ripamonti 2004; Ripamonti 2006; Ripamonti et al. 2008b; Ripamonti 2017).

There is a "continuum" between the soluble and solid states of the newly formed bone as in the skeleton (Reddi 1997) in which the "continuum" is regulated by signals in solution interacting with the insoluble extracellular matrix (Reddi 1997). Embedded secreted osteogenic proteins within the concavities of the biomimetic matrices initiate the induction of bone formation as a secondary response (Ripamonti 2004; Ripamonti 2017; Ripamonti 2018).

2.5 DIFFUSIBLE SIGNALS, MICROENVIRONMENTS AND THE ESTABLISHMENT OF MORPHOGENETIC GRADIENTS

Finally, it is important to conclude this chapter with why and when the induction of bone forms and sometime repairs and regenerates by presenting the last three iconographic plates that morphologically describe the induction of morphogenetic gradients across boundaries of different matrices and substrata, initiating selected tissue induction and morphogenesis. Tissue induction recapitulates embryonic development more so, however, possibly recapitulating the molecular and morphogenetic matrix that *ab initio* set into motion the cascade of cellular proliferation, differentiation and finally the induction of complex highly specialized tissues and matrices such as cartilage and bone.

Figure 2.6 shows the induction of morphogenetic microenvironments across the substratum of coral-derived bioreactors with the initiation of morphogenetic gradients as established by the induction of collagenous condensations across the substrata (Figs. 2.7a,d). Synchronously, there is also the establishment of the morphogenetic, osteogenetic and differentiating "osteogenetic vessels" of Trueta's definition (Trueta 1963) in close relationship with the inductive substratum (Figs. 2.7b,c). Such morphogenetic inductive gradients finally lead to the spontaneous and/or intrinsic induction of bone formation tightly adhered to the inductive substratum by day 60 after heterotopic implantation (Fig. 2.7e) (Ripamonti et al. 1993).

We believe that the induction and the remodelling of the collagenous condensations make up a fundamental morphogenetic module across animal phyla, morphogenizing tissue induction and morphogenesis across the mammalian skeletons.

Kusumbe et al. (2014) re-visited the osteogenetic vessels of Trueta's definition (Trueta 1963), identifying a specific vessel sub-type, type H endothelial cells, in bone to couple angiogenesis to osteogenesis, maintaining perivascular osteoprogenitors. Additional studies also indicated the critical role of Notch expression and activity promoting angiogenesis and osteogenesis in bone (Ramasamy et al. 2014).

The critical role of angiogenesis in tissue morphogenesis and regeneration has been shown by a series of communications describing the vast plasticity of the endothelial cells to construct morphogenesis *via* secreted signals during regeneration (Ramasamy et al. 2015; Gomez-Salinero and Raffi 2018; Kovacic and Boehm 2009; Benjamin et al. 1988).

Other studies have shown the plasticity of the endothelium to convert into multipotent stem-like cells (Medici et al. 2011). Finally, recent studies have indicated that the "archetypal multipotent" (Crisan et al. 2008) mesenchymal stem cells (MSCs) are of perivascular origin, as the pericytes, indicating that the vessel's walls harbour

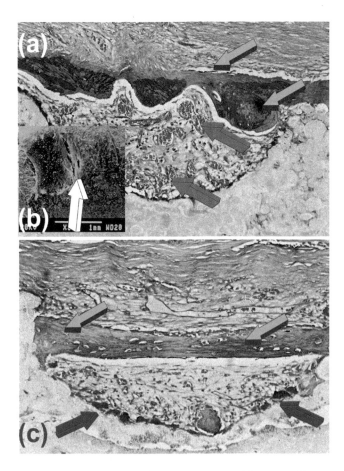

FIGURE 2.6 Morphogenetic microenvironments generated by concavities carved in solid discs of crystalline sintered hydroxyapatites on both planar surfaces (Ripamonti et al. 1999). (a,c) Bridges of mineralized bone form across the margins of concavities of crystalline sintered hydroxyapatites (Ripamonti et al. 1999; Ripamonti et al. 2012b). (b) Collagenous condensations first form across the edges of the concavities (*white* arrow) 5 days after heterotopic implantation (Ripamonti et al. 2012b). The established microenvironment and prominent angiogenesis (*pink* arrows a) surrounding the newly formed bone (*light blue* arrows) induce constructs bridging the edges of the concavities 90 days after heterotopic *rectus abdominis* implantation. Osteoclastic activity (*dark blue* arrows in c) initiates resorption with Ca⁺⁺ release, angiogenesis, cell differentiation and the induction of bone formation as a secondary response (Klar et al. 2013; Ripamonti et al. 2014; Ripamonti 2017).

progenitor pericytic cells capable of differentiation including the osteogenic phenotype (Crisan et al. 2008; Chen et al. 2009).

Tractional fields across the concavity microenvironments are shown in Figure 2.6, highlighting the induction of tractional forces across the margins of the concavities and the induction of bone formation by day 90 after heterotopic implantation of crystalline sintered hydroxyapatite discs with concavities on both planar surfaces (Fig. 2.6). Inset b

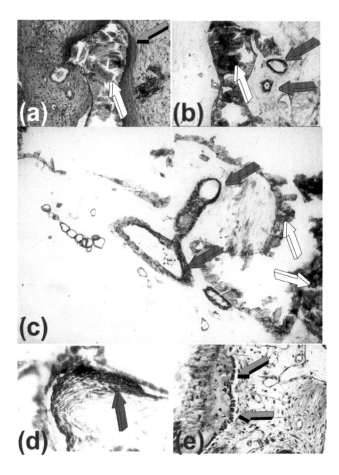

FIGURE 2.7 Early morphogenetic inductive events initiating the spontaneous and/or intrinsic induction of bone formation by coral-derived macroporous bioreactors when implanted in the *rectus abdominis* muscle of the chacma baboon *Papio ursinus* and harvested on days 30 and 60 (Ripamonti et al. 1993a; Ripamonti 2017). (a) Induction of mesenchymal collagenous condensations (*light blue* arrow) on day 30 along the profile of the implanted calcium carbonate/hydroxyapatite replica. (b) Serial section and specular image of (a) showing alkaline phosphatase staining of invading capillaries surrounding the implanted coral-derived construct (*magenta* arrows). White arrows in (a) and (b) indicate the heterotopically implanted coral-derived bioreactors. (c) Invading capillaries into the macroporous spaces of the coral-derived bioreactors (*white* arrows) showing prominent alkaline phosphatase staining on the capillary tunics (*magenta* arrows). (d) Prominent alkaline phosphatase staining (*magenta* arrow) of collagenous condensations against the coral-derived biomatrix 60 days after heterotopic implantation. (e) Induction of bone formation, mineralization of the newly formed bone and osteoid synthesis by multiple contiguous osteoblasts (*light blue* arrows) 60 days after heterotopic implantation. Note the tight adhesion of the newly formed mineralized bone attached to the coral-derived mineralized substratum. Undecalcified sections from tissue blocks embedded in K-Plast methyl-methacrylate with section cut at 3 to 5 μm stained with toluidine blue in 30% ethanol or with Goldner's trichrome stain (Ripamonti et al. 1993a).

shows collagen bundles extending between the edges of a concavity (Fig. 2.6b) as early as 5 days after intramuscular heterotopic implantation of geometric hydroxyapatite-coated titanium implants (Ripamonti et al. 2012a and Chapter 7 of this volume).

The established tractional and morphogenetic fields across the concavities of the substratum remarkably induce the differentiation of bone across the collagenous condensations with newly formed membranous bone bridging the edges of the concavities 90 days after heterotopic implantation in the *rectus abdominis* muscle of *Papio ursinus* (Figs. 2.6a,c).

Perhaps however the most remarkable morphogenetic microenvironment is shown in Figure 2.8. Two adjacent coral-derived bioreactors preloaded with doses of 125 μg recombinant human transforming growth factor-β_3 (hTGF-β_3) implanted in the *rectus abdominis* muscle show the differentiation and the induction of heterotopic endochondral ossicles (Fig. 2.8a *white* arrow). There is prominent induction of cartilage as a precursor anlage (Fig. 2.8a *magenta* arrow) just above the *rectus abdominis* fascia and between the two intramuscularly implanted super-activated bioreactors (Fig. 2.8).

There is no doubt that the extended range of biological activity is due to the establishment of a diffusion gradient and to the induction of a sequential chain of cellular and molecular induction that results in the initiation of a cartilage anlage with bone differentiation 2–3 cm away from the implanted carriers.

The establishment of a rapid morphogenetic hTGF-β_3 gradient, so far unreported as yet by any gene products of the TGF-β supergene family members, is possibly further potentiated by cell-to-cell receptor activation and by the establishment of a TGF-β_3 morphogen gradient by long-range activation.

The induction of an endochondral ossicle with clear-cut cartilage differentiation is certainly not due to release of the TGF-β_3 signal from the previously coated calcium phosphate-based constructs. Rather the endochondral ossicle formed after the establishment of a long-range activity of the hTGF-β_3 isoform attached to the calcium phosphate-based bioreactors.

Remarkably, the induction of bone forms *via* endochondral bone formation within the *rectus abdominis* muscle of the heterotopically implanted *Papio ursinus*. Figure 2.8 once again shows that the induction of a cartilage anlage (Fig. 2.8a) is a *conditio sine qua non* for the induction of bone formation even after the establishment of a diffusion gradient across boundaries of the calcium phosphate-based bioreactors and the *rectus abdominis* muscle.

The cartilage anlage thus retains a memory of developmental events by the expression of the *TGF-β_3* gene and gene products masterminding the origin of the cartilaginous matrices. It is noteworthy that high-power analyses of the cartilage anlage generating the induction of bone formation between the implanted coral-derived bioreactors show the presence of columnar elements, recapitulating the basic features of the mammalian growth plate (Fig. 2.8b).

Phylogenetically ancient genes and gene products are continuously deployed through evolution and differentiation of animal phyla *via* speciating events obeying the overall plan of Nature's parsimony in controlling tissue induction and morphogenesis.

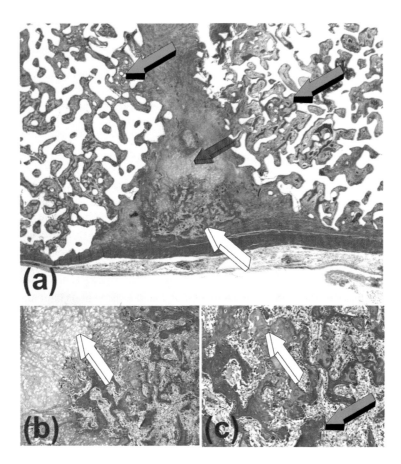

FIGURE 2.8 Induction of heterotopic endochondral ossicles spontaneously initiating within the *rectus abdominis* muscle of the chacma baboon *Papio ursinus* after implantation of coral-derived constructs super activated by 125 µg recombinant human transforming growth factor-β_3 (hTGF-β_3). (a) The heterotopic ossicle (*white* arrow) forms after induction of a cartilage anlage replicating the organizational structure of the cartilage growth plate. The inductive microenvironment is established by diffusible signals from the super-activated bioreactors previously combined with doses of hTGF-β_3. (b, c) Details of the newly generated cartilage anlage with columns of proliferating chondrocytes (*white* arrow in b). (c) Prominent induction of trabecular bone just above the peritoneal fascia (*white* arrow) with multiple osteoblastic cells secreting osteoid matrix along the trabeculae of the newly formed bone (*light blue* arrow).

Nature's creation of master genes including the *TGF-β_3* gene and gene product is to continuously deploy pleiotropic multifaceted genes for fundamental morphogenetic and differentiating pathways (Ripamonti 2019).

Indeed, we showed that tissue folding and arching of the continuously erupting dental lamina of the selachian *Carcharinus obscurus* dusky shark is characterized by *TGF-β_3* expression (Ripamonti et al. 2018; Ripamonti 2019).

Tissue bending by tractional forces and more particularly tissue folding is a developmental morphogenetic module evolutionarily shared by choanoflagellates with a myosin-mediated morphogenesis. This has been conserved through evolution, sharing myosin-mediated morphogenetic events including tissue bending as an hallmark feature of animal morphogenesis (Brunet et al. 2019; Tomancak 2019).

We conclude this chapter on "When and Why Bone Forms and Sometimes Repairs and Regenerates" by once again stating that the pleiotropism of the *TGF-β₃* gene controls tissue induction and morphogenesis across animal phyla and speciation from phylogenetically ancient animals to the human primate *Homo sapiens*. TGF-β also regulates the mechanical properties and composition of the bone matrix (Balooch et al. 2005). Morphogenetic pathways include mineralized tesserae in extant sharks *Carcharinus obscurus*, and folding of the continuously erupting dental lamina in dusky sharks with *TGF-β₃* expression during folding (Ripamonti et al. 2018).

The TGF-β₃ master gene and gene product mastermind osteogenesis in mammals with prominent induction of endochondral bone in the non-human primate *Papio ursinus*. *TGF-β₃* further controls the induction of cementogenesis from Diadectidae, stem mammals and mosasaurs to the chacma baboon *Papio ursinus* (Ripamonti 2019), *via* the induction of prominent cementogenesis coupled with angiogenetic signals expressed by the newly formed almost trabecular-like induced cementum in extant primates (Ripamonti et al. 2017). TGF-β₃ masterminds the induction of the vascularized tri-dimensional structure of the periodontal ligament spaces in primates with the induction of stemness from multifunctional vascular and perivascular stem cell niches (Ripamonti 2019).

REFERENCES

Badylak, S.; Rosenthal, N. Regenerative Medicine: Are We There Yet? *NPJ Regen. Med.* 2017, *201*(2), 2. doi:10.1038/s41536-016-0005-9.

Balooch, G.; Balooch, M.; Nalla, R.K.; Schilling, S.; Filvaroff, E.H.; Marshall, G.W.; Marshall, S.J.; Ritchie, R.O.; Derynck, R.; Alliston, T. TGF-Beta Regulates the Mechanical Properties and Composition of the Bone Matrix. *Proc. Natl. Acad. Sci. U.S.A.* 2005, *102*(52), 18813–18.

Baptista, P.M.; Atala, A. Regenerative Medicine: The Hurdles of Hope. *Transl. Res.* 2014, *163*(4), 255–58.

Bely, A.E. Evolutionary Loss of Animal Regeneration: Patterns and Process. *Integr. Comp. Biol.* 2010, *50*(4), 515–27.

Bely, A.E.; Nyberg, K.G. Evolution of Animal Regeneration: Re-Emergence of a Field. *Trends Ecol. Evol. (Amst.)* 2010, *25*(3), 161–70.

Benjamin, L.E.; Hemo, I.; Keshet, E. A Plasticity Window for Blood Vessels Remodelling Is Defined by Pericyte Coverage of the Preformed Endothelial Network and Is Regulated by PDGF-B and VEGF. *Development* 1988, *125*(9), 1591–98.

Brunet, T.; Larson, B.T.; Linden, T.A.; Vermeij, M.J.A.; McDonald, K.; King, N. Light-Regulated Collective Contractility in a |Multicellular Choanoflagellate. *Science* 2019, *366*(6463), 326–34.

Burchardt, H. Biology of Bone Transplantation. *Orthop. Clin. North Am.* 1987, *18*(2), 187–96.

Celeste, A.J.; Iannazzi, J.M.; Taylor, J.A.; Hewick, R.C.; Rosen, V.; Wang, E.A.; Wozney, J.M. Identification of Transforming Growth Factor-β Family Members Present in Bone-Inductive Protein Purified from Bovine Bone. *Proc. Natl. Acad. Sci. U.S.A.* 1990, *87*(24), 9843–47.

Cell Editorial. Pulling It All Together. *Cell* 2014, *157*(1), 1–2. doi:10.1016/j.cell.2014.03.22.

Chen, C.-W.; Montelatici, E.; Crisan, M.; Corselli, M.; Huard, J.; Lazzari, L.; Péault, B. Perivascular Multi-Lineage Progenitor Cella in Humans Organs: Regenerative Units, Cytokine Sources or Both? *Cytokine Growth Factor Rev.* 2009, *20*(5–6), 429–34.

Cooper, G.M.; Mooney, M.P.; Gosain, A.K.; Campbell, P.G.; Losee, J.; Huard, J.E. Testing the Critical Size in Calvarial Bone Defects; Revisiting the Concept of a Critical-Size Defect. *Plast. Reconstr. Surg.* 2011, *6*, 1685–92.

Crisan, M.; Yap, S.; Casteilla, L. et al. A Perivascular Origin for Mesenchymal Stem Cells in Multiple Human Organs. *Cell Stem Cell* 2008, *3*(3), 301–13.

Ferretti, C.; Ripamonti, U.; Tsiridis, E.; Kerawala, C.J.; Mantalaris, A.; Heliotis, M. Osteoinduction: Translating Preclinical Promises into Reality. *Br. J. Oral Maxillofac. Surg.* 2010, *48*(7), 536–39.

Ferretti, C.; Muthray, E.; Rikhotso, E.; Reyneke, J.; Ripamonti, U. Reconstruction of 56 Mandibular Defects with Autologous Compressed Particulate Cortico-Cancellous Bone Grafts. *Br. J. Oral Maxillofac. Surg.* 2016, *54*(3), 322–26. doi:10.1016/j.bjoms.2015.12.014.

Ferretti, C.; Ripamonti, U. Long Term Follow-Up of Pediatric Mandibular Reconstruction with Human Recombinant Transforming Growth Factor-β_3. *J. Craniofac. Surg.* 2020, *31*(5), 1424–1429. doi:10.1097/SCS.0000000000006568.

Friedenstein, A.Y. The Humeral Nature of the Osteogenic Activity of Transitional Epithelium. *Exp. Biol.* 1962, *54*, 1385–87.

Friedenstein, A.Y. Induction of Bone Tissue by Transitional Epithelium. *Clin. Orthop. Rel. Res.* 1968, *59*, 21–37.

Gomez-Salinero, J.M.; Rafii, S. Endothelial Cell Adaptation in Regeneration. *Science* 2018, *362*(6419), 1116–17.

Gosain, A.K.; Song, L.S.; Yu, P.; Mehrara, B.J.; Maeda, C.Y.; Gold, L.I.; Longaker, M.T. Osteogenesis in Cranial Defects: Reassessmentof the Concept of Critical Size and the Expression of TGF-β Isoforms. Plast. Recontr. *Surg.* 2000, *106*(2), 360–72.

Gosain, A.K.; Santoro, T.D.; Song, L.S.; Capel, C.C.; Sudhakar, P.V.; Matloub, H.S. Osteogenesis in Calvarial Defects: Contribution of the Dura, the Pericranium, and the Surrounding Bone in Adult *versus* Infant Animals. *Plast. Recontr. Surg.* 2003, *112*(2), 515–27.

Gosain, A.K.; Gosain, S.A.; Sweeney, W.M.; Song, L.S.; Amarante, M.T.J. Regulation of Osteogenesis and Survival within Bone Grafts to the Calvaria: The Effect of the Dura *versus* the Pericranium *Plast. Recontr. Surg.* 2011, *128*(1), 85–94.

Habal, M.B. Bone Grafting in Craniofacial Surgery. *Clin. Plast. Surg.* 1994, *21*(3), 349–63.

Huggins, C.B. The Formation of Bone under the Influence of Epithelium of the Urinary Tract. *Arch. Surg.* 1931, *22*, 377–408.

Khouri, R.K.; Koudsi, B.; Reddi, H. Tissue Transformation In Vivo. A Potential Practical Application. *JAMA* 1991, *266*(14), 1953–5.

King, R.S.; Newmark, P.A. The Cell Biology of Regeneration. *J. Cell. Biol.* 2012, *196*(5), 553–62.

Klar, R.M.; Duarte, R.; Dix-Peek, T.; Dickens, C.; Ferretti, C.; Ripamonti, U. Calcium Ions and Osteoclastogenesis Initiate the Induction of Bone Formation by Coral-Derived Macroporous Constructs. *J. Cell. Mol. Med.* 2013, *17*(11), 1444–57.

Klar, M.R.; Duarte, R.; Dix-Peek, T.; Ripamonti, U. The Induction of Bone Formation by the Recombinant Human Transforming Growth Factor-β3. *Biomaterials* 2014, *35*(9), 2773–88. doi:10.1016/j.biomaterials.2013.12.062.

Kovacic, J.C.; Boehm, M. Resident Vascular Progenitor Cells: An Emerging Role for Non-Terminally Differentiated Vessel-Resident Cells in Vascular Biology. *Stem Cell Res.* 2009, *2*(1), 2–15.

Kusumbe, A.P.; Ramasamy, S.K.; Adams, R.H. Coupling of Angiogenesis and Osteogenesis by a Speficic Vessel Subtype in Bone. *Nature* 2014, *507*(7492), 323–28.

Kwan, M.D.; Wan, D.C.; Wang, Z.; Gupta, D.M.; Slater, B.J.; Longaker, M.T. Microarray Analysis of the Role of Regional Dura Mater in Cranial Suture Fate. *Plast. Reconstr. Surg.* 2008, *2*(2), 389–99.

Lacroix, P. Recent Investigations on the Growth of Bone. *Nature* 1945, *156*(3967), 576–77.

Langer, R.; Vacanti, J. Tissue Engineering. *Science* 1993, *260*(5110), 920–26.

Langer, R. Chemical and Biological Approaches to Regenerative Medicine and Tissue Engineering. *Mol. Front. J.* 2019, *3*(2), 1–7. doi:10.1142/S25229732519400091.

Levander, G. Tissue Induction. *Nature* 1945, *155*(3927), 148–49.

Levander, G.; Willestaedt, H. Alcohol-Soluble Osteogenic Substance from Bone Marrow. *Nature* 1946, *3992*, 587.

Lismaa, S.E.; Kaidonis, X.; Nicks, M.; Bogush, N.; Kikuchi, K.; Naqvi, N.; Harvery, R.P.; Husain, A.; Grahm, R.M. Comparative Regenerative Mechanisms across Different Mammalian Tissues. *NPJ Reg. Med.* 2018, *3*(1), 1–20. doi:10.1038/s41536-018-0044-5.

Luyten, F.P.; Cunningham, N.S.; Ma, S.; Muthukumaran, N.; Hammonds, R.G.; Nevins, W.B.; Wood, W.I.; Reddi, A.H. Purification and Partial Amino Acid Sequence of Osteogenin, a Protein Initiating Bone Differentiation. *J. Biol. Chem.* 1999, *264*(23), 13377–80.

Manolagas, S.C.; Kronenberg, M. Reproducibility of Results in Preclinical Studies: A Perspective from the Bone Field. *J. Bone Miner. Res.* 2014, *29*(10), 2131–40.

Martin, I. Engineered Tissues as Customized Organ Gems. *Tissue Eng. A* 2014, *20*(7–8), 1132–33.

Medici, D.; Shore, E.M.; Lounev, V.Y.; Kaplan, F.D.; Kalluri, R.; Olsen, B.R. Conversion of Vascular Endothelial Cells into Multipotent Stem-Like Cells. *Nat. Med.* 2011, *16*(12), 1400–06.

Özkaynak, E.; Rueger, D.C.; Drier, E.A.; Corbett, C.; Ridge, R.J.; Sampath, T.K.; Oppermann, H. OP-1 cDNA Encodes an Osteogenic Protein in the TGF-Beta Family. *EMBO J.* 1990, *9*(7), 2085–93.

Paralkar, V.M.; Nandedkar, A.K.N.; Pointer, R.H.; Kleinman, H.K.; Reddi, A.H. Interaction of Osteogenin, a Heparin Binding Bone Morphogenetic Protein, with Type IV Collagen. *J. Biol. Chem.* 1990, *265*(28), 17281–84.

Paralkar, V.M.; Vukicevic, S.; Reddi, A.H. Transforming Growth Factor β Type 1 Binds to Collagen Type IV of Basement Membrane Matrix: Implications for Development. *Dev. Biol.* 1991, *143*(2), 303–10.

Ramasamy, S.K.; Kusumbe, A.P.; Wang, L.; Adams, R.H. Endothelial Notch Activity Promotes Angiogenesis and Osteogenesis in Bone. *Nature* 2014, *507*(7492), 376–80.

Ramasamy, S.K.; Kusumbe, A.P.; Adams, R.H. Regulation of Tissue Morphogenesis by Endothelial Cell-Derived Signals. *Trends Cell Biol.* 2015, *25*(3), 148–57.

Reddi, A.H.; Huggins, C. Biochemical Sequences in the Transformation of Normal Fibroblasts in Adolescent Rats. *Proc. Natl. Acad. Sci. U.S.A.* 1972, *69*(6), 1601–5.

Reddi, A.H. Cell Biology and Biochemistry of Endochondral Bone Development. *Coll. Rel. Res.* 1981, *1*(2), 209–26.

Reddi, A.H. Bone Morphogenesis and Modeling: Soluble Signals Sculpt Osteosomes in the Solid State. *Cell* 1997, *89*(2), 159–61.

Reddi, A.H. Morphogenesis and Tissue Engineering of Bone and Cartilage: Inductive Signals, Stem Cells, and Biomimetic Matrices. *Tissue Eng.* 2000, *6*(4), 351–59.

Reichman, O.J. Evolution of Regenerative Capabilities. *Am. Nat.* 1984, *123*(6), 752–63.

Ripamonti, U.; van den Heever, B.; van Wyk, J. Expression of the Osteogenic Phenotype in Porous Hydroxyapatite Implanted Extraskeletally in Baboons. *Matrix* 1993, *13*(6), 491–502.

Ripamonti, U.; Reddi, A.H. Bone Morphogenetic Proteins: Applications in Plastic and Reconstructive Surgery. Adv. Plast. Reconstr. Surg. 1995, *11*, 47–65.

Ripamonti, U.; Duneas, N.; van den Heever, B.; Bosch, C.; Crooks, J. Recombinant Transforming Growth Factor-β_1 Induces Endochondral Bone in the Baboon and Synergize with Recombinant Human Osteogenic protein-1 (Bone Morphogenetic protein-7) to Initiate Rapid Bone Formation. *J. Bone Miner. Res.* 1997, *12*, 1584–95.

Ripamonti, U.; Crooks, J.; Kirkbride, A.N. Sintered Porous Hydroxyapatites with Intrinsic Osteoinductive Activity: Geometric Induction of Bone Formation. *S. Afr. J. Sci.* 1999, *95*, 335–43.

Ripamonti, U.; Crooks, J.; Matsaba, T.; Tasker, J. Induction of Endochondral Bone Formation by Recombinant Human Transforming Growth Factor-β_2 in the Baboon (*Papio ursinus*). *Growth Factors* 2000, *17*(4), 269–85.

Ripamonti,U.; Ramoshebi, L.N.; Matsaba, T.; Tasker, J.; Crooks, J.; Teare, J. Bone Induction by BMPs/OPs and Related Family Members. The Critical Role of Delivery Systems. *J. Bone Joint Surg.* [A] 2001, 83-A(Suppl. 1 (Pt 2)), S116–127.

Ripamonti, U. Osteogenic Proteins of the Transforming Growth Factor-ß Superfamily. In: H.L. Henry, A.W. Norman (eds.), *Encyclopedia of Hormones*, Academic Press, San Diego, USA 2003, 80–6.

Ripamonti, U. Soluble, Insoluble and Geometric Signals Sculpt the Architecture of Mineralized Tissues. *J. Cell. Mol. Med.* 2004, *8*(2), 169–80.

Ripamonti, U.; Ramoshebi, L.N.; Patton, J.; Matsaba, T.; Teare, J.; Renton, L. Soluble Signals and Insoluble Substrata: Novel Molecular Cues Instructing the Induction of Bone. In: E.J. Massaro, J.M. Rogers (eds.), *The Skeleton*, Humana Press, New Jersey, USA 2004, Chapter 15, 217–27.

Ripamonti, U. Soluble Osteogenic Molecular Signals and the Induction of Bone Formation. Biomaterials Leading Opinion Paper 2006, *27*(6), 807–22.

Ripamonti, U.; Ferretti, C.; Heliotis, M. Soluble and Insoluble Signals and the Induction of Bone Formation: Molecular Therapeutics Recapitulating Development. *J. Anat.* 2006, *209*(4), 447–68.

Ripamonti, U.; Heliotis, M.; Ferretti, C. Bone Morphogenetic Proteins and the Induction of Bone Formation: From Laboratory to Patients. *Oral Maxillofac. Surg. Clin. North Am.* 2007, *19*(4), 575–89.

Ripamonti, U.; Ramoshebi, L.N.; Teare, J.; Renton, L.; Ferretti, C. The Induction of Bone Formation by Transforming Growth Factor-β_3: Experimental Studies in the Non-Human Primate *Papio ursinus*. *J. Cell. Mol. Med.* 2008a, *12*(3), 1029–48.

Ripamonti, U.; Richter, P.W.; Nilen, R.W.N.; Renton, L. The Induction of Bone Formation by *Smart* Biphasic Hydroxyapatite Tricalcium Phosphate Biomimetic Matrices in the Non-Human Primate *Papio ursinus*. *J. Cell. Mol. Med.* 2008b, *12*(6B), 2609–21.

Ripamonti, U.; Crooks, J.; Khoali, L.; Roden, L. The Induction of Bone Formation by Coral-Derived Calcium Carbonate/Hydroxyapatite Constructs. *Biomaterials* 2009, *30*(7), 1428–39.

Ripamonti, U. Biomimetism, Biomimetic Matrices and the Induction of Bone Formation. *J. Cell. Mol. Med.* 2009, *13*(9B), 2953–72.

Ripamonti, U.; Roden, L.C. The Induction of Bone Formation by the Transforming Growth Factor-β_2 in the Non-Human Primate *Papio ursinus* and Its Modulation by Skeletal Muscle Responding Stem Cells. *Cell. Prolif.* 2010, *43*(3), 207–18.

Ripamonti, U.; Klar, R.M.; Renton, L.F.; Ferretti, C. Synergistic Induction of Bone Formation by hOP- 1, hTGF-β_3 and Inhibition by Zoledronate in Macroporous Coral-Derived Hydroxyapatites. *Biomaterials* 2010, *31*(25), 6400–410.

Ripamonti, U.; Teare, J.; Ferretti, C. A Macroporous Bioreactor Superactivated by the Recombinant Human Transforming Growth Factor-β_3. *Front. Physiol.* 2012a, *3*, 172. doi:10.3389/phys.2012.00172.

Ripamonti, U.; Roden, L.C.; Renton, L.F. Osteoinductive Hydroxyapatite-Coated Titanium Implants. *Biomaterials* 2012b, *33*(15), 3813–23.

Ripamonti, U.; Duarte, R.; Ferretti, C. Re-Evaluating the Induction of Bone Formation in Primates. *Biomaterials* 2014, *35*(35), 9407–22.

Ripamonti, U.; Dix-Peek, T.; Parak, R.; Milner, B.; Duarte, R. Profiling Bone Morphogenetic Proteins and Transforming Growth Factor-Bs by hTGF-β_3 Pre-Treated Coral-Derived Macroporous Constructs: The Power of One. *Biomaterials* 2015, *49*, 90–102.

Ripamonti, U. Induction of Bone Formation in Primates. In: U. Ripamonti (ed.) *The Transforming Growth Factor-beta3*, CRC Press Tylor & Francis Group, Boca Raton, 2016, 1–197.

Ripamonti, U.; Ferretti, C. Regeneration of Mandibular Defects in Non-Human Primates by the Transforming Growth Factor-β_3 and Translational Research in Clinical Contexts. In: U Ripamonti (ed.) *Induction of Bone Formation in Primates. The Transforming Growth Factor–beta3*, CRC Press, Boca Raton, Chapter 5, 2016, 105–21.

Ripamonti, U.; Parak, R.; Klar, R.M.; Dickens, C.; Dix-Peek, T.; Duarte, R. The Synergistic Induction of Bone Formation by the Osteogenic Proteins of the TGF-β Supergene Family. *Biomaterials* 2016a, *104*, 279–96. doi:10.1016/j.biomaterials.2016.07.018.

Ripamonti, U.; Klar, M.R.; Parak, R.; Dickens, C.; Dix-Peek, T.; Duarte, R. Tissue Segregation Restores the Induction of Bone Formation by the Mammalian Transforming Growth Factor-β_3 in Calvarial Defects of the Non-Human Primate *Papio ursinus*. *Biomaterials* *86*, 2016b, 21–32. doi:10.1016/j.biomaterials2016.01.071.

Ripamonti, U. Biomimetic Functionalized Surfaces and the Induction of Bone Formation. *Tissue Eng.* 2017, *23*(21,22), 1197–209.

Ripamonti, U.; Parak, R.; Klar, R.M.; Dickens, C.; Dix-Peek, T.; Duarte, R. Cementogenesis and Osteogenesis in Periodontal Tissue Regeneration by Recombinant Human Transforming Growth Factor-β3; a Pilot Study in Papio ursinus. *J. Clin. Periodontol.* 2017, *44*(1), 83–95.

Ripamonti, U.; Roden, L.; van den Heever, B. Sharks, Sharks Cartilages and Shark Teeth: A Collaborative Africa-USA Study to Attempt to Induce "*Bone: Formation by Autoinduction*" in Cartilaginous Fishes. *S. Afr. Dent. J.* 2018, *73*(1), 11–21.

Ripamonti, U. Functionalized Surface Geometries Induce: "*Bone: Formation by Autoinduction*". *Front. Physiol.* 2018a, *8*, 1084. doi:10.3389/fphys.2017.01084.

Ripamonti, U. Introductory Remarks on Cartilages, Bones and on "*Bone: Formation by Autoinduction*". *S. Afr. Dent. J.* 2018b, *73*(1), 6–10.

Ripamonti, U. Developmental Patterns of Periodontal Tissue Regeneration. Developmental Diversities of Tooth Morphogenesis Do Also Map Capacity of Periodontal Tissue Regeneration? *J. Periodont. Res.* 2019, *54*(1), 10–26. doi:10.1111/jre.12596.

Roberts, A.B.; Sporn, M.B.; Assoian, R.K.; Smith, J.M.; Roche, N.S.; Wakefield, L.M.; Heine, L.L.; Liotta, L.A.; Falanga, V.; Kehrl, J.H.; Fauici, A.S. Transforming Growth Factor Type β: Rapid Induction of Fibrosis and Angiogenesis *In Vivo* and Stimulation of Collagen Formation *In Vitro*. *Proc. Natl. Acad. Sci. U.S.A.* 1986, *83*(12), 4167–71.

Sacerdotti, C.; Frattin, G. Sulla Produzione Eteroplastica dell' Osso. *Riv. Accad. Med. Torino* 1901, *27*, 825–36.

Sampath, T.K., Coughlin, J.E., Whetstone, R.M., Banach, D., Corbett, C., Ridge, R.J., Ozkaynak, E., Oppermann, H., Rueger, D.C. Bovine Osteogenic Protein is Composed of Dimerrs of OP-1 and BMP-2A, two Members of the Transforming Growth Factor-beta Superfamily. *J. Biol. Chem.* 1990, **265**, 13198–205.

Seeherman, H.J.; Berasi, S.P.; Brown, C.T.; Martinez, R.X.; Jou, Z.S.; Jelinsky, S.; Cain, M.J.; Grode, J.; Tumelty, K.; Bohner, M.; Grinberg, O.; Orr, N.; Shoseyov, O.; Eyckmans, J.; Chen, C.; Moralrs, P.; Wilson, C.G.; Vanderploeg, E.J.; Wozney, J.M. A BMP/Activin A Chimera Is Superiorvto Native BMPs and Induces Bone Repair in Non-Human Primates Whwn Delivered in a Composite Matrix. *Sci. Transl. Med.* 2019, *11*(489), eaar4953.

Shinozaki, M.; Kawara, S.; Hayashi, N.; Kakinuma, T.; Igarashi, A.; Takehara, K. Induction of Subcutaneous Tissue Fibrosis in Newborn Mice by Transforming Growth Factor Beta – Simultaneous Application with Basic Fibroblast Growth Factor Causes Persistent Fibrosis. *Biochem. Biophys. Res. Commun.* 1997, *240*(2), 292–97.

Takagi, K.; Urist, M.R. The Reaction of the Dura to Bone Morphogenetic Protein (BMP) in Repair of Skull Defects. *Ann. Surg.* 1982, *1*(1), 100–09.

Takahashi, K.; Yamanaka, S. Induction of Pluripotent Stem Cells from Mouse Embryonic and Adult Fibroblast Cultures by Defined Factors. *Cell* 2006, *126*(4), 663–76.

Takahashi, K.; Tanabe, K.; Ohnuki, M.; Narita, M.; Ichisata, t.; Tomoda, K.; Yamanaka, S. Induction of Pluripotent Stem Cells from Adult Human Fibroblasts by Defined Factors. *Cell* 2007, *131*(5), 1–12.

Takahashi, K.; Yamanaka, S. Induced Pluripotent Stem Cells in Medicine and Biology. *Development* 2013, *140*(12), 2457–61.

Tanaka, E.M.; Reddien, P.W. The Cellular Basis for Animal Regeneration. *Dev. Cell* 2011, *21*(1), 172–85.

Tomancak, P. Evolutionary History of Tissue Bending. *Science* 2019, *366*(6463), 300–01. doi:10.1126/science.aaz1289.

Trueta, J. The Role of the Vessels in Osteogenesis. *J. Bone Joint Surg.* 1963, *45B*, 402–18.

Urist, M.R. Bone: Formation by Autoinduction. *Science* 1965, *220*(3698), 893–99. doi:10.1126/science.150.3698.893.

Urist, M.R.; Silverman, F.; Büring, K.; Dubuc, F.L.; Rosenberg, J.M. The Bone Induction Principle. *Clin. Orthop. Rel. Res.* 1967, *53*, 243–83.

Urist, M.R. The Reality of a Nebulous Enigmatic Myth. *Clin. Orthop. Relat. Res.* 1968, *59*, 3–5.

Wang, E.A.; Rosen, V.; Cordes, P.; Hewick, R.M.; Kriz, M.J.; Luxenberg, D.P.; Sibley, B.S.; Wozney, J.M. Purification and Characterization of Other Distinct Bone-Inducing Factors. *Proc. Natl. Acad. Sci. U.S.A.* 1988, *85*(24), 9484–88.

Williams, D.F. Tissue Engineering: The Multidisciplinary Epitome of Hope and Despair. In: R. Paton, L. McNamara (eds.) *Studies on Mutidisciplinary*, Elsevier BV, Amsterdam, 2006, 483–524.

Wozney, J.M.; Rosen, V.; Celeste, A.J.; Mitsock, L.M.; Whitters, M.J.; Kriz, R.W.; Hewick, R.M.; Wang, E.A. Novel Regulators of Bone Formation: Molecular Clones and Activities. *Science* 1988, *242*(4885), 1528–34.

3 The Induction of Bone Formation and the Osteogenic Proteins of the Transforming Growth Factor-β Supergene Family
Pleiotropism and Redundancy

Ugo Ripamonti

3.1 WHY BONE?

The grand multifaceted "history of the evolutionary development of the vertebrate skeletal tissue" is lucidly presented by Romer (1963) who, in his far-reaching essay "The 'ancient history' of bone", asks the seemingly simple question: "Why bone?" (Romer 1963).

Romer lists the function of storage of ions, particularly Ca^{++}, and proteins, though later states that this function would appear to be of secondary importance (Romer 1963). "Why then bone?" Romer suggests that the induction of bone formation or osteogenesis, and thus the development of the dermal skeletal armour, is why bone formed. Bone as a skeletal armour developed as a biological and physical defence against the predaceous habits of the eurypterids, largely feeding on the early vertebrate "bony fishes" at the evolutionary beginning of the vertebrates (Romer 1963). It seems likely that the eurypterids' aggressive feeding on the evolutionary beginning of the vertebrates "was primarily responsible for the development of the vertebrate bony armour" (Romer 1963).

It is worth reporting *verbatim* the final statement of Romer's essay on "the ancient history of bone":

It thus seems highly probable that the bony skeleton without which the evolution of the higher vertebrates could never have taken place, owes its origin, close to half a billion years ago, to the threat of invertebrates predation on our feeble primitive fish ancestors.
(Romer 1963)

Indeed, figuratively, Romer suggested that, in contrast to later times, the "vertebrates were the underdogs" (Romer 1963), and the predators were invertebrates, the euryp-terids, much larger than the contemporary vertebrates, "with effective biting mouth parts in addition to grasping claws" (Romer 1963).

There is of course no direct correlation between being massively predated and the evolutionary appearances of the vertebrates' bony armour and later of the skeleton. Rather, positive gene mutations selectively at that point in time set into motion different calcium homeostasis profiles (Venkatesh et al. 2014) with later the appearance of the osteogenic proteins of the TGF-β supergene family (Ripamonti 2003). The development and appearance of specific secretory calcium binding phosphoproteins (SCPPs), arising from the Sparc-like1 gene (Sparcl1), would have been critical for the evolutionary beginning of bone formation (Venkatesh et al. 2014).

The work of Romer (1963) grandly shows that, evolutionarily, bone was an ancestral character, and that previously labelled cartilaginous fishes like sharks, skates and rays were not primitive, and that the cartilaginous skeleton formed following a "degenerative slump from bone bearing ancestors" (Romer 1963). Indeed, "sharks and rays are placoderm descendants which have degenerated in their skeletal structure from an ancestral condition in which bone was present" (Romer 1963).

Romer summarizes by stating that the origin of the vertebrates' skeletal evolution is "the reverse of the truth" (Romer 1963). Instead of evolutionarily starting with a cartilaginous anlage, defined as the ancestral guiding developmental matrix for the late induction of endochondral bone formation, the early vertebrates had a high degree of bone development. Romer reports that several genera and species displayed "a slump towards a cartilaginous condition" and that "bone is an ancient, rather than a relatively new, skeletal material in the history of vertebrates" (Romer 1963).

We have recently reviewed and expanded these observations by reporting unique histological data from tissue specimens harvested from the dorsal musculature of *Carcharinus obscurus* adolescent sharks (Ripamonti 2017; Ripamonti 2018; Ripamonti et al. 2018). Our experimentation then at the Dental Research Institute of the University and at the Bone Cell Biology Section of the NIH, Bethesda, MD, USA, was focused on trying to force the selachian elasmobranch *C. obscurus* to initiate the induction of bone formation by implanting a variety of osteoinductive matrices and proteins in heterotopic dorsal muscular sites of *C. obscurus* sharks, cartilaginous fishes, class *Chondrichthyes* (Ripamonti 2018; Ripamonti et al. 2018). Heterotopic implantation and tissue harvest were done at the Oceanographic Research Institute, Marine Parade, Durban (Ripamonti 2018; Ripamonti et al. 2018).

Because of the spectacular induction of chondrogenesis in the macroporous spaces of a coral-derived bioreactor harvested from the dorsal musculature of the selachian

C. obscurus shark, the writer would like to present the images of the induction of chondrogenesis as the first figure of this chapter reviewing the formation of bone, osteogenesis and the induction of bone formation by the osteogenic proteins of the TGF-β supergene family (Ripamonti 2003). Figure 3.1 presents unique images of the induction of chondrogenesis by macroporous coral-derived bioreactors harvested from the dorsal musculature of the dusky shark *C. obscurus* (Ripamonti et al. 2018).

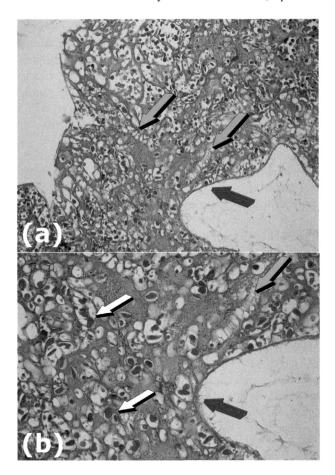

FIGURE 3.1 Induction of chondrogenesis within the macroporous spaces of a coral-derived construct implanted in the dorsal musculature of a *Carcharinus obscurus* shark and harvested on day 21 from the selachian fish housed in the salty ponds of the Oceanographic Research Institute, Marine Parade, Durban, South Africa (Ripamonti 2018; Ripamonti et al. 2018). (a) Chondroblastic cells attached to the coral-derived macroporous construct (*dark blue* arrow) close to progressively differentiating chondroblasts (*light blue* arrow) patterning the newly formed cartilage as in the mammalian growth plate but without any evidence of vascular invasion and chondrolysis (Ripamonti et al. 2018). (b) High-power view illustrating chondrogenic material, chondroblastic cells (*white* arrows) with detail of progressively differentiating chondroblasts (*light blue* arrow). Undecalcified sections from resin-embedded specimens stained with modified Goldner's trichrome.

Chondrogenesis, as shown in Figure 3.1, also requires us to ask: why the induction of chondrogenesis in a macroporous calcium phosphate-based construct after implantation in heterotopic intramuscular sites of a selachian fish? *C. obscurus* responding cells, migrating into the macroporous spaces and in contact with the calcium phosphate substratum with variable surface nanotopographies, differentiate into chondroblastic cells with matrix deposition (Fig. 3.1). Though the bioactive calcium phosphate-based bioreactor in contact with responding cells is conducive and inducive to developmental events resulting in the induction of chondrogenesis, there is a lack of bone differentiation (Ripamonti et al. 2018). We have previously stated that the DNA of *C. obscurus* does not retain the genetic memory of the developmental cascade of the induction of bone formation following selected gene mutations ablating the osteogenetic programme as well as lacking capillary sprouting, angiogenesis and vascular invasion of the cartilage leading to chondrolysis. In mammals, the latter is followed by the induction of osteoblastogenesis and endochondral osteogenesis (Reddi 1981; Ripamonti 2017; Ripamonti et al. 2018). Capillary sprouting and osteoblastogenesis are not shown in the cartilaginous endoskeleton of *C. obscurus* nor during chondrogenesis by coral-derived bioreactors (Fig. 3.1) (Ripamonti et al. 2018).

Before attempting to address these seemingly complicated pleotropic evolutionary and biologically relevant questions, we need once again to review the images presented in Figure 3.1, dissecting the evolutionary and biological significance of columns of progressively differentiating chondroblasts patterning the newly formed cartilage within the macroporous spaces of the coral-derived bioreactor heterotopically implanted in *C. obscurus* (Ripamonti et al. 2018).

Columnar condensations of chondroblastic cells are to be found in the pleiotropic uniqueness of the cartilaginous growth plate, a fundamental compartmentalized biological system of growth that masterminds the tri-dimensional growth of the mammalian axial skeleton that forms through endochondral osteogenesis *via* the cartilage anlage. The cartilaginous anlage is the bioreactor that engineers vascular invasion with capillary sprouting, chondrolysis, the induction of osteoblastic cells, early bone matrix deposition and, finally, the induction of bone formation (Reddi 1981; Reddi and Kuettner 1981; Ripamonti 2017).

Masterfully Romer shows that "cartilage is among vertebrates, essentially an embryonic adaptation" (Romer 1963), and that to properly grow and tri-dimensionally expand bones of the axial skeleton, it would have been necessary to "invent" the growth plate to "achieve an orderly embryological development of a vertebrate" (Romer 1963).

The development of the axial skeleton in mammals was only possible following the development of the ancestral cartilaginous or cartilage matrix that formed, as per Romer's statement, following a "slump towards a cartilaginous condition" (Romer 1963).

Such a cartilaginous condition inherited by sharks, skates and rays evolved into cartilaginous skeletons after genetic mutations and ablation of angiogenic mechanisms controlling the evolution and development of the cartilaginous anlage, the mechanism that Nature proposed for the induction of endochondral osteogenesis as opposed to the relatively simple appositional growth of membranous osteogenesis (Fig. 3.2) (Romer 1963; Ripamonti 2006; Ripamonti et al. 2006; Ripamonti 2017).

We did previously dwell on the uniqueness of the cartilaginous growth plate phylogenetically invocated to initiate endochondral osteogenesis and the development of the axial skeleton (Ripamonti 2006; Ripamonti 2017). The induction or generation of the primordial cartilaginous anlage masterminded the molecular and structural organization of the cartilage anlage, the pleiotropic bioreactor that is essential for the induction of endochondral osteogenesis (Fig. 3.3).

The induction of the cartilage anlage, differentiating angiogenesis, capillary sprouting, vascular invasion initiate chondrolysis of the mineralized hypertrophic chondrocytes and the induction of osteoblastogenesis in a tight molecular, cellular and morphological cross talk are illustrated in Figures 3.3, 3.4, and 3.5.

The selected figures show the induction of developmental aggregating condensations of highly differentiating cells in human embryonic limbs with later the induction of cartilage anlages of the humerus and radius (Figs. 3.3a,b) as well as an embryonic femur (Fig. 3.4a). The high-power view shows the tight morphological and molecular cross talk between hypertrophic mineralized chondrocytes, with the exquisite details of the invading and sprouting capillaries (Fig. 3.4b). Angiogenesis and capillary invasion (Fig. 3.4b) initiate chondrolysis that brings about the differentiation of osteoblastic cells, early matrix deposition and the growth of the long bones *via* endochondral osteogenesis (Fig. 3.4b).

It is significant that endochondral osteogenesis as initiated in embryonic development (Figs. 3.3b, 3.4a,b) is recapitulated in postnatal osteogenesis during the induction of bone formation (Fig. 3.4c), whereby the induction of endochondral bone by demineralized bone matrices or by soluble osteogenic molecular signals of the TGF-β supergene family (Fig. 3.4c) (Ripamonti et al. 1992; Klar et al. 2014 Ripamonti 2003) reconstituted with different delivery systems firstly initiate chondrogenesis followed by vascular invasion, heralding chondrolysis and the induction of bone formation (Fig. 3.4c).

Primarily, membranous osteogenesis originates after the induction of mesenchymal condensations populated by contiguous osteoblasts that later continuously secrete bone matrix as yet to be mineralized, or osteoid, around invading and supporting central blood vessels (Fig. 3.2). These "osteogenetic vessels" (Trueta 1963) are responsible for the induction of "osteogenesis in angiogenesis" (Fig. 3.2) (Ripamonti 2006; Ripamonti et al. 2006; Ripamonti et al. 2007). Membranous or dermal bones that form by mesenchymal condensations are present in mammals only in plate-like mineralized condensations of the craniofacial bones and clavicle, "without intimate connections with surrounding organs" (Romer 1963).

Why then bone and the growth mechanisms of the embryonic cartilaginous plate for the longitudinal growth of the mammalian axial skeleton? (Figs. 3.3, 3.4)? After millions of years of evolution, Nature has had the capacity to nucleate and evolve highly sophisticated tissues and organs. A classic example is the evolution of the skeleton, thus providing the emergence of the vertebrates. *Par force* deambulation and body erection needed to develop, thus freeing the upper limbs for superior foraging and more industrious *Homo*-like activities including the use of tools for hunting,

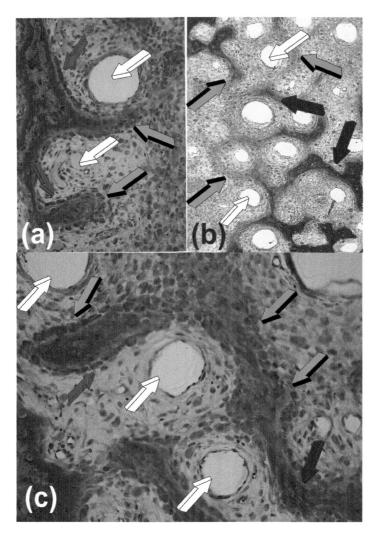

FIGURE 3.2 Osteogenesis in angiogenesis (Ripamonti 2004; Ripamonti 2006; Ripamonti et al. 2006; Ripamonti et al. 2007) in heterotopic intramuscular sites of the Chacma baboon *Papio ursinus*. Digital images show the critical role of the vessels and angiogenesis in the induction of bone formation. (a) Central blood vessels/capillaries (*white* arrows) surrounded by mesenchymal condensations (*light blue* arrows) enveloping the central blood vessels. The mechano-physical process surrounding the central morphogenetic capillary initiates osteoblastogenesis along the mesenchymal condensation facing the central blood vessel (*magenta* arrows). (b,c) Low-power view depicting several central blood vessels (*white* arrows), each surrounded by enveloping mesenchymal condensations at different stage of differentiation and with (*dark blue* arrows) or without (*light blue* arrows) induction of mineralization within the newly formed mesenchymal condensations. (c) The central blood vessels patterning the induction of bone formation are acting as a frame to induce the Haversian system of the lamellar osteonic bone of the primate lamellar bone (Ripamonti 2006; Ripamonti 2010).

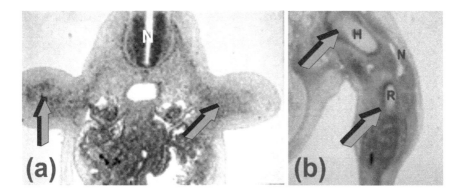

FIGURE 3.3 Induction of tissue morphogenesis and the induction of endochondral bone formation in human embryos highlighting embryonic differentiation of the upper limbs, mesenchymal condensation, cellular aggregation and differentiation during limb budding in human embryo (a) with later differentiation and morphogenesis of cartilage anlages (b) of the humerus (H) and radius (R). (a) Upper limb buds (*light blue* arrows) show condensation and aggregation of mesenchymal cells for later induction of the upper limb and endochondral bone formation of the humerus and radius as depicted in (b). N indicates the embryonic development of the neuronal crest (N). (b) Histological preparation of a human embryonic upper limb depicting the basic biology of endochondral bone formation *via* the synthesis and maturation of cartilaginous anlages critical for bone development to occur. The anlages (*light blue* arrows) after vascular invasion and capillary sprouting within the cartilaginous matrices will guide the induction of bone differentiation with formation of the cartilaginous growth plates, essential bioreactors for the induction of endochondral bone formation and the growth and elongation of the long bones of the axial skeleton.

foraging and above all, however, for maternal care, contributing thus to the speciation of the genus *Homo* ultimately directing the emergence of *Homo sapiens*. The emergence of osteogenesis and later the evolution of the vertebrates, of the skeleton and *Homo sapiens* after a billion years from *Drosophila melanogaster* have permitted the precision self-assembly of the bone unit or osteosome (Reddi 1997) and the remodelling processes of the skeleton (Parfitt 1994; Manolagas and Jilka 1995). The concavities as sculpted in the bio-inspired biomimetic bioceramics biomimetize the remodelling cycle of the osteonic cortico-cancellous bone (Ripamonti 2006; Ripamonti 2009; Klar et al. 2013; Ripamonti et al. 2014).

Nature has masterminded the evolution of the skeleton, deploying common yet limited mechanisms to direct the emergence not only of the skeleton but of several specialized tissue and organs (Ripamonti 2006). The TGF-β and BMPs' families reflect Nature's parsimony in controlling multiple specialized functions or pleiotropy, deploying several osteogenic molecular signals with minor variation in amino acid motifs within highly conserved carboxy terminal domains (Ripamonti 2003; Ripamonti et al. 2004; Ripamonti 2006; Ripamonti et al. 2014). The evolutionary conservation of the TGF-β supergene family is superbly demonstrated by the remarkable observation that recombinant decapentaplegic and 60A proteins, gene

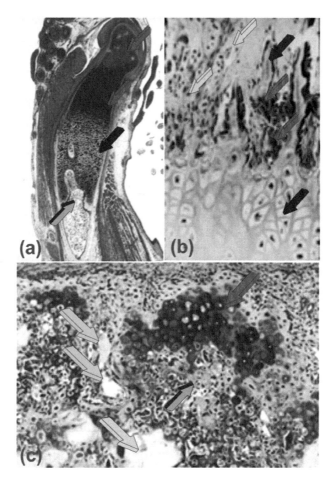

FIGURE 3.4 Digital images from human embryos highlighting endochondral bone formation of a femur detailing vascular invasion, chondrolysis and the induction of bone formation. (a) The cartilage anlage of the femur (*magenta* arrows) develops into hypertrophic chondrocytes (*dark blue* arrow) with vascular invasion and chondrolysis (*light blue* arrow) that set the stage for osteoblastic cell differentiation, matrix synthesis and the induction of endochondral bone formation. (b) Cell to cell and matrix signalling cross talk is shown in (b) through the phases of induction of hypertrophic chondrocytes (*dark blue* arrow), chondrolysis and angiogenesis and vascular invasion (white arrows) that bring about osteoblastic cell differentiation and osteoid matrix synthesis (*magenta* arrows). (c) Recapitulation of embryonic development in postnatal tissue induction and morphogenesis by combining 2.5 μg recombinant human osteogenic protein-1 (hOP-1) with 25 mg of rat allogeneic inactive and insoluble collagenous bone matrix (ICBM) (*grey* arrow). The induction of heterotopic induction of endochondral bone forms by recapitulating embryonic development inducing cartilaginous anlages (*magenta* arrow) at the periphery of the implanted bioreactor. The induction of chondrogenesis and of the cartilage anlage is followed by vascular invasion and chondrolysis (*cyan* arrows) followed by the induction of osteoblastic-like cells with matrix secretion and the induction of bone formation (*light blue* arrow).

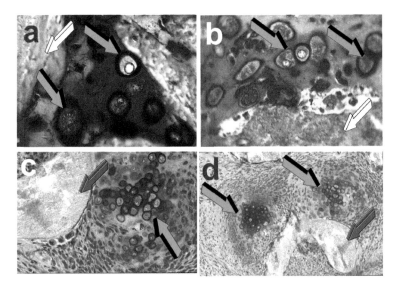

FIGURE 3.5 Tissue induction and morphogenesis recombining soluble osteogenic molecular signals with insoluble signals or substrata initiating the ripple-like cascade of "Bone: Formation by Induction" (Urist 1965). (a,b) Induction of chondrogenesis with differentiation of hypertrophic chondrocytes (*light blue* arrows) by 0.1–2.5 µg highly purified naturally derived osteogenic fractions, osteogenin (Ripamonti et al. 1992), reconstituted with 25 mg of rat inactive and insoluble collagenous bone matrix (*white* arrows). Note chondrogenic material and hypertrophic chondrocytes in tight association with the matrix used as a carrier for the biological activity of osteogenin (Ripamonti et al. 1992). (c,d) Combinatorial molecular protocols combining highly purified osteogenin after adsorption, affinity and molecular sieve gel filtration chromatography to coral-derived calcium phosphate bioreactors (*magenta* arrows) implanted in the subcutaneous space of the rodent bioassay (Ripamonti et al. 1992). Induction of cartilaginous islands (*light blue* arrows) recapitulating the generation of cartilage anlages in embryonic differentiation in close proximity to the coral-derived bioreactors implanted subcutaneously in the rodent bioassay (Ripamonti et al. 1992).

products of the fruit fly *Drosophila melanogaster*, induce bone formation in mammals (Sampath et al. 1993).

> Nature has thus usurped phylogenetically ancient amino acid motifs and sequences deployed for dorso-ventral patterning in *Drosophila melanogaster* to set the unique vertebrate trait of the induction of bone formation rather than evolving new genes and gene products for the induction of bone and the emergence of the skeleton, and thus of the vertebrates.
>
> **(Ripamonti 2003; Ripamonti 2004; Ripamonti 2006; Ripamonti 2009)**

With the induction of skeletogenesis and body erection, locomotion was *par force* the next evolutionary step characterizing the emergence of the vertebrates and vertebrates' speciation. Bipedalism and deambulation were hallmark steps in hominid's evolution, controlling the emergence of *Homo habilis* and later of *Homo erectus* walking out of Africa to colonize the planet landscape. Studies showed conserved gene families, including the remarkably conserved Hox transcription factor-dependent programme in ancient primitive marine vertebrates, the skate *Leucoraja erinacea*, at about 420,000 years before the present (Jung et al. 2018). Such skates share highly conserved neuronal circuitry that is essential for land deambulation (Jung et al. 2018). The communication by Jung et al. (2018) has indicated that circuitry that is essential for walking "evolved through adaptation of a genetic regulatory network shared by all vertebrates with paired appendages" capable thus of locomotion (Jung et al. 2018).

The genesis of bone has further invocated "bone in the solid state" (Reddi 1997), a solid mineralized matter with the supramolecular assembly of structural proteins, collagens and vascular structures permeating the osteonic bone in contact with bone marrow. The precision self-assembly of the bone unit or osteosome (Reddi 1997), with the remodelling processes of the osteonic bone, has been a superior example of Nature's creativity, design and architecture (Reddi 1997). The remodelling of the skeleton, the formation of bone by osteoblasts and the resorption of bone by osteoclasts make up a closely integrated homeostatic system (Reddi 1997) and, at the same time, provide the geometric pattern of the remodelling osteonic cycle, the skeleton thus acting as "a giant molecular machine" (Reddi 2001, personal communication).

It is noteworthy that selected and phylogenetically few ancient circuitries masterminded the emergence of the skeleton, the "bone induction principle" (Urist et al. 1967), of the vertebrates, and thus of the *Homo* clade. *Drosophila* genes and expressed proteins more than 800 million years before the present were operational to construct the unique vertebrate trait of the induction of bone formation, and later of skeletogenesis (Ripamonti 2006; Ripamonti 2010). Similarly, highly conserved circuitry has been shown between primitive marine vertebrates of the genus *Leucoraja* and other vertebrates, highlighting how *Hox* gene expression controlled the evolutionary transitions between undulatory and ambulatory motor circuit connectivity programmes (Jung et al. 2014).

These observations, together with the prominent evolutionary conservation of TGF-β family members as shown by the induction of bone formation by phylogenetically ancient genes and gene products such as decapentaplegic (*dpp*) and *60A* of the fruit fly *Drosophila melanogaster*, indicate the apparently simple construction design of our common ancestors or Urbilateria having "most of the developmental gene pathways from which animals are built" (De Robertis and Sasai 1996).

The conservation of the *Hox* genes between the fruit fly *Drosophila melanogaster* and vertebrates is highly remarkable, and once again reflects Nature's parsimony in deploying genes and gene products responsible for the generation of the pleiotropic diversity of the multiple species of animal life on earth, key questions of evo-devo (De Robertis 2008).

The highly conserved circuitries of genes, gene products and gene pathways developing the pleiotropy ancestry of animals' phyla reflect the "awesome diversity and beauty of life" (De Robertis 2008); the ancestry of the concavity, the "shape of life" (Ripamonti 2004; Ripamonti 2006; Ripamonti 2010; Ripamonti 2012), biomimetizes the ancestral repetitive multi-million-years-tested designs and topographies of Nature. The concavity, as cut into biomimetic matrices or biomimetized along titania planar surfaces plasma-sprayed by sintered crystalline hydroxyapatites is thus the geometric signal that initiates the induction of bone formation. Concavities are endowed with smart functional shape memory geometric cues in which soluble signals induce morphogenesis, and physical forces, imparted by the geometric topography of the substratum, dictate biological patterns, constructing the induction of bone formation and regulating the expression of gene products as a function of the structure.

Why bone then? Possibly Nature's whole developmental biological and evolutionary plan was simply to provide deambulation and body erection, freeing the upper limbs for superior foraging and more industrious *Homo*-like activities including the use of tools for hunting and above all, however, for maternal care, contributing thus to the speciation of the Homo clade, ultimately directing the emergence of *Homo sapiens*. In other words, the induction of bone and osteogenesis was only to finalize the evolution of *Homo sapiens* on the planet earth.

3.2 MORPHOGENS AND THE INDUCTION OF BONE FORMATION

In a series of seminal contributions to the induction of bone formation, M.R. Urist provides some insights into the fascinating phenomenon of "Bone: formation by autoinduction" (1965). Together with other exceptional experimentalists even before his time on the induction of bone formation (Huggins 1931; Levander 1938; Levander 1945; Levander and Willestaedt 1946; Bridges and Prichard 1958; Moss 1958; Trueta 1963), great strides were made to at least hypothesize the presence of substances within the bone matrix capable of cellular induction, additionally attempting to extract this substance from bone matrices (Levander 1945). Levander's great insights further allowed him to state that "the same chemical substances are active both during the embryonal differentiation and during post-foetal growth. Regeneration of tissues is, in other words, a repetition of embryonal development" (Levander 1945).

It is the beginning of the grand quest of tissue induction and morphogenesis, a quest that will finally identify the rules that govern pattern formation, tissue induction and morphogenesis. Last-century experimentalists started to address how the fundamental problems of tissue induction and morphogenesis develop *ab initio* during embryonic development, setting the stage for the fabrication of a vast multitude of organs and tissues of all the animal kingdom. After so many molecular breakthroughs that have pinpointed cellular life to its minute details (*Cell* Editorial 2014), the question that now arises is, what next beyond morphogens and stem cells (Ripamonti and Duarte 2019)?

The fundamental tenet of tissue engineering is to combine soluble molecular signals with insoluble signals or substrata to erect scaffolds of smart biomaterial matrices that mimic the supramolecular assembly of the extracellular matrix of bone (Reddi 1994; Ripamonti and Reddi 1995; Ripamonti and Duneas 1996; Reddi 2000).

In the bone matrix, molecular signals are in solution, interacting with the insoluble signal of the extracellular matrix (Reddi 1997). Soluble molecular signals have been pursued, searched, later identified, purified to homogeneity and cloned (Wozney et al. 1988; Özkaynak et al. 1990; Reddi 1997; Reddi 2000; Ripamonti 2006). This has *par force* shown that the bone matrix is a rich repository of several morphogens and cytokines which are endowed with the striking capacity not only to promote and guide tissue induction and morphogenesis but also to initiate the ripple-like cascade of the induction of bone formation in heterotopic sites, that is, in extraskeletal heterotopic sites where there is no bone (Ripamonti 2006).

As Marshall Urist stated in the pages of the *Journal of Bone and Joint Surgery* [A] (1952), "The nature of the osteogenetic principle which initiates and sustains the process of bone formation is a fundamental problem of osteogenesis" (Urist and McLean 1952). The grand challenge of investigating the "osteogenetic principle", later defined as the "bone induction principle" (Urist et al. 1967), has resulted in the identification of a number of initiators that control tissue induction and morphogenesis (Sampath et al. 1987; Wozney et al. 1988; Özkaynak et al. 1990; Reddi 1997; Reddi 2000; Ripamonti 2006; Ripamonti et al. 2006; Ripamonti et al. 2007). These initiators have been grandly postulated by Turing (1952), who indicated that mammalian tissues contain "morphogens, or forms generating substances", thus initiating tissue induction and morphogenesis (Turing 1952).

Last-century research has identified several mammalian extracellular matrices including dentine and bone that, upon demineralization and implantation in heterotopic extraskeletal sites of animal models, would result in the *de novo* induction of bone formation in a variety of animal models including rodents, lagomorphs, canines and non-human and human primates (Fig. 3.4) (Ripamonti 2006; Ripamonti et al. 2006; Ripamonti et al. 2007; Ferretti et al. 2010; Ferretti and Ripamonti 2020).

The nature of the osteogenetic principle and of "the bone induction principle" of Urist's definition (Urist et al. 1967) was, however, difficult to characterize, not least because "bone is in the solid state" (Reddi 1997).

Solubilization of the putative bone-inductive proteins from the "bone matrix in the solid state" (Reddi 1997) was elegantly disclosed by dissociatively extracting the demineralized bone matrix in an insoluble signal, mainly collagenous bone matrix and solubilized molecular signals in solution after chaotropic dissociative extraction of the bone matrix in 4 M guanidinium hydrochloride (Gdn-HCl) or 6 M urea (Sampath and Reddi 1981; Sampath and Reddi 1983). These were key experiments that firstly identified solubilized morphogens within the bone matrix, and secondly provided chromatographic procedures for the final purification of the osteoinductive signals (Sampath et al. 1987; Wang et al. 1988; Luyten et al. 1999; Ripamonti et al. 1992).

The chaotropic dissociative extraction experiments (Sampath and Reddi 1981; Sampath and Reddi 1983) provided the operational reconstitution of the soluble molecular signal with an insoluble signal or substratum that triggers the bone induction cascade (Figs. 3.4c, 3.5) (Ripamonti et al. 1992; Ripamonti and Reddi 1995; Reddi 1997).

3.3 THE OSTEOGENIC PROTEINS OF THE TRANSFORMING GROWTH FACTOR-β SUPERGENE FAMILY

This operational reconstitution further provided evidence of the homology of the putative osteoinductive molecular signals, whereby several homologous signals purified from different mammalian matrices would also induce bone formation, provided that the xenogeneic protein extracts were reconstituted, for example, with rat insoluble collagenous bone matrix (ICBM) when implanted in the rat subcutaneous assay (Sampath and Reddi 1983; Ripamonti and Reddi 1995).

Xenogeneic highly purified bovine osteogenic fractions induce bone when reconstituted with allogeneic baboon insoluble collagenous bone matrices and implanted in heterotopic intramuscular *rectus abdominis* sites of the Chacma baboon *Papio ursinus* (Ripamonti et al. 1991). It was noteworthy that highly purified naturally derived bovine osteogenic fractions induced alveolar bone, the *de novo* initiation of Sharpey's fibres inserting into newly generated cementoid deposited along newly formed mineralized cementum when using baboon insoluble collagenous bone matrices when implanted in Class II furcation defects of the Chacma baboon *Papio ursinus* (Ripamonti et al. 1994).

This was the forerunner of large-scale purifications of the putative osteogenetic proteins from large quantities of bovine bone matrices. This yielding purified proteins to homogeneity (Sampath et al. 1987; Wang et al. 1988; Luyten et al. 1999; Ripamonti et al. 1992). Purified proteins were subjected to amino acid sequencing, followed by molecular cloning and expression of the recombinant human proteins, and collectively defined as bone morphogenetic proteins (BMPs) (Wozney et al. 1988; Ozkäyanak et al. 1990; Celeste et al. 1990; Sampath et al. 1990).

Purification and cloning of the BMP family set the stage for testing of the recombinant human proteins in pre-clinical and clinical studies (Ripamonti and Reddi 1995; Ripamonti 2006).

Research studies have shown that osteogenic members of the TGF-β supergene family (Ripamonti 2003) are conserved and phylogenetically expressed several million years before the emergence of vertebrate animals including the speciation of the *Homo* clade. TGF-β family members are expressed and synthesized in a variety of animal forms as different as the fruit fly *Drosophila melanogaster*, the nematode *Caenorhabditis elegans*, the fresh-water zebrafish *Danio rerio* and the common platanna African clawed frog, *Xenopus laevis*.

The induction of bone formation, and of the skeleton, was then possible firstly by recapitulating embryonic development, and secondly by synthesizing a large set of multifunctional pleiotropic morphogens capable of the induction of bone formation (Wozney et al. 1988; Ozkäyanak et al. 1990; Celeste et al. 1990; Sampath et al. 1990; Reddi 2000; Ripamonti 2006). Such morphogens individually directly, synergistically and synchronously (Ripamonti et al. 2015 Ripamonti et al. 2016) generated the induction of bone formation and thus skeletogenesis. The generation of the skeleton, "a giant molecular machine" (Reddi 2001, personal communication), controlled morphologically, mechanically, endocrinologically, ionically and molecularly the emergence of the vertebrates and the induction of the *Homo* clade.

We alluded above to the pleiotropic vastness of members of the TGF-β supergene family, whereby gene products of the fruit fly *Drosophila melanogaster*, when implanted in recombinant form in the heterotopic subcutaneous space of mammals, initiate the induction of endochondral bone formation (Sampath et al. 1993).

Interesting data have shown how opposing cellular responses of fibrosis and new myocardium formation are spatially and temporally coordinated during heart regeneration in the zebrafish *Danio rerio* through Smad3-dependent TGF-β signalling. Such signalling orchestrates the interplay between scar-based repair and cardiomyocyte-based regeneration (Chablais and Jaźwińska 2012).

Members of the TGF-β superfamily of proteins are involved in dorso-ventral patterning in the nematode *Caenorhabditis elegans* and the African clawed frog *Xenopus laevis* (De Robertis 2008). We have indicated that Nature's plans were to usurp already operating DNA base pairs and selected gene products to control

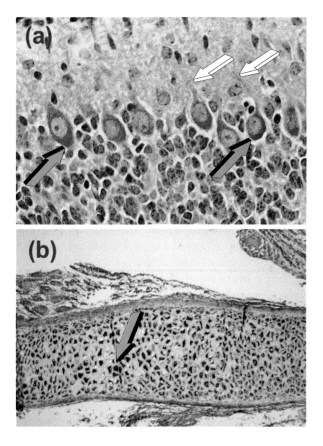

FIGURE 3.6 Pleiotropism and apparent redundancy of soluble molecular signals of the transforming growth factor-β (TGF-β) supergene family controlling the induction of several organs and tissues of the mammalian body. (a) BMP-3 immunolocalization in the cytoplasm of Purkinjie cells (*light blue* arrows) in the cerebellum of 13-day-old pups (Thomadakis et al. 1999). Immunolocalization of BMP-3 is also detected in the axonal neurite of the cells (*white* arrows). Note the prominent stain within the cytoplasm of Purkinjie cells (*light blue* arrows). What is the rationale and biological function of the BMP-3 gene product within Purkinjie cells cytoplasm and axonal neurite? (b) *In situ* hybridization of osteogenic protein-1 (OP-1/BMP-7) mRNA in the developing femur of a human embryo (Ripamonti et al. 2006).

the emergence of bone as an organ, and later of the skeleton. The emergence of the soluble osteogenic molecular signals of the TGF-β supergene family was then masterminded with a comparatively limited amino acid sequence variation in the operational, highly conserved carboxy-terminal domains (Wozney et al. 1988; Reddi 2000; Ripamonti 2006; Ripamonti et al. 2015).

This resulted in the synthesis and expression of several multifaceted pleiotropic gene products, controlling not just the induction of bone formation and skeletogenesis but the splanchnic mass, the peripheral and central nervous systems, the kidneys and the lungs, as well as the mammalian masticatory apparatus masterminding tooth development, periodontal tissue induction and cementogenesis (Ripamonti 2007; Ripamonti 2019).

Perhaps it is worth recording as the final image for this chapter on the osteogenic proteins of the TGF-β supergene family (Fig. 3.6a) (Ripamonti 2003) the as yet unknown multiple pleiotropic roles of the *BMP-3* gene and gene product and the biological rationale of BMP-3 expression in the cerebellum and in the cytoplasm of its Purkinje's cells (Fig. 3.6a) (Thomadakis et al. 1999).

REFERENCES

Bridges, J.B.; Pritchard, J.J. Bone and Cartilage Induction in the Rabbit. *J. Anat.* 1958, *92*(1), 28–38.

Celeste, A.J.; Iannazzi, J.M.; Taylor, J.A.; Hewick, R.C.; Rosen, V.; Wang, E.A.; Wozney, J.M. Identification of Transforming Growth Factor-β Family Members Present in Bone-Inductive Protein Purified from Bovine Bone. *Proc. Natl. Acad. Sci. U.S.A.* 1990, *87*(24), 9843–47.

Cell Editorial. Pulling It All Together. *Cell* 2014, *157*(1), 1–2. doi:10.1016/j.cell.2014.03.22.

Chablais, F.; Jaźwińska, A. The Regenerative Capacity of the Zebrafish Heart Is Dependent on TGFβ Signalling. *Development* 2012, *139*(11), 1921–30. doi:10.1242/dev.078543.

De Robertis, E.M.; Sasai, Y.; Common, A. A Common Plan for Dorsoventral Patterning in Bilateria. *Nature* 1996, *380*(6569), 37–40.

De Robertis, E.M. Evo-Devo: Variations on Ancestral Themes. *Cell* 2008, *132*(2), 185–95. doi:10.1016/j.cell.2008.01.003.

Ferretti, C.; Ripamonti, U.; Tsiridis, E.; Kerawala, C.J.; Mantalaris, A.; Heliotis, M. Osteoinduction: Translating Preclinical Promises into Reality. *Br. J. Oral Maxillofac. Surg.* 2010, *48*(7), 536–39.

Ferretti, C.; Ripamonti, U. Long-Term Follow-Up of Paediatric Mandibular Reconstruction with Human Transforming Growth Factor-β₃. *J. Craniofac. Surg.* 2020, *31*(5), 1424–29.

Huggins, C.B. The Formation of Bone under the Influence of Epithelium of Urinary Tract. *Arch. Surg.* 1931, *32*, 915–31.

Jung, H. Mazzoni, E.O. Soshnikova, N. Hanley, O. Venkatesh, B. Duboule, D. Dasen, J. Evolving *Hox* Activity Profiles Govern Diversity in Locomotor Systems. *Developmental Cell* 2014, **29**, 171–87.

Jung, H.; Baek, M.; D'Elia, K.; Boisvert, C.; Currie, P.D.; Tay, B.-H.; Venkatesh, B.; Brown, S.T.; Heguy, A.; Schoppik, D.; Dasen, J.S. The Ancient Origins of Neural Substrates for Land Walking. *Cell* 2018, *172*(4), 667–82. doi:10.1016/j.cell.2018.01.013.

Klar, M.R.; Duarte, R.; Dix-Peek, T.; Dickens, C.; Ferretti, C.; Ripamonti, U. Calcium Ions and Osteoclastogenesis Initiate the Induction of Bone Formation by Coral-Derived Macroporous Constructs. *J. Cell. Mol. Med.* 2013, *17*(11), 1444–57.

Klar, M.R.; Duarte, R.; Dix-Peek, T.; Ripamonti, U. The Induction of Bone Formation by the Recombinant Human Transforming Growth Factor-β₃. *Biomaterials* 2014, *35*(9), 2773–88. doi:10.1016/j.biomaterials.2013.12.062.

Levander, G. A Study on Bone Regeneration. *Surg. Obstetr.* 1938, *67*, 705–14.

Levander, G. Tissue Induction. *Nature* 1945, *155*(3927), 148–49.

Levander, G.; Willestaedt, H. Alcohol-Soluble Osteogenetic Substance from Bone Marrow. *Nature* 1946, *3992*, 587.

Luyten, F.P.; Cunningham, N.S.; Ma, S.; Muthukumaran, N.; Hammonds, R.G.; Nevins, W.B.; Wood, W.I.; Reddi, A.H. Purification and Partial Amino Acid Sequence of Osteogenin, a Protein Initiating Bone Differentiation. *J. Biol. Chem.* 1999, *264*(23), 13377–80.

Manolagas, S.C.; Jilka, R.L.; Marrow, Bone Bone Marrow, Cytokines, and Bone Remodeling. Emerging Insights into the Pathophysiology of Osteoporosis. *N. Engl. J. Med.* 1995, *332*(5), 305–11.

Moss, M.L. Extraction of an Osteogenic Inductor Factor from Bone. *Science* 1958, *127*(3301), 755–56.

Özkaynak, E.; Rueger, D.C.; Drier, E.A.; Corbett, C.; Ridge, R.J.; Sampath, T.K.; Oppermann, H. OP-1 cDNA Encodes an Osteogenic Protein in the TGF-Beta Family. *E.M.B.O. J.* 1990, *9*(7), 2085–93.

Parfitt, A.M. Osreonal and Hemi-Osteonal Remodeling: The Spatial and Temporal Framework for Signal Traffic in Adult Human Bone. *J. Cell. Biochem.* 1994, **55**, 273–86.

Reddi, A.H. Cell Biology and Biochemistry of Endochondral Bone Development. *Coll. Rel. Res.* 1981, *1*(2), 209–26.

Reddi, A.H.; Kuettner, K.E. Vascular Invasion of Cartilage: Correlation of Morphology with Lysozyme, Glycosaminoglycans, Protease, and Protease-Inhibitory Activity During Endochondral Bone Development. *Dev. Biol.* 1981, *82*(2), 217–23.

Reddi, A.H. Symbiosis of biotechnology and biomaterials: Applications in tissue engineering of bone. *J. Cell. Biochem.* 1994, *56*(2), 192–95.

Reddi, A.H. Bone Morphogenesis and Modeling: Soluble Signals Sculpt Osteosomes in the Solid State. *Cell* 1997, *89*(2), 159–61.

Reddi, A.H. Morphogenesis and Tissue Engineering of Bone and Cartilage: Inductive Signals, Stem Cells, and Biomimetic Matrices. *Tissue Eng.* 2000, *6*(4), 351–59.

Ripamonti, U.; Magan, A.; Ma, S.; van den Heever, B.; Moehl, T.; Reddi, A.H. Xenogeneic Osteogenin, a Bone Morphogenetic Protein, and Demineralized Bone Matrices, Including Human, Induce Bone Differentiation in Athymic Rats and Baboons. *Matrix* 1991, *11*(6), 404–11.

Ripamonti, U.; Ma, S.; Cunningham, N.; Yeates, L.; Reddi, A.H. Initiation of Bone Regeneration in Adult Baboons by Osteogenin, a Bone Morphogenetic Protein. *Matrix* 1992, *12*(5), 369–80.

Ripamonti, U.; Ma, S.; Reddi, A.H. The Critical Role of Geometry of Porous Hydroxyapatite Delivery System in Induction of Bone by Osteogenin, a Bone Morphogenetic Protein. *Matrix* 1992, *12*(3), 202–12.

Ripamonti, U.; Heliotis, M.; van den Heever, B.; Reddi, A.H. Bone morphogenetic proteins induce periodontal regeneration in the baboon. *J. Periodontal Res.* 1994, *29*(6), 439–45.

Ripamonti, U.; Reddi, A.H. Bone Morphogenetic Proteins: Applications in Plastic and Reconstructive Surgery. Adv. *Plast. Reconstr. Surg.* 1995, *11*, 47–73.

Ripamonti, U.; Duneas, N. Tissue Engineering of Bone by Osteoinductive Biomaterials. *MRS Bulletin* November 1996, 36–9.

Ripamonti, U. Osteogenic Proteins of the Transforming Growth Factor-ß Superfamily. In: H.L. Henry, A.W. Norman (eds.) *Encyclopedia of Hormones*, Academic Press, San Diego, CA 2003, 80–6.

Ripamonti, U. Soluble, Insoluble and Geometric Signals Sculpt the Architecture of Mineralized Tissues. *J. Cell. Mol. Med.* 2004, *8*(2), 169–80.

Ripamonti, U.; Ramoshebi, L.N.; Patton, J.; Matsaba, T.; Teare, J.; Renton, L. Soluble Signals and Insoluble Substrata: Novel Molecular Cues Instructing the Induction of Bone. In: E.J. Massaro, J.M. Rogers (eds.) *The Skeleton*, Humana Press, Totowa, New Jersey 2004, Chapter 15, 217–27.

Ripamonti, U. Soluble Osteogenic Molecular Signals and the Induction of Bone Formation. Biomaterials Leading Opinion Paper 2006, *27*(6), 807–22.

Ripamonti, U.; Ferretti, C.; Heliotis, M. Soluble and Insoluble Signals and the Induction of Bone Formation: Molecular Therapeutics Recapitulating Development. *J. Anat.* 2006, *209*(4), 447–68.

Ripamonti, U.; Heliotis, M.; Ferretti, C. Bone Morphogenetic Proteins and the Induction of Bone Formation: From Laboratory to Patients. *Oral Maxillofac. Surg. Clin. North Am.* 2007, *19*(4), 575–89.

Ripamonti, U. Biomimetism, Biomimetic Matrices and the Induction of Bone Formation. *J. Cell. Mol. Med.* 2009, *13*(9B), 2953–72.

Ripamonti, U. Soluble and Insoluble Signals Sculpt Osteogenesis in Angiogenesis. *World J. Biol. Chem.* 2010, *26*(5), 109–32.

Ripamonti, U. The Concavity: The "Shape of Life"and the Control of Bone Differentiation – Feature Paper – *Science in Africa* **May** 2012.

Ripamonti, U.; Roden, L.; Renton, L.; Klar, R.M.; Petit, J.-C. The Influence of Geometry on Bone: Formation by Autoinduction. *Science in Africa* 2012. http://www.scienceinafric a.co.za/2012/Ripamonti_bone.htm.

Ripamonti, U.; Duarte, R.; Ferretti, C. Re-Evaluating the Induction of Bone Formation in Primates. *Biomaterials* 2014, *35*(35), 9407–22.

Ripamonti, U.; Dix-Peek, T.; Parak, R.; Milner, B.; Duarte, R. Profiling Bone Morphogenetic Proteins and Transforming Growth Factor-Bs by hTGF-β_3 Pre-Treated Coral-Derived Macroporous Constructs: The Power of One. *Biomaterials* 2015, *49*, 90–102.

Ripamonti, U.; Parak, R.; Klar, R.M.; Dickens, C.; Dix-Peek, T.; Duarte, R. The Synergistic Induction of Bone Formation by the Osteogenic Proteins of the TGF-β Supergene Family. *Biomaterials* 2016, *104*, 279–96.

Ripamonti, U. Biomimetic Functionalized Surfaces and the Induction of Bone Formation. *Tissue Eng.* 2017, *23*(21,22), 1197–209.

Ripamonti, U. Introductory Remarks on Cartilages, Bones and on "*Bone: Formation by Autoinduction*". *S. Afr. Dent. J.* 2018, *73*(1), 6–10.

Ripamonti, U.; Roden, L.; van den Heever, B. Sharks, Sharks Cartilages and Shark Teeth: A Collaborative Africa-USA Study to Attempt to Induce "*Bone: Formation by Autoinduction*" in Cartilaginous Fishes. *S. Afr. Dent. J.* 2018, *73*(1), 11–21.

Ripamonti, U.; Duarte, R. Inductive Surface Geometries: Beyond Morphogens and Stem Cells. *S. Afr. Dent. J.* 2019, *74*(8), 421–44.

Romer, A.S. The "Ancient History" of Bone. *Ann. N. Y. Acad. Sci.* 1963, *109*, 168–76.

Sampath, T.K.; Reddi, A.H. Dissociative Extraction and Reconstitution of Extracellular Matrix Components Involved in Local Bone Differentiation. *Proc. Natl. Acad. Sci. U.S.A.* 1981, *78*(12), 7599–603.

Sampath, T.K.; Reddi, A.H. Homology of Bone-Inductive Proteins from Human, Monkey, Bovine, and Rat Extracellular Matrix. *Proc. Natl. Acad. Sci. U.S.A.* 1983, *80*(21), 6591–95.

Sampath, T.K.; Muthukumaran, N.; Reddi, A.H. Isolation of Osteogenin, an Extracellular Matrix-Associated Bone-Inductive Protein, by Heparin Affinity Chromatography. *Proc. Natl. Acad. Sci. U.S.A.* 1987, *84*(20), 7109–13.

Sampath, T.K.; Coughlin, J.E.; Whetstone, R.M.; Banach, D.; Corbett, C.; Ridge, R.J.; Özkaynak, E.; Oppermann, H.; Rueger, D.C. Bovine Osteogenic Protein Is Composed of Dimers of OP-1 and BMP-2A, Two Members of the Transforming Growth Factor-Beta Superfamily. *J. Biol. Chem.* 1990, *265*(22), 13198–205.

Sampath, T.K.; Rashka, K.E.; Doctor, J.S.; Tucker, R.F.; Hoffmann, F.M. Drosophila Transforming Growth Factor-β Superfamily of Proteins Induce Endochondral Bone Formation in Mammals. *Proc. Natl. Acad. Sci. U.S.A.* 1993, *90*(13), 6004–08.

Thomadakis, G.; Crooks, J.; Rueger, D.; Ripamonti, U. Immunolocalization of Bone Morphogenetic Protein-2, –3 and Osteogenic Protein-1 During Murine Tooth Morphogenesis and Other Craniofacial Structures. *Eur. J. Oral Sci.* 1999, *107*, 368–77.

Trueta, J. The Role of the Vessels in Osteogenesis. *J. Bone Joint Surg.* 1963, *45B*, 402–18.

Turing, A.M. The Chemical Basis of Morphogenesis. *Philos. Trans. R. Soc. Lond.* 1952, *237*, 27–41.

Urist, M.R.; McLean, F.C. Osteogenic Potency and New Bone Formation by Induction in Transplants to the Anterior Chamber of the Eye. *J. Bone Joint Surg. Am* 1952, *34A*, 443–76.

Urist, M.R. Bone: Formation by Autoinduction. *Science* 1965, *220*(3698), 893–99. doi:10.1126/science.150.3698.893.

Urist, M.R.; Silverman, F.; Büring, K.; Dubuc, F.L.; Rosenberg, J.M. The Bone Induction Principle. *Clin. Orthop. Rel. Res.* 1967, *53*, 243–83.

Venkatesh, B.; Lee, A.P.; Ravi, V.; Maurya, A.K.; Lian, M.M.; Swarm, J.B.; Ohta, Y.; Flajnik, M.F.; Sutoh, Y.; Kasahara, M.; Hoon, S.; Gangu, V.; Roy, S.W.; Irimia, M.; Korzh, V.; Kondrychyn, I.; Lim, Z.W.; Tay, B.H.; Tohari, S.; Kong, K.W.; Ho, S.; Lorente-Galdos, B.; Quilez, J.; Marques-Bonet, T.; Raney, B.J.; Ingham, P.W.; Tay, A.; Hillier, L.W.; Minx, P.; Boehm, T.; Wilson, R.K.; Brenner, S.; Warren, W.C. Elephant Shark Genome Provides Unique Insights into Gnathostome Evolution. *Nature* 2014, *505*(7482), 174–79.

Wang, E.A.; Rosen, V.; Cordes, P.; Hewick, R.M.; Kriz, M.J.; Luxenberg, D.P.; Sibley, B.S.; Wozney, J.M. Purification and Characterization of Other Distinct Bone-Inducing Factors. *Proc. Natl. Acad. Sci. U.S.A.* 1988, *85*(24), 9484–88.

Wozney, J.M.; Rosen, V.; Celeste, A.J.; Mitsock, L.M.; Whitters, M.J.; Kriz, R.W.; Hewick, R.M.; Wang, E.A. Novel Regulators of Bone Formation: Molecular Clones and Activities. *Science* 1988, *242*(4885), 1528–34.

4 Coral-Derived Hydroxyapatite-Based Macroporous Bioreactors Initiate the Spontaneous Induction of Bone Formation in Heterotopic Extraskeletal Sites
Morphological Time Studies

Ugo Ripamonti

4.1 SELF-INDUCING OSTEOINDUCTIVE BIOMIMETIC MATRICES

The central question in developmental biology, and thus regenerative medicine and tissue engineering alike, is the molecular basis of pattern formation (Reddi 1984; Lander 2007; De Robertis 2008). Developmental strides in understanding tissue induction and morphogenesis eventually discovered morphogenetic soluble molecular signals, or morphogens, that initiate pattern formation, tissue induction and morphogenesis (Turing 1952; Reddi 1984; Reddi 1997; Lander 2007; De Robertis 2008; Kerzberg and Wolpert 2007; Wolpert 1996).

Regenerative medicine is the grand multidisciplinary challenge of molecular, cellular and evolutionary biology requiring the integration of tissue biology, tissue engineering, developmental biology and experimental surgery to explore how to trigger *de novo* and *ex novo* tissues and organs in man. We previously stated in a Special Topic in *Plastic and Reconstructive Surgery* (Ripamonti and Duneas 1998) that, as a prelude to morphogenesis, the genesis of form and function, the generation of cellular diversity, or differentiation, must first occur (Reddi 1981; Reddi 1984; Lander 2007; De Robertis 2008; Kerzberg and Wolpert 2007; Wolpert 1996; Reddi 2000). As a result, there must exist several signalling molecules, or "morphogens", which were first defined by Turing as "forms generating substances" (Turing 1952).

Morphogens are expressed and secreted by a variety of differentiating cells that control and guide differentiating pathways acting on progenitor and responding cells, including stem cells, initiating pattern formation and the attainment of tissue form and function (Lander 2007; De Robertis 2008; Kerzberg and Wolpert 2007; Wolpert 1996; Reddi 2000; Ripamonti and Duneas 1998; Ripamonti 2006).

As briefly discussed in Chapters 1 and 2, the emerging novel question of bio-materials science is whether biomaterials scientists, developmental and molecular biologists alike can assemble self-initiating biomimetic matrices that *per se* induce desired and specific morphogenetic responses from the host tissues without the addi-tion of exogenously applied soluble molecular signals of the transforming growth factor-β (TGF-β) supergene family (Ripamonti 2003), powerful initiators of tissue induction and morphogenesis (Wozney et al. 1988; Reddi 2000; Ripamonti 2006; Ripamonti 2010; Ripamonti et al. 2007; Ripamonti 2012).

What next beyond morphogens and stem cells? The novel and exciting concept of bone tissue engineering is to develop *smart* functionally bioinspired biomimetic matrices (Ripamonti et al. 1993a; Ripamonti et al. 1999; Ripamonti 2012; Ripamonti et al. 2008; Ripamonti 2014; Ripamonti 2017; Ripamonti 2018; Ripamonti and Duarte 2019) as finally reported in *Science* (Christman 2019). Such matrices *per se* initiate the spontaneous and/or intrinsic induction of specialized tissue formation without the exogenous application of soluble molecular signals of the TGF-β supergene fam-ily (Ripamonti et al. 1993a; Ripamonti 1996; Ripamonti al. 1999).

Biomaterials for tissue repair that could "promote exogenous healing without delivering cells or therapeutics" were reported, however (Christman 2019), without describing or referencing the paradigmatic shift from carrier matrices combined with soluble molecular signals to biomimetic matrices *solo* as already reported in the last century for the induction of bone formation by *smart* biomimetic matrices (Ripamonti 1991; Ripamonti et al. 1993a; Ripamonti 1996; Ripamonti et al. 1999).

The communication in *Science* (Christman 2019) regretfully ended by stating "biomaterial scaffolds can recreate the microenvironment and influence the immune system and tissue regeneration, they are already poised to have immediate patient impact and represent an alternative paradigm for regenerative medicine" without quoting prior publications having stated that macroporous *smart* biomimetic matrices are changing the paradigm of bone tissue engineering by creating matrices that *per se* initiate the induction of bone differentiation without the exogenous applications of soluble osteogenic molecular signals of the TGF-β supergene family (Ripamonti et al. 1993a; Ripamonti 1996; Ripamonti et al. 1999; Ripamonti 2017; Ripamonti 2018).

Smart functionalized biomimetic surfaces in their own right can initiate the ripple-like cascade of bone differentiation by induction (Ripamonti 1990; Ripamonti 1991; Ripamonti et al. 1993a; Ripamonti 1996; Ripamonti and Duneas 1996; Ripamonti et al. 1999; Ripamonti et al. 2001; Klar et al. 2013; Ripamonti 2004; Ripamonti 2017; Ripamonti 2018).

A number of systematic studies in non-human primates have firmly established that certain macroporous hydroxyapatites obtained from the calcium carbon-ate exoskeletons of coral *Goniopora* are endowed with the striking prerogative of spontaneously inducing bone within the macroporous bioreactors. Remarkably, the induction of bone within the macroporous spaces of the coral exoskeleton initiates after implantation in heterotopic *rectus abdominis* sites of the Chacma baboon *Papio ursinus*, where there is no bone (Ripamonti 1990; Ripamonti 1991; Ripamonti et al. 1993a; Ripamonti 1996; Ripamonti 2000; Klar et al. 2013; Ripamonti 2013; Ripamonti 2017; Ripamonti 2018; Ripamonti and Duarte 2019).

Sequential systematic studies investigated the induction of tissue patterning and organization by coral-derived bioreactors harvested from the *rectus abdominis* of *Papio ursinus* on days 15, 30, 60, 90, 180, 270 and 365 (Klar et al. 2013; Ripamonti 1991; Ripamonti et al. 1993a; Ripamonti 1996; Ripamonti et al. 2009; Ripamonti et al. 2010).

The molecular studies resolving the spontaneous and/or intrinsic induction of bone formation are reported in other chapters mechanistically assigning the induction of bone formation by coral-derived bioreactors when implanted intramuscularly in *Papio ursinus* (Ripamonti et al. 1993a; Klar et al. 2013; Ripamonti et al. 2014; Ripamonti et al. 2015; Ripamonti 2017).

This chapter focuses on the time study of the induction of bone formation from days 15 to 365, providing morphological highlights on the induction of bone formation by the heterotopically implanted bioreactors. This complete set of data was prepared from several specimens implanted in the Chacma baboon *Papio ursinus* across the century from 1988 to 2018 spanning more than 30 years of focused research investigations on the spontaneous induction of bone formation by calcium phosphate-based bioreactors (Ripamonti 2017; Ripamonti 2018).

Data were generated and implemented by the writer from the very beginning with histological techniques always modified for superior clarity initiated and masterminded by Barbara van den Heever who managed to train several histologists on K-Plast resin-embedded undecalcified blocks for superior sectioning on rotary sledge heavy duty Leica microtomes (Ripamonti et al. 1993a). Later, Ruqayya Parak added the third dimension of the undecalcified sectioning cut on the Exakt diamond saw grinding and polishing equipment. Both undecalcified histological techniques provided significant insights into the induction of bone formation as initiated by coral-derived bioreactors (Ripamonti et al. 1993a) which were later mechanistically assigned (Klar et al. 2013; Ripamonti et al. 2015; Ripamonti 2017).

The spontaneous induction of bone formation that initiates bone morphogenesis in calcium phosphate-based bioreactors when implanted in heterotopic *rectus abdominis* sites of *Papio ursinus* (Ripamonti 1990; Ripamonti 1991) soon developed as a multi-centred research effort across the globe (Yuan et al. 2010; Barradas et al. 2011; Ripamonti 2017; Ripamonti 2018). Of note, Yamasaki reported heterotopic bone formation around granular porous hydroxyapatites subcutaneously in mongrel dogs by day 90 after implantation (Yamasaki 1990). In contrast, there was no evidence of bone formation but only of fibrous encapsulation after implantation of dense hydroxyapatite granules (Yamasaki 1990; Yamasaki and Sakai 1992). The studies suggested that unspecified "physical and chemical factors" in the microenvironment provided by the granular hydroxyapatites might have initiated the induction of bone formation (Yamasaki 1990; Yamasaki 1992).

Our first demonstration of the morphogenesis of bone in the macroporous spaces of coral-derived bioreactors harvested on day 90 from the *rectus abdominis* muscle of *Papio ursinus* proposed the induction of patterned collagenous condensations as a prerequisite for the later induction of bone formation (Ripamonti 1990). Our communications suggested that bone-differentiating factors or osteogenins, then recently isolated and partially purified from bovine bone matrices (Luyten et al. 1999), could be absorbed onto the hydroxyapatite substratum (Ripamonti 1990; Ripamonti 1991).

The morphogenesis of bone in replicas of porous hydroxyapatite obtained from the conversion of calcium carbonate exoskeletons of coral was presented by a large morphological time study in the *rectus abdominis* muscle of *Papio ursinus* on days 90, 180 and 270 (Ripamonti 1991). The morphogenesis of bone was intimately associated with differentiation of connective tissue condensations at the hydroxyapatite interface (Ripamonti 1991). The induction of bone formation was characterized by a prominent vascular component (Ripamonti 1991). As reported by Yamasaki in canines (1990), the induction of bone in the Chacma baboon *Papio ursinus* also formed by direct intramembranous ossification without cartilage differentiation (Ripamonti 1990; Ripamonti 1991). Our interpretation then was that circulating osteogenins (Luyten et al. 1999) could have been adsorbed onto the hydroxyapatite substratum, initiating the induction of bone formation as a secondary response (Ripamonti 1991; Ripamonti et al. 1993a). Indeed, our interpretation was substantiated by the observation that several schemes for growth factor purification involved hydroxyapatite chromatography (Urist et al. 1984; Luyten et al. 1999). Adsorption of mammalian osteogenins onto hydroxyapatite gels is a fundamental step for their purification (Urist et al. 1984; Luyten et al. 1999; Ripamonti et al. 1992a).

The adsorption of bone morphogenetic proteins onto the substratum of hydroxyapatite was then exploited to construct an osteogenic delivery system after chromatographic adsorption of highly purified osteogenic fractions onto coral-derived macroporous bioreactors (Figs. 4.1, 4.2) (Ripamonti et al. 1993b). Protein fractions after gel filtration chromatography onto tandem Sephacryl S-200 columns (Ripamonti et al. 1992a) (Fig. 4.1a inset) with biological activity in the subcutaneous space of the rat, as determined by alkaline phosphatase activity of harvested and homogenized heterotopic implants (Fig. 4.1a), were loaded onto macroporous constructs, 25 mm in diameter, implanted into calvarial defects of the Chacma baboon *Papio ursinus* (Ripamonti et al. 1992b). Histological analyses showed the induction of bone formation across the macroporous spaces by day 30 after implantation (Fig. 4.1b) (Ripamonti et al. 1992b).

An additional batch of baboon demineralized bone matrix (2 kg starting material) was chaotropically extracted in 4 M guanidine hydrochloride (Gdn-HCl), 50 Mm TRIS HCl, pH 7.4, containing protease inhibitors as described (Ripamonti et al. 1993b). Extracted proteins were exchanged with 6 M urea and sequentially chromatographed onto hydroxyapatite-Ultrogel (IBF Biotechnics) adsorption and heparin-Sepharose (Pharmacia LKB) affinity chromatography columns, eluted and washed as described (Luyten et al. 1999; Ripamonti et al. 1992a). Osteogenic eluted fractions with osteogenic activity in the rodent subcutaneous bioassay were separated from high molecular weight components in a single peak (Figs. 4.1c,d). Protein fractions with osteogenic activity in the subcutaneous space of the rat (Fig. 4.1c, inset) (Ripamonti et al. 1993b) were pooled and exchanged with 8 volumes of 5 mM

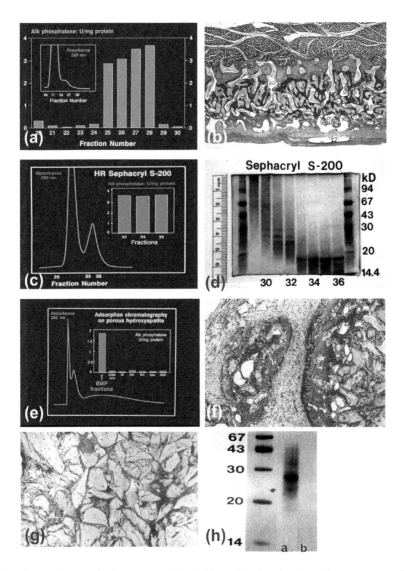

FIGURE 4.1 Reconstituting osteogenic soluble molecular signals with macroporous biomimetic matrices by hydroxyapatite adsorption chromatography for later surgical implantation in the Chacma baboon *Papio ursinus*. (a) Alkaline phosphatase activity on day 11 of 25 mg of rat insoluble collagenous bone matrix (ICBM) reconstituted with baboon Sephacryl S-200 gel filtration fractions (inset) bioassayed in the subcutaneous space of Long–Evans rats. The osteogenic activity is confined to fractions 25 to 28 with an apparent molecular mass range of 26–42 kDa on sodium dodecyl sulphate polyacrylamide gel electrophoresis (not shown). (b) Osteogenic fractions in 5 mM HCl were combined with coral-derived macroporous constructs (Ripamonti et al. 1992b). Tissue specimens were harvested on day 30 and showed the induction of bone formation (b) across the central regions of the macroporous coral-derived constructs (Ripamonti et al. 1992b). (c) Three lots of 2–2.3 kg of baboon bone matrix starting material were extracted with 4 M guanidinium HCl, purified

FIGURE 4.1 (CONTINUED)
by heparin-Sepharose, hydroxyapatite-Ultrogel and gel filtration chromatography on tandem S-200 Sephacryl columns (Ripamonti et al. 1992a; Ripamonti et al. 1993b). Protein fractions with osteogenic activity in the rat subcutaneous bioassay were pooled and loaded onto tandem S-200 Sephacryl gel filtration chromatography columns. The osteogenic activity was isolated in a single peak (c) with osteogenic activity confined to fractions 33 to 35 with high alkaline phosphatase activity (c inset). Sodium dodecyl sulphate polyacrylamide gel electrophoresis (SDS-PAGE) (d) shows that protein fractions were separated by high molecular weight contaminants in a single peak with high osteogenic activity in the subcutaneous space of Long–Evans rats as determined by alkaline phosphates activity (inset c) and histological analyses (Ripamonti et al. 1993b). (e) Highly purified osteogenic factions were pooled and exchanged with 5 mM HCl acid and loaded, in a final concentration of 4.2 mg in 4 ml of 5 mM HCl, onto a Pharmacia LKB chromatography column, 25 mm of internal diameter, containing discs of coral-derived macroporous hydroxyapatite substrata, 25 mm in diameter (Fig. 4.2a). Coral-derived discs, acting as hydroxyapatite adsorption chromatography substratum, were equilibrated in 5 mM HCl (20 ml of column volume) (Ripamonti et al. 1993b). Adsorption of the highly purified osteogenic fractions was monitored at absorbance of 280 nm (e). Before adsorption chromatography on the coral-derived biomatrices packed onto the Pharmacia column, pooled osteogenic fractions were bioassayed in the subcutaneous tissue of the rat. Harvested ossicles showed alkaline phosphatase activity (e inset BMPs' fractions) with newly formed bone (f) populated by contiguous osteoblasts. After loading, the coral-derived hydroxyapatite adsorption chromatography column was run in close circuit, recirculating the hydroxyapatite eluate. The column was left to dry, and the eluate (20 ml) was collected and bioassayed for potential residual osteogenic activity. Protein fractions in 50, 200, 500, 1000 and 2000 μl eluate were reconstituted with rat ICBM and bioassayed in the subcutaneous space of the rat (Ripamonti et al. 1993b). There was lack of alkaline phosphatase activity (e inset) which was confirmed by lack of bone formation by induction on harvested implants on day 11 (g). (h) 15% SDS silver stained gel under non-reducing conditions (Ripamonti et al. 1993b). Lane a, proteins fractions before adsorption chromatography on coral-derived macroporous constructs (Fig. 4.2a). Lane b: apparent lack of proteins in the hydroxyapatite eluate. Molecular mass markers are given in kDa.

HCl using diafiltration membranes (Amicon YM-10). Pooled fractions were chromatographed onto macroporous coral-derived hydroxyapatite constructs in disc configuration, 25 mm in diameter, inserted into a Pharmacia LKD chromatography column, 25 mm internal diameter (Fig. 4.2a). Confirmation of the osteogenic activity before adsorption chromatography of the loaded protein fractions was obtained by the rodent subcutaneous assay (Fig. 4.1f) (Reddi and Huggins 1972; Ripamonti et al. 1993b). Histological results showed the prominent induction of bone formation by pooled osteogenic fractions (Fig. 4.1f) before chromatography onto coral-derived hydroxyapatite bioreactors. These were made in macroporous discs configuration 25 mm in diameter for proper adaptation into a Pharmacia LKD column, 25 mm internal diameter (Fig. 4.2a) (Ripamonti et al. 1993b).

The chromatographic profile of highly purified pooled BMPs' fractions loaded onto coral-derived macroporous hydroxyapatite constructs inserted into a Pharmacia LKD chromatography column (Fig. 4.2a) shows the osteogenic activity of the pooled osteogenic fractions (Fig. 4.1e,f) separated by a single chromatographic peak during adsorption chromatography (Fig. 4.1e). There was a lack of osteogenic activity

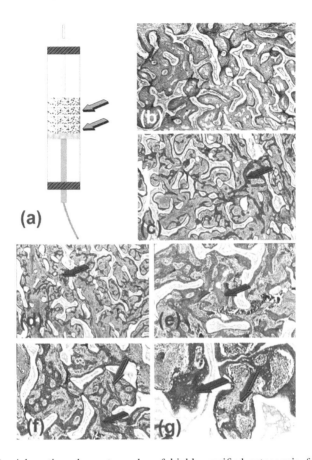

FIGURE 4.2 Adsorption chromatography of highly purified osteogenic fractions onto coral-derived calcium phosphate discs. (a) Coral-derived macroporous discs of crystalline hydroxyapatite packed into a Pharmacia LKD chromatography column acting as the column substratum for adsorption chromatography (Ripamonti et al. 1993b). (b) Control coral-derived construct without adsorption of highly purified osteogenic fractions showing lack of bone formation 30 days after heterotopic intramuscular implantation. Fibrovascular tissue ingrowth within the macroporous spaces but lack of bone differentiation. (c,d,e,f,g) Series of morphological images of coral-derived macroporous bioreactors loaded with highly purified naturally derived osteogenic fractions from baboon bone matrices (Ripamonti et al. 1993b). Induction of bone formation (*dark blue* arrows) across the macroporous spaces in coral-derived constructs after adsorption chromatography (Fig. 4.2a). Substantial induction of bone formation by day 30 after heterotopic *rectus abdominis* implantation. The induction of bone forms across the macroporous spaces together with substantial vascular invasion and angiogenesis (e). Newly formed woven bone invades the macroporous spaces three-dimensionally, networking across the macroporous bioreactors. (f,g) Woven bone generates within the macroporous spaces with plumped osteoblastic cells secreting new matrix across the trabeculae of the newly formed bone (*magenta* arrows in f,g). The induced newly formed bone engineers trabeculations with a three-dimensional network of fine trabeculae (*magenta* arrows) surrounded by sprouting capillaries filling the available macroporous spaces. Decalcified sections cut at 6 μm from paraffin-embedded blocks stained with a modified Goldner's trichrome.

of collagenous bone matrix reconstituted with non-osteogenic eluates after adsorption chromatography (Fig. 4.1g), demonstrating that pooled osteogenic fractions were chromatographically adsorbed onto the coral-derived macroporous substrata (Ripamonti et al. 1993b). The electrophoretic profile on a 15% acrylamide SDS silver stained gel shows the apparent lack of protein in the hydroxyapatite eluates (Fig. 4.1b, lane b), with the corresponding lack of biological activity in the heterotopic subcutaneous rodent bioassay (Fig. 4.1g) (Ripamonti et al. 1993b).

Figure 4.2 highlights the adsorption chromatography on macroporous discs of coral-derived constructs inserted into a Pharmacia LKD column, 25 mm internal diameter (Fig. 4.2a). The implantation of coral-derived macroporous constructs *solo* without chromatographic adsorption of pooled osteogenic fractions showed the lack of bone formation by day 30 after implantation (Fig. 4.2b). In marked contrast, the reconstitution and biological activity of the soluble molecular signals chromatographed and adsorbed onto the coral-derived bioreactors show a remarkable and extensive induction of bone differentiation by day 30 after heterotopic *rectus abdominis* intramuscular implantation (Figs. 4.2c,d,e,f,g). There is substantial angiogenesis and capillary sprouting as previously described in coral-derived constructs implanted in the *rectus abdominis* muscle (Ripamonti 1991; Ripamonti et al. 1993a). Newly formed bone by induction is seen across the macroporous spaces together with prominent angiogenesis (Figs. 4.2e,f,g) with the induction of fine highly cellular trabeculations covered by secreting osteoblasts forming elongating highly cellular matrices into the macroporous spaces (Figs. 4.2f,g).

The two-dimensional morphological digital images of Figures 4.2f and g directly illustrate complex morphogenetic three-dimensional interlacing newly formed highly cellular trabeculations spanning across the macroporous spaces amongst sprouting capillaries (Fig. 4.2f,g). Trabeculae, whilst depositing newly secreted osteoid matrix, elongate into the macroporous spaces and merge with other trabeculae enriched by osteoid seams to form a tri-dimensional engineered construct continuously sustained by capillary sprouting and invasion (Fig. 4.2f,g).

A short paragraph is needed to highlight the selection of the first sub-heading of this chapter; i.e. "Self-Inducing Osteoinductive Biomimetic Matrices". In 2011, the then editor-in-chief of *Biomaterials*, David Williams, published a focused editorial titled "The Continuous Evolution of Biomaterials", setting novel requirements for the journal, also outlining six sections under which submitted manuscripts should find a home when published after a further tightening of the review process requesting mechanistic data *vs.* no longer acceptable descriptive studies (Williams 2011).

In personal e-mail correspondence (October 25 2010), I suggested to the then editor-in-chief that he

> could still consider a section for biomaterials that per se induce molecular and cellular cascades of events leading to de novo engineering of new constructs, primarily osteogenesis and angiogenesis. Since I have found that the concavity does initiate the ripple-like cascade of bone formation by induction, I have always tried to define or find the proper definition for a biomaterial that *per se* initiates the induction of angiogenesis followed by osteogenesis, i.e. *osteogenesis in angiogenesis*. I have used the term *per se*

or *on its own right* but the terminology for a biomaterial that by virtue of its composition and/or geometry starts cellular differentiation and gene expression has still eluded my scientific imagination; I have thought of self-inducing or self-assembling biomaterials for the induction of bone formation.

Our research efforts and systematic data in *Papio ursinus* indicate that such bioactive *smart* biomimetic biomaterials should not be listed under "Biomaterials and the stem cell niche" or "Biomaterials and regenerative medicine". The suggestion was that such *smart* biomimetic matrices should perhaps deserve to have a specific section on their own as self-inducing and/or self-assembling biomimetic matrices (Ripamonti, personal communication, October 25, 2010).

When mailing David Williams for his agreement to quote the above e-mails, David indicated (May 2019) that a term was introduced when chairing a consensus conference on definitions of biomaterials as "Tissue-inducing biomaterials". Such a biomaterial was defined as "A biomaterial designed to induce the regeneration of damaged or missing tissues or organs without the addition of cells and/or bioactive factors" (Zhang and Williams 2019).

This chapter thus discusses *tissue-inducing biomaterials* or *smart* biomaterial biomimetic matrices that in their own right set into motion cell attachment, cell differentiation and proliferation. There is osteoclast and osteoblast cell differentiation; osteoclastic priming of the crystalline hydroxyapatite surfaces results in nanotopographical geometric configurations with Ca^{++} release within the concavities microenvironment. Ca^{++} release induces angiogenic cell differentiation together with the induction of the osteogenic phenotype on cells attached to the substratum. This is followed by *bone morphogenetic proteins'* expression and synthesis, and finally secretion and embedding of the osteogenic gene products into the *smart* geometrically nanotopographically modified concavities that self-initiate the induction of bone formation as a secondary response (Ripamonti et al. 1993a; Ripamonti 1996; Ripamonti et al. 1999; Ripamonti et al. 2001; Ripamonti 2004; Klar et al. 2013; Ripamonti 2017; Ripamonti 2018).

4.2 TISSUE INDUCTION AND MORPHOGENESIS BY CORAL-DERIVED HYDROXYAPATITE CONSTRUCTS ON DAY 15 AFTER HETEROTOPIC *RECTUS ABDOMINIS* IMPLANTATION

The earlier time studies analyzed by our laboratories were 15-day time studies of coral-derived bioreactors heterotopically implanted in the *rectus abdominis* muscle (Klar et al. 2013; Klar et al. 2014). At tissue harvest, the *rectus abdominis* muscle tightly envelopes the implanted coral-derived constructs supplying fibrovascular tissue invasion with capillary sprouting penetrating the peripheral macroporous spaces (Figs. 4.3, 4.4). Images show pronounced fibrovascular invasion with the early induction of mesenchymal collagenous condensations tightly generated against the hydroxyapatite surfaces (Fig. 4.4). Of interest, some but not all cut sections on day 15 show the induction of fibrin-fibronectin rings expanding within the macroporous

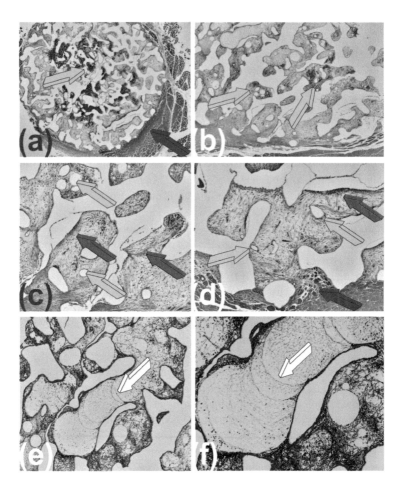

FIGURE 4.3 Early extracellular matrix and vascular morphogenetic events on day 15 in macroporous coral-derived bioreactors implanted in the *rectus abdominis* muscle of the Chacma baboon *Papio ursinus*. (a) The *rectus abdominis* muscle (*dark blue* arrows) tightly surrounds the implanted bioreactors, and fibrovascular mesenchymal tissue penetrates within the external macroporous spaces blending with haemorrhagic areas within the central regions of the heterotopically implanted bioreactor (*white* arrow) (b) Magnified peripheral and internal regions showing vascular invasion (*white* arrows) with the differentiation of mesenchymal tissue enveloping and holding the invading vasculature across the macroporous spaces (*white* arrows). (c,d) Peripheral and internal regions showing pronounced vascular invasion and capillary sprouting (*white* arrows) with early developmental events responsible for the induction of collagenous condensations at the hydroxyapatite interface (*magenta* arrows). (e,f) Induction of fibrin-fibronectin rings (*white* arrows) within the macroporous spaces as highlighted in coral-derived constructs super activated by doses of the hTGF-β3 isoform (Klar et al. 2014; Ripamonti et al. 2015; Ripamonti et al. 2016).

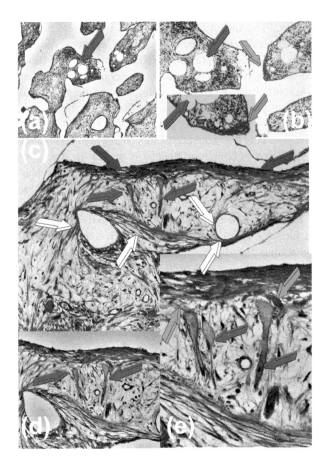

FIGURE 4.4 Developmental steps on day 15 differentiating collagenous condensations at the hydroxyapatite interface, angiogenesis, capillary sprouting and the early development of bone matrix at the hydroxyapatite interface. (a,b) Angiogenesis and capillary sprouting are the hallmark of the spontaneous and/or intrinsic induction of bone formation by macroporous coral-derived bioreactors (*magenta* arrows). The images show pronounced capillary invasion by day 15 after heterotopic implantation within the macroporous spaces in close relationship with corpuscolate haemorrhagic material after the surgical implantation (*magenta* arrows). (b and inset) Within a concavity of the macroporous bioreactor there is the early differentiation of bone matrix against the hydroxyapatite interface (*light blue* arrows and inset) showing the early differentiation of bone matrix with embedded osteocytes against the interface of the crystalline substratum. (c) Low-power view detailing the induction of collagenous condensations at the hydroxyapatite interface (*white* arrows) with capillary invasion within the organizing fibrovascular stroma (*light blue* arrows). Longitudinally cut proliferating capillaries (*magenta* arrows) show the close relationship with the remodelling collagenous condensations. Capillary sprouting feeds the mesenchymal condensations also with perivascular pericytic cells. The tight relationship between newly induced condensations and capillary sprouting and invasion is shown in (d) and (e); high-power views show the large nucleated hyperchromatic endothelial cells (*magenta* arrows) in a tight relationship with mesenchymal cells of the collagenous condensation migrating and almost touching the invading capillaries as per Trueta's definition of the "osteogenetic vessels" (Trueta 1963).

spaces (Fig. 4.3e,f). Pronounced fibrin-fibronectin rings were previously shown in macroporous bioreactors super activated by hTGF-β_3 when harvested on day 15 after heterotopic *rectus abdominis* implantation (Klar et al. 2014; Ripamonti et al. 2014; Ripamonti et al. 2015). On day 15 in coral-derived constructs *solo*, vascular invasion with capillary sprouting was foremostly represented at the peripheral spaces of the implanted bioreactors sustained by pronounced fibrocellular invasion supporting the induction of mesenchymal collagenous condensations (Figs. 4.3, 4.4).

Figure 4.4 reconstructs the sequence of morphological, cellular and ionic events leading to the induction of bone morphogenesis within the macroporous spaces of the implanted coral-derived bioreactors. There is substantial vascular invasion with capillary sprouting sustained by an organizing fibrocellular network (Figs. 4.4a,b) surrounded by corpuscolated red blood cells as a result of the surgical implantation (Figs. 4.4.a,b). There is differentiation of osteoclastic cells priming the coral-derived hydroxyapatite surfaces. Light blue arrows in Figure 4.4b point to the beginning of bone formation against the substratum with embedded osteocytes (Figs. 4.4b and inset *light blue* arrows) together with pronounced capillaries invasion (Figs. 4.4b and inset, *magenta* arrows).

More revealing, are Figures 4.4c and related high-power views (Figs. 4.4d,e). The digital images show the induction of collagen condensations on day 15 against the substratum primed by osteoclastic migration and topographical alterations. Mesenchymal cells are embedded along the collagenic pattern supported by capillary sprouting along the macroporous spaces (*white* arrows) (Fig. 4.4c).

Fortuitously, the digital images show capillaries not only cut transversely but also longitudinally, highlighting thus the copious interconnecting capillary sprouting and invasion generated by the substratum with tightly attached embedded collagenous condensations (Figs. 4.4c *dark blue* arrows). High-power views (Figs. 4.4d,e) show the capillary invasion exquisitely touching the capillary microenvironment of the collagenous condensations. The highest magnification (Fig. 4e) shows capillaries' endothelial cells with hyperchromatic nuclei morphologically fulfilling the "angiogenetic vessels" of Trueta's definition (Fig. 4e *light blue* arrow) (Trueta 1963). Vessels not only bring nutrient supply for the induction of the collagenous condensations but more likely pericytes and perivascular cells to continuously support the growth and the inductive capacities of the collagenous condensations (Figs. 4.4d,e).

Note in Figure 4.4e the exquisite relationships of the sprouting capillaries with mesenchymal cells seemingly migrating from the vascular cellular network of the collagenous condensations (Fig. 4.4e *light blue* arrows). Attached to the hydroxyapatite interface (Fig. 4.4e), mesenchymal cells are differentiating into osteoblastic-like cells, later expressing, secreting and embedding osteogenic gene products onto the biomimetic matrix, initiating the induction of bone formation as a secondary response (Ripamonti et al. 1993a; Ripamonti 2004; Klar et al. 2013; Ripamonti et al. 2014; Ripamonti 2017; Ripamonti 2018).

Which are the responding, invading and differentiating mesenchymal cells, stem or not stem cells, pericytic and perivascular somatic cells de-differentiating whilst attached to the calcium phosphate-based biomatrix? Invading progenitors and differentiating cells are exposed to Ca^{++} release within the protective microenvironment of the concavities of the substratum, and to angiogenetic and morphogenetic stimuli that set into motion the induction of bone formation (Vlodavsky et al. 1987; Folkman et al. 1988; Paralkar et al. 1990; Paralkar et al. 1991).

Identification of precursor and responding cells has been an impossible task in experiments in the non-human primate *Papio ursinus*. The nature of experimentation in primates calls for a precise balance between achievable results and the number of animals in experimentation, which, *par force*, needs to adhere to the three "r"s of primate experimentation, i.e. refine, reduce and replace. The last "r", however, is a difficult if not impossible task, since our systematic experimentation using identical coral-derived bioreactors across two centuries has been continuously surgically implanted by the same operator, in the same animal species, and prepared by the same histological techniques, to which only later the undecalcified third dimension of the Exakt cutting and grinding method has been added (Ripamonti 2017; Ripamonti 2018 for reviews).

It is noteworthy that these systematic studies showed reproducible and highly comparable morphological and molecular results across several experiments in *Papio ursinus* (Ripamonti 1990; Ripamonti 1991; Ripamonti et al. 1993a; Ripamonti 1996; Ripamonti 2004; Ripamonti 2009; Ripamonti et al. 2009; Ripamonti et al. 2010; Klar et al. 2013; Ripamonti et al. 2015).

Figure 4.5 presents digital images of fragmenting *rectus abdominis* muscle penetrating the peripheral macroporous spaces of the heterotopically implanted bioreactors (*light blue* arrows in a,b) highlighting fibrocellular muscular fragmentation. White arrows throughout the presented digital images indicate the tight relationship of the fragmented muscle fibres and cells with the substratum. Fragmentation with possible degeneration of the muscle fibres releases several phenotypes including myofibroblasts, undifferentiated mesenchymal cells and pericytes from intramuscular capillaries, as well as myoendothelial cells as described by Zheng et al. in *Nature Biotechnology* (2007). Such myoendothelial cells, isolated from human *rectus abdominis* striated muscle, are endowed with the prerogative of differentiating into myoblastic and/or osteogenic phenotypes according to the microenvironment affecting the recipient stem cell niches (Zheng et al. 2007).

Morphological analyses show the often tight relationships of de-differentiating muscle cells or differentiating myoblasts attached to the calcium phosphate-based substratum (Fig. 4.5e *light blue* arrow), possibly in contact with capillary elements just beneath the cell and against the coral-derived biomatrix hypertrophic nuclei (Fig. 4.5e *magenta* arrows). In previous studies, we stated that myoendothelial cells may respond to bone morphogenetic proteins previously bound to type IV collagen and other extracellular matrix components including laminin domains (Ripamonti et al. 2009).

The chemical composition of the calcium phosphate-based biomimetic matrix itself, particularly the bioavailability of Ca^{++} *per se*, initiates the induction of bone formation (Klar et al. 2013). Circulating monocytes, stimulated by continuous releases of Ca^{++} into the microenvironment, fuse to form macrophages. The continuous release of Ca^{++} by the resorbed biomimetic matrices prompts the phenotypic change of macrophages to undergo further transformation into osteoclasts (Kanatani

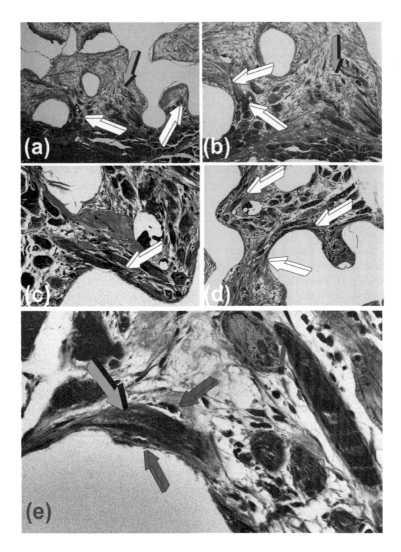

FIGURE 4.5 Myofibroblast invasion, fragmentation of the muscle *rectus abdominis* fibre bundles with differentiation and de-differentiation of pericytic perivascular and/or myoendo-thelial cells at the substratum interface (*white* arrows). (a,b) Fragmentation of muscle fibres of the invading *rectus abdominis* striated muscle (*white* arrows). (c,d) High-power views to further highlight fragmentation and invasion of myoblastic cellular elements within the mac-roporous spaces in tight contact with the crystalline substratum (*white* arrows). (e) Tight and exclusive relationship of a myoblastic cell (*light blue* arrows) with the crystalline substratum possibly in contact with capillaries (*magenta* arrows) surrounding the differentiating cell towards the osteoblastic phenotype. Decalcified sections cut at 6 μm from paraffin-embedded blocks stained with a modified Goldner's trichrome.

et al. 1991; Müller et al. 2008). Once osteoclastic cells are activated, the biomimetic matrix is resorbed subjacent to the ruffle border of the multinucleated osteoclastic cells cell by acid hydrolysis. The imprint in the biomimetic calcium phosphate-based matrix is a negative impression of the cell itself, its size and shape directed by the sealing zone of the osteoclasts (Ripamonti et al. 2012). Such impressions are either single concavities, or more commonly, grooves, lacunae and/or pits that form as a succession of individual pits that coalesce into concavities as the osteoclasts trek along the surface of the biomimetic matrix (Ripamonti et al. 2012).

qRT-PCR of homogenized bioreactors, on day 15, shows upregulation of *Type IV Collagen* (Klar et al. 2013) correlating with the morphological evidence of angiogenesis invading the macroporous spaces. In our first communications describing the induction of bone formation by coral-derived macroporous bioreactors, we described the peculiar pattern of alignment of condensed mesenchymal collagenous fibres in direct apposition to the coral-derived hydroxyapatite crystalline surfaces (Fig. 4.4) (Ripamonti 1990; Ripamonti 1991). These studies showed that such collagenous condensations precede the formation of bone by supporting the induction of the osteogenic phenotype of cellular elements within the condensations attached to the crystalline hydroxyapatite biomatrix (Ripamonti 1990; Ripamonti 1991).

We later showed that tissue patterning and the induction of collagenous condensations aligned against the coral-derived macroporous surfaces are critical steps for the spontaneous and/or intrinsic induction of bone formation (Ripamonti et al. 1993a; Klar et al. 2013). Of note, and as a *conditio sine qua non*, the induction and tissue patterning of mesenchymal condensations are preceded by osteoclastic activity priming the macroporous surfaces of the heterotopically implanted biomimetic matrices (Ripamonti et al. 2010; Klar et al. 2013). The generation of osteoclasts and osteoclastic priming of the surfaces of the heterotopically implanted bioreactors lead to nanotopographical geometric alterations of the hydroxyapatite substratum. Cell surface osteoclastic geometric modifications result in the generation of self-inductive nanopatterned geometric configurations with Ca^{++} release within the confined spaces of the macroporous cavities and concavities of the substratum (Ripamonti et al. 2010; Klar et al. 2013; Ripamonti 2017; Ripamonti 2018).

Elegantly, the critical role of osteoclastic activity in priming the macroporous calcium phosphate-based surfaces is indirectly shown by the lack of bone induction by coral-derived bioreactors when pre-treated with the osteoclastic inhibitor bisphosphonate zoledronate zometa (Fig. 4.6) (Ripamonti et al. 2010; Klar et al. 2013).

Further revealing the critical role of osteoclastic differentiation and priming of the hydroxyapatite patterned surfaces, experiments with coral-derived bioreactors preloaded with the Ca^{++} channel blocker verapamil hydrochloride strongly inhibited the induction of collagenous condensations and mesenchymal tissue patterning against the hydroxyapatite interface (Fig. 4.7) (Klar et al. 2013). Of note, qRT-PCR showed a direct correlation between the lack of bone formation as well as the induction of mesenchymal collagenous condensations with downregulation of *bone morphogenetic protein-2* (*BMP-2*) together with upregulation of the BMPs' inhibitor *Noggin* gene.

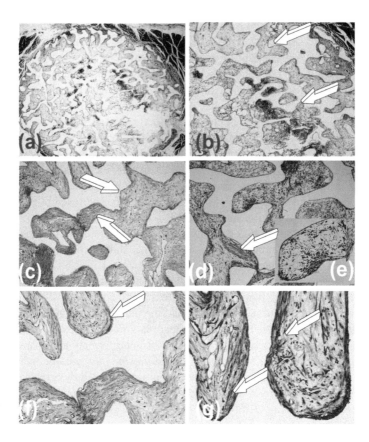

FIGURE 4.6 Inhibition of the induction of mesenchymal collagenous condensations on day 15 with lack of bone formation by day 90 after heterotopic intramuscular implantation of coral-derived constructs pre-treated with the osteoclastic inhibitor zoledronate zometa (Ripamonti et al. 2010; Klar et al. 2013). (a,b) Low-power views of coral-derived bioreactors showing limited and disorganized fibrovascular tissue invasion within the macroporous constructs (*white* arrows). (c,d) Digital images depicting lack of collagenous condensations with limited fibrovascular invasion and modelling (*white* arrows). Topographically geometrically modified surfaces by osteoclastogenesis serve to align tissue patterning and cellular differentiation against the geometrically modified surfaces. Lack of early induction of collagenous condensations against the hydroxyapatite substratum in zoledronate zometa-treated bioreactors (c,d inset e). Limited surface priming by osteoclastogenesis inhibited by preloaded doses of the bisphosphonate zoledronate zometa (Ripamonti et al. 2010; Klar et al. 2013) causes osteogenetic ineffective mesenchymal aggregations (*white* arrows) (f,g) with lack of cellular differentiation within non-conducive and non-inducive condensations.

The data indicated that the induction of bone formation by coral-derived macroporous bioreactors is *via* the *bone morphogenetic proteins* pathway. The induction of bone is initiated by a local peak of Ca^{++} activating stem cell differentiation into the osteoblastic phenotype (Klar et al. 2013). Cellular differentiation is followed by the expression and synthesis of angiogenic and osteogenic gene products. These are

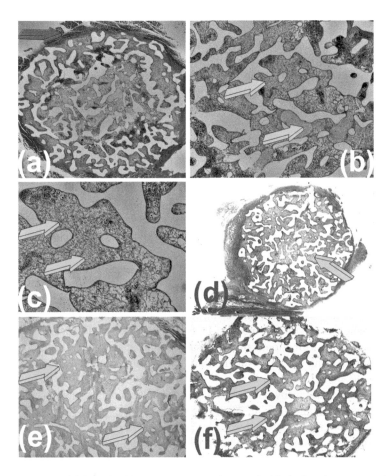

FIGURE 4.7 Inhibition of collagenous condensations and of inducive tissue patterning by coral-derived bioreactors preloaded with the Ca^{++} channel blocker verapamil hydrochloride. (a,b,c) Disorganized tissue patterning and collagenous condensation in Ca^{++} channel blocker verapamil hydrochloride pre-treated macroporous surfaces on day 15 after heterotopic intramuscular implantation (Klar et al. 2013). (c) Limited tissue patterning and organization of collagenous condensations with a loose fibrovascular tissue matrix expanding within the macroporous spaces (*light blue* arrows). (d,e) Limited and disorganized tissue patterning of collagenous condensation after Ca^{++} channel blocker verapamil hydrochloride on day 60 after implantation. (f) Evaluation of histological sections on day 90 showed lack of or limited bone formation by induction with upregulation of *Noggin*, also shown on day 15 together with downregulation of the *bone morphogenetic protein-2* (BMP-2) gene (Klar et al. 2013).

embedded into the biomimetic matrices with the induction of bone formation as a secondary response (Ripamonti et al. 1993a; Klar et al. 2013; Ripamonti et al. 2014; Ripamonti et al. 2015; Ripamonti 2017; Ripamonti 2018).

Pre-treatment of the coral-derived bioreactors with doses of 125 or 150 µg recombinant human Noggin (hNoggin), a BMPs' inhibitor (Zimmerman et al. 1996), blocks the spontaneous and/or intrinsic induction of bone formation by the coral-derived bioreactors (Fig. 4.8). The lack of bone differentiation is by blocking tissue induction and morphogenesis of patterned collagenous condensations at the hydroxyapatite interface (Figs. 4.8a,b,c,d) (Klar et al. 2014; Ripamonti et al. 2015).

We reported that the addition of hNoggin to the bioreactors *solo* profoundly inhibited the sequential morphogenetic cascades of tissue patterning and alignment of collagenous condensations so critical for the induction of bone formation when using coral-derived bioreactors (Klar et al. 2014; Ripamonti et al. 2015). It is noteworthy to highlight that the inhibitory effect of hNoggin pre-treatment indicates that the spontaneous and intrinsic osteoinductivity of the coral-derived macroporous bioreactors is initiated *via* the BMPs' pathway during alignment and patterning of the collagenous condensations, before osteoblastic-like cells differentiation. Inhibition of tissue patterning and lack of proper functionality of collagenous condensations block the bone induction cascade showing that *BMPs'* expression synthesis and embedding within the collagenous condensations are pre-requisites for the spontaneous osteoinductivity of coral-derived macroporous bioreactors (Fig. 4.8).

Pre-treatment of the coral-derived bioreactors with the osteoclast inhibitor bisphosphonate zoledronate zometa abolishes the induction of bone formation as evaluated on day 90 (Ripamonti et al. 2010) (Fig. 4.9a,b). Preloading the coral-derived bioreactors with recombinant hOP-1 (also known as BMP-7) results in the induction of bone formation across the macroporous spaces as evaluated on day 90 (Fig. 4.9c). Loading 125 μg hNoggin onto hOP-1 pre-treated constructs blocks the induction of bone formation across the macroporous spaces, which are now infiltrated by a loose mesenchymal tissue without the induction of functional collagenous condensations (Ripamonti et al. 2010) (Fig. 4.9d).

Elegantly, the induction pathways by *bone morphogenetic proteins* expression and synthesis are shown by the substantial induction of bone formation by a coral-derived bioreactor on day 90 (Fig. 4.9e) whilst bioreactors *solo* pre-treated with 125 μg hNoggin show lack of bone differentiation (Fig. 4.9f). In hNoggin pre-treated bioreactors, tissue patterning and alignment of collagenous condensations are substantially inhibited and delayed (Fig. 4.9f) when compared to untreated macroporous constructs *solo* (Klar et al. 2014; Ripamonti et al. 2015). The data gathered so far in our systematic studies in *Papio ursinus* show that blocking BMPs by preloading hNoggin onto the coral-derived constructs blocks the induction of bone differentiation not by blocking osteoblastic cell differentiation but by aborting the functional induction of mesenchymal collagenous condensations at the interface of the macroporous bioreactors.

Similarly, pre-treatment of coral-derived bioreactors with hNoggin on calcium phosphate-based constructs preloaded with doses of recombinant hTGF-β_3 blocks the induction of bone formation (Klar et al. 2014; Ripamonti et al. 2014; Ripamonti et al. 2015). The induction of bone formation is blocked by blocking the induction of collagenous condensations with proper tissue patterning and alignment (Figs. 4.8e,f, 4.10). This has shown that the induction of bone formation as initiated by hTGF-β_3 in the *rectus abdominis* of *Papio ursinus* is *via* the BMPs' pathway. hTGF-β_3 controls the induction of bone formation by regulating the expression of *BMPs via Noggin*

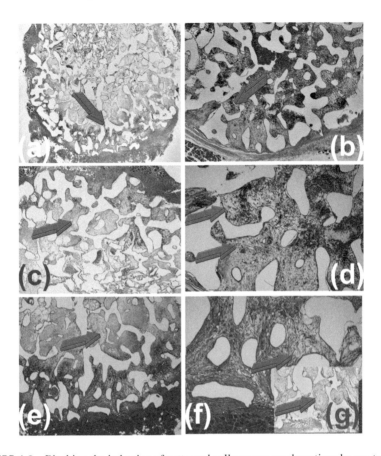

FIGURE 4.8 Blocking the induction of patterned collagenous condensations by pre-treating coral-derived bioreactors with 125 or 150 μg recombinant hNoggin, a BMPs' inhibitor. The lack of bone differentiation by hNoggin/pre-treated macroporous bioreactors is by blocking the induction of patterned inductive mesenchymal condensations at the hydroxyapatite macroporous surfaces. Profound inhibition of collagenous tissue patterning blocks the induction of bone differentiation. The digital images shown in (a,b,c,d) reveal the formation of a loose fibrovascular matrix with lack of tissue patterning and alignment (*magenta* arrows). The lack of proper tissue patterning of collagenous condensation in hNoggin-treated macroporous bioreactors indicates the critical role of BMPs and of *bone morphogenetic proteins* expression and synthesis long before osteoblastic cell differentiation. This highlights the critical role of the assemblage of collagenous condensations at the hydroxyapatite interface for bone formation to occur (Ripamonti 1991; Ripamonti et al. 1993a; Klar et al. 2014; Ripamonti et al. 2014; Ripamonti et al. 2015). (e,f) Limited or loose fibrovascular tissue invasion in macroporous spaces of coral-derived bioreactors pre-treated with 125 μg doses of recombinant human transforming growth factor-β_3 (hTGF-β_3) additionally preloaded with 125 or 150 μg hNoggin (Klar et al. 2014; Ripamonti et al. 2015). The rapid induction of bone by the hTGF-β_3 isoform (Ripamonti et al. 1998; Ripamonti et al. 2015) is blocked by blocking the induction of collagenous condensations at the hydroxyapatite interface (*magenta* arrows). The data show that the induction of bone formation by the hTGF-β_3 isoform is *via* the BMPs' pathway regulating the expression of *BMPs* and it is blocked by hNoggin (inset g *magenta* arrow) (Ripamonti et al. 2015).

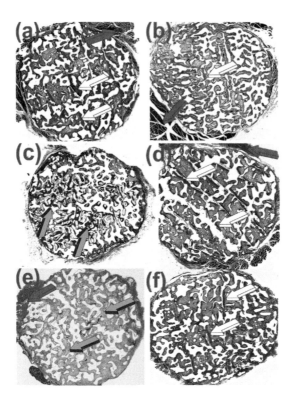

FIGURE 4.9 Modulation and inhibition of the spontaneous induction of bone formation by doses of the bisphosphonate zoledronate zometa and recombinant human Noggin (hNoggin). Both compounds block the morphological and molecular pathways of the spontaneous and/ or intrinsic induction of bone formation by coral-derived macroporous bioreactors when heterotopically implanted in the *rectus abdominis* muscle (*magenta* arrows), where there is no bone, of the Chacma baboon *Papio ursinus*. (a,b) Low-power images showing the induction of poorly patterned collagenous condensations (*white* arrows) at the macroporous surfaces of the coral-derived bioreactors pre-treated with 240 µg zoledronate zometa with lack of bone differentiation 90 days after heterotopic implantation (Ripamonti et al. 2010; Klar et al. 2013). (c) Preloading macroporous bioreactors with 125 µg doses of human recombinant osteogenic protein-1 (hOP-1) (also known as BMP-7) sets into motion the induction of bone formation across the macroporous spaces (*light blue* arrows). Pre-treating coral-derived bioreactors with 125 (Klar et al. 2014) or 150 µg (Ripamonti et al. 2015) recombinant hNoggin, a BMPs' inhibitor, onto coral-derived bioreactors preloaded with 125 µg hOP-1 blocks the induction of bone formation (d) with the induction of poorly patterned assembled collagenous condensations (*white* arrows d). (e) Substantial induction of bone formation (*light blue* arrows) by coral-derived bioreactors implanted in the *rectus abdominis* muscle (*magenta* arrow). (f) The induction of bone formation by coral-derived bioreactors *solo* is blocked by preloading the calcium phosphate-based macroporous constructs with 150 µg hNoggin (Ripamonti et al. 2015). The images show how dramatically the BMPs' pathway controls and masterminds the spontaneous and/or intrinsic induction of bone formation by coral-derived bioreactors. Decalcified sections cut at 6 µm from paraffin-embedded blocks stained with a modified Goldner's trichrome.

expression (Klar et al. 2014; Ripamonti et al. 2014; Ripamonti et al. 2015). The induction of bone formation is by upregulating the expression of several profiled *BMPs* and it is blocked by hNoggin (Klar et al. 2014; Ripamonti et al. 2015) (Figs. 4.8e,f, Fig. 4.10).

Our present interpretation focuses on both morphological and molecular evidence of selected gene expression' pathways that shows that preloading macroporous bioreactors with doses of recombinant hNoggin results in the formation of haphazardly patterned collagenous condensations (Figs. 4.9d,f) (Ripamonti et al. 2015). The poorly patterned condensations lack capillary sprouting and angiogenesis with lack of cellular differentiation without induction of bone formation (Ripamonti et al. 2015) (Figs. 4.8, 4.9).

Somehow mechanistically identically, the lack of patterned induction of collagenous condensations blocks the induction of bone formation when super activated hTGF-β_3 bioreactors are implanted in binary applications with 125 or 150 μg hNoggin (Figs. 4.8e,f, 4.10) (Klar et al. 2014; Ripamonti et al. 2015). The lack of bone differentiation by 125 μg hTGF-β_3 preloaded bioreactors with 125 or 150 μg hNoggin more vividly perhaps shows the direct role of BMPs in setting into motion the induction of bone differentiation *via* the induction of properly organized and patterned collagenous condensations (Fig. 4.10) (Ripamonti et al. 1993a; Klar et al. 2014; Ripamonti et al. 2015; Ripamonti 2017).

Invading sprouting capillaries are brought into the macroporous spaces across the open interconnected spaces, or septa, of the coral-derived bioreactor. Invading capillaries bring about bundles of responding cells migrating along newly formed collagenous fibres. It is likely that myoendothelial cells within the *rectus abdominis* striated muscle of primates (Zheng et al. 2007) migrate to the hydroxyapatite surfaces and microenvironment to differentiate into osteoblastic-like cells by high Ca++, capillary invasion and angiogenesis.

Based on the fact that osteoclastic primed nano-patterned geometries and Ca++ release with capillary sprouting result in the differentiation of osteoblastic-like cells attached to the geometrically conditioned substrata, we propose that either myoblastic cells, myoendothelial cells, perivascular and/or pericytic cells de-differentiate into *de novo* differentiating somatic stem cells in contact with the inductive geometric microenvironment (Fig. 4.4). In the presence of high Ca++, invading responding cells directly differentiate into osteoblastic cells expressing and secreting a battery of osteogenic genes and gene products initiating the induction of bone formation in tight contact with the geometrically primed and modified surfaces by osteoclastogenesis (Fig. 4.4b and inset) (Ripamonti et al. 2010; Klar et al. 2013; Ripamonti et al. 2015; Ripamonti 2017; Ripamonti 2018).

On day 30, there is remodelling of the collagenous condensations, with further alignment against the micro patterned surface with vascular invasion (Fig. 4.11). So far, however, the induction of bone formation has not been reported on day 30 when studying coral-derived macroporous constructs. Back in the 1990s, we postulated that the lack of bone formation on day 30 suggested a critical concentration lag time of putative BMPs' synthesis that is required for the initiation of bone formation (Ripamonti 1991; Ripamonti et al. 1993a). Several years later, and now into the

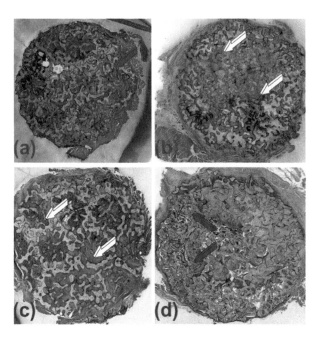

FIGURE 4.10 Low-power digital macrophotography highlighting the mechanisms of the induction of bone formation by the recombinant human transforming growth factor-β_3 (hTGF-β_3) when pre-combined *solo* or in binary application with recombinant human Noggin (hNoggin) to macroporous coral-derived bioreactors and implanted in heterotopic *rectus abdominis* sites of the Chacma baboon *Papio ursinus* (Klar et al. 2014; Ripamonti et al. 2015). (a) Prominent induction of bone formation (arrowed in *red-orange*) within the macroporous spaces (*blue* arrow) by 125 µg hTGF-β_3 applied singly to the macroporous bioreactor. (b,c,d) Preloading the coral-derived bioreactors with binary application of 125 µg hTGF-β_3 or 150 µg hNoggin (Ripamonti et al. 2015) blocks the induction of bone formation throughout the macroporous spaces (*white* arrows b,c). The lack of bone formation is by blocking the assemblage of patterned collagenous condensations by blocking the expressed *BMPs*' genes and gene products (Klar et al. 2014; Ripamonti et al. 2015). Note the lack of patterned collagenous condensations so critical for bone induction to occur (Ripamonti et al. 1993a). Indirectly, the morphological experiments show that the induction of bone formation by the hTGF-β_3 isoform is *via* the *bone morphogenetic proteins* (*BMPs*) pathway. hTGF-β_3 elicits the induction of bone formation by expressing *BMPs* with the secretion of related gene products that initiate the induction of bone formation within the macroporous spaces of the coral-derived bioreactors. Doses of hNoggin, affecting the BMPs' receptor binding, block the induction of bone formation, indicating thus that the induction of bone formation by the hTGF-β_3 isoform is *via* the *BMPs*' pathway, and that the spontaneous and/or intrinsic induction of bone formation is also *via* the BMPs' pathway, since 125 or 150 µg hNoggin blocks the induction of bone formation by the macroporous bioreactors implanted *solo* in the *rectus abdominis* of *Papio ursinus* (Klar et al. 2014; Ripamonti et al. 2015; Ripamonti 2017; Ripamonti 2018). In other specimens (d), the induction of bone formation is sometimes seen at the periphery of the macroporous constructs (*blue* arrows d), indicating the limited tri-dimensional diffusion of the preloaded hNoggin across the macroporous spaces of the treated bioreactors. (a,b,c and d) Undecalcified sections cut at 37 µm cut and polished with the Exakt diamond saw grinding and polishing equipment (Klar et al. 2014; Ripamonti et al. 2015).

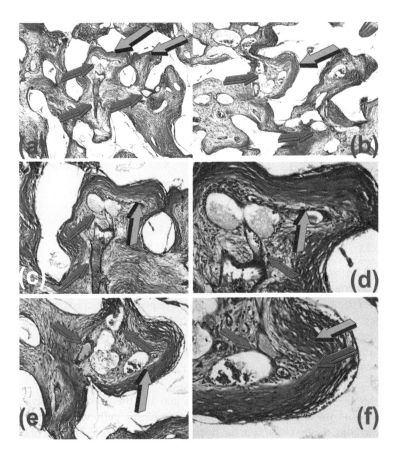

FIGURE 4.11 Induction of patterned tissue morphogenesis with maturational gradients of collagenous condensations at the interface of the coral-derived constructs with angiogenesis and capillary sprouting within the macroporous spaces on day 30 after intramuscular implantation of the coral-derived bioreactors. Bioreactors were implanted without the exogenous applications of osteogenic soluble molecular signals. (a,b) Low-power views showing mesenchymal tissue condensations at the interface of the heterotopically implanted bioreactors (*light blue* arrows), with capillary invasion and angiogenesis (*magenta* arrows). (c,d) Morphological details of collagenous condensations (*light blue* arrows) showing patterning with mesenchymal cells invasion in close contact with capillary sprouting (*magenta* arrows). (e,f) Details of patterned collagenous condensations with exquisite relationships with the supporting angiogenesis with capillary sprouting and invasion (*magenta* arrows) penetrating the collagenous condensations (*red* arrow in f) supplying nutrients as well as pericytic perivascular and/or myoendothelial cells for the continuous induction of bone formation. The symbiosis of mesenchymal tissue patterning and collagenous condensations with capillary invasion, angiogenesis and capillary sprouting is the most significant event in the spontaneous and/or intrinsic induction of bone formation by coral-derived macroporous bioreactors. Nothing is known about the molecular cross talk between invading endothelial, perivascular and/or myoendothelial cells and the patterning collagenous condensations prior to or during the induction of bone formation between days 30 and 60. Sections cut at 6 μm prepared from paraffin-embedded decalcified blocks stained with Goldner's trichrome.

21st century, the available data indicate that the lack of bone formation by induction by coral-derived macroporous bioreactors is rather time-dependent by osteoclastic priming, Ca^{++} release and available concentrations within anatomically different macroporous spaces, induction and remodelling of mesenchymal collagenous condensations together with de-differentiation of inductive stem cells from invading myoblastic, myoendothelial, perivascular and/or pericytic cells later expressing, secreting and embedding bone morphogenetic proteins gene products onto the substratum (Ripamonti et al. 2010; Klar et al. 2013; Ripamonti et al. 2015; Ripamonti 2017; Ripamonti 2018).

4.3 TISSUE INDUCTION AND MORPHOGENESIS BY CORAL-DERIVED HYDROXYAPATITE CONSTRUCTS ON DAYS 60 AND 90 AFTER HETEROTOPIC *RECTUS ABDOMINIS* IMPLANTATION

Bone thus forms between days 30 and 60 after heterotopic implantation of the coral-derived macroporous bioreactors (Figs. 4.11, 4.12) (Ripamonti et al. 1993a; van Eeden and Ripamonti 1994). As discussed in Chapter 1, osteoclastic priming of the implanted macroporous surfaces with Ca^{++} release, together with the induction of angiogenesis and capillary sprouting, are the primary events leading to the differentiation of the osteogenic phenotype by the geometrically modified surfaces after osteoclastic priming of the macroporous surfaces (Figs. 4.11, 4.12) (Ripamonti et al. 2010; Klar et al. 2013).

On day 60, there is cell differentiation into the osteogenic phenotype at the hydroxyapatite interface with proliferating osteoblastic cells with hyperchromatic nuclei secreting the first matrix at the hydroxyapatite interface (Figs. 4.12a,b,c). Cell differentiation within patterned collagenous condensations between days 30 and 60 initiates the cascade of the spontaneous and/or intrinsic induction of bone formation by coral-derived bioreactors (Fig. 4.12d). Cell differentiation within collagenous condensations at the hydroxyapatite interface (Fig. 4.12b,c) is the critical event for bone differentiation to occur (Ripamonti et al. 1993a).

Gene expression analyses on day 60 after *rectus abdominis* implantation of coral-derived bioreactors showed the relative expression of *BMP-3* as determined from the homogenized tissues of the implanted bioreactors, however with limited *BMP-3* expression in the surrounding *rectus abdominis* muscle (Ripamonti et al. 2015). *BMP-3* expression values within the bioreactors, two- to three-fold higher than in the surrounding *rectus abdominis* muscle, were very similar to both hTGF-β_3-treated and non-treated bioreactors (Ripamonti et al. 2015). On day 90, there was over-expression of *BMP-3* in both hTGF-β_3-treated and non-treated control bioreactors as well as in the surrounding *rectus abdominis* muscle (Ripamonti et al. 2015). This indicates that the expression of the *BMP-3* gene is a critical step for the spontaneous and/or intrinsic induction of bone formation. Studies have shown that BMP-3 is a negative regulator of bone mass *in vivo* (Daluiski et al. 2001). *BMP-3* over-expression after intramuscular implantation of coral-derived bioreactors may initiate signalling to recruit monocyte osteoclast precursor. This will in turn lead to an

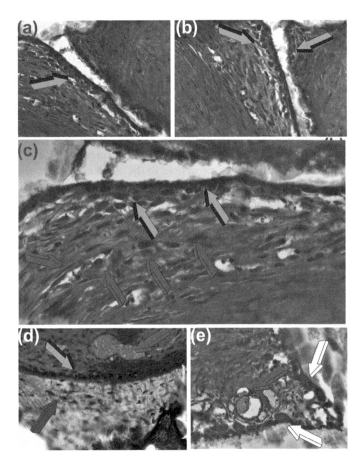

FIGURE 4.12 Morphological cellular events *de novo* initiating the spontaneous and/ or intrinsic induction of bone formation by coral-derived macroporous bioreactors when implanted in the *rectus abdominis* muscle of the Chacma baboon *Papio ursinus* and harvested on day 60 after intramuscular implantation. (a,b) Cellular differentiation at the coral-derived hydroxyapatite surface with generation of the osteoblastic phenotype (*light blue* arrows) on day 60 after *rectus abdominis* implantation. (c) High-power view highlighting osteoblast-like cell differentiating at the hydroxyapatite interface (*light blue* arrows) with migrating perivascular and/or myoendothelial cells from the vascular compartment (*magenta* arrows) to the osteogenetic compartment (*light blue* arrows) with secreting osteoblasts directly against the substratum. (d) Spontaneous induction of bone formation within a concavity of the coral-derived substratum with osteoblastic cells (*light blue* arrow) populating the newly formed bone with embedded osteocytes (*dark blue* arrow). Note capillary invasion (*magenta* arrow) close to the induction of bone formation (*light blue* arrow). (e) Recapitulating morphological steps critical for the induction of bone formation to occur: osteoclastic cell differentiation at the hydroxyapatite interface (*white* arrows) together with capillary sprouting and invasion (*magenta* arrow). Note the exquisite relationships between angiogenesis and capillary sprouting (*magenta* arrow) and the osteoclasts resorbing the calcium phosphate-based construct. Undecalcified sections cut at 3 to 6 μm from historesin embedded undecalcified blocks harvested on day 60 after intramuscular implantation (Ripamonti et al. 1993a).

increase of topographical geometric modifications with higher Ca^{++} release acceler-
ating the ripple-like cascade of the induction of bone formation as morphologically
and molecularly described (Klar et al. 2013; Ripamonti et al. 2015).

BMP-3 over-expression is seen on day 60 after *rectus abdominis* implantation of
both hTGF-β_3-treated and non-treated control bioreactors (Ripamonti et al. 2015).
This over-expression in the homogenized bioreactors with limited expression in the
surrounding *rectus abdominis* muscle correlates to the initiation of bone formation
after differentiation of osteoblastic-like cells tightly attached to the differentiating
biomimetic bioreactor surface (Fig. 4.12). *BMP-3* over-expression both in treated and
untreated coral-derived constructs as well as in the surrounding *rectus abdominis*
muscle correlates with significant induction of bone formation across the macroporous spaces as seen 90 days after heterotopic implantation (Fig. 4.14).

Available sections show the critical role of the substratum geometry in controlling the induction of bone formation. Bone initiates within concavities of the
macroporous substratum on day 60 after *rectus abdominis* implantation (Fig. 4.13).
Bone matrix is surfaced by contiguous rows of active osteoblastic cells continuously
secreting bone matrix with embedded osteocytes (Figs. 4.13c,d,e,f).

There is an exquisite spatial and topographical relationship between invading
sprouting capillaries and secreting osteoblastic cells (Figs. 4.12, 4.13). Figure 4.13
elegantly shows the critical role of the geometry of the substratum in controlling
the spontaneous and/or intrinsic induction of bone formation without exogenously
applied osteogenic proteins of the TGF-β supergene family (Figs. 4.13c,d,e,f)
(Ripamonti et al. 1999; Ripamonti et al. 2001; Ripamonti 2004; Klar et al. 2013;
Ripamonti 2017; Ripamonti 2018). The concavity of the calcium phosphate-based
construct is an optimal protected microenvironment that maintains Ca^{++} within the
macroporous spaces, enabling capillary sprouting and invasion and supporting the
differentiation of the osteogenic phenotype in mesenchymal cells, stem cells and
progenitors attached the substratum concavities (Figs. 4.13c,d,e,f). Ripamonti et al.
1993a; Ripamonti 2017; Klar et al. 2013; Ripamonti 2018).

By day 90, bone formation within the macroporous spaces is often substantial
(Fig. 4.14). The newly formed bone is highly vascularized and tightly attached to
the substratum. Across the serially cut specimens, there are always transitional morphological features that temporo-spatially reconstruct the spontaneous induction
of bone formation. The images show the induction of bone morphogenesis within
the macroporous spaces (Figs. 4.14a,b) together with osteoclastic resorption of both
matrices and the newly formed bone. There is bone remodelling with the induction
of lamellar bone (Figs. 4.14c,d), angiogenesis and capillary sprouting. Morphology
across the macroporous spaces recapitulates a series of events across spatio-temporal
time periods as also shown on day 60 after heterotopic implantation (Figs. 4.14e,f).
Osteoclastic activity with remarkable capillary invasion is followed by the induction
of bone formation and remodelling within concavities of the coral-derived substratum on day 90 (Figs. 4.14c,d). Osteoclastogenesis, capillary sprouting and the induction of bone formation are recapitulated on day 60 across serially cut macroporous
constructs (Figs. 4.14e,f).

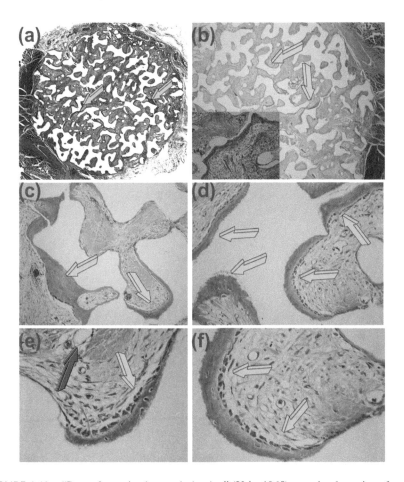

FIGURE 4.13 "Bone: formation by autoinduction" (Urist 1965) upon implantation of coral-derived calcium phosphate-based constructs in the *rectus abdominis* muscle of *Papio ursinus* and harvested on day 60 after intramuscular implantation (Klar et al. 2013). (a) Low-power view depicting the whole mounted slide of the bioreactor spontaneously initiating the induction of bone formation (*light blue* arrows) on day 60 after *rectus abdominis* implantation. (b) Low-power view highlighting the induction of bone formation (*light blue* arrows) tightly bound to the hydroxyapatite substratum. Inset in (b) shows the induction of bone formation against the substratum in close relationship with vascular invasion and capillary sprouting. (c,d,e,f) The critical role of geometry in the induction of bone formation by coral-derived macroporous constructs. Bone (*light blue* arrows) initiates within concavities of the macroporous spaces. Such concavities retain Ca^{++} inducing angiogenesis and cell differentiation towards the osteoblastic phenotype. Newly formed bone (*light blue* arrows) is covered by contiguous osteoblasts in close relationship with capillary sprouting and invasion. Newly formed bone matrix has embedded osteocytes, and it is tightly attached into the coral-derived calcium phosphate-based substratum. Decalcified sections cut at 4 μm stained with Goldner's trichrome (Klar et al. 2013).

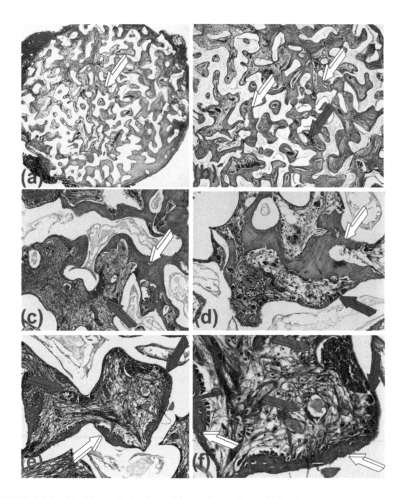

FIGURE 4.14 Significant induction of bone formation within the macroporous spaces of coral-derived bioreactors *solo* harvested from the *rectus abdominis* muscle of *Papio ursinus* on day 90 after intramuscular implantation. (a) Low-power view showing substantial induction of bone formation (*light blue* arrow) across the macroporous spaces. (b) Detail of (a) highlighting the induction of bone formation (*white* arrows) supported by prominent angiogenesis and capillary sprouting (*magenta* arrow). (c,d,e,f) Recapitulating morphological and molecular steps leading to the spontaneous and/or intrinsic induction of bone formation (Klar et al. 2013; Ripamonti et al. 2014; Ripamonti 2017; Ripamonti 2018). The four digital images show the induction of newly formed bone (*white* arrows), bone remodelling with osteoclastic activity, pronounced angiogenesis and capillary sprouting (*magenta* arrows) and osteoclastic cell differentiation and activity (*blue* arrows), priming the substratum generating topographical modifications targeting responding mesenchymal cells to differentiate into osteoblastic-like cells on days 90 (c,d) and 60 (e,f) (Klar et al. 2013; Ripamonti 2017; Ripamonti 2018). From days 15 to 90 there is thus continuous recapitulation of the mechanisms responsible for the induction of bone formation, so that all available macroporous spaces may be generating bone across the bioreactor. Decalcified sections cut at 4 μm stained with Goldner's trichrome (Ripamonti 1990; Ripamonti 1991; Ripamonti et al. 1993a).

On day 90, bone progresses from the central to the internal more peripheral macroporous spaces (Figs. 4.15a,b) (Ripamonti 1990; Ripamonti 1991; Ripamonti et al. 1993a). Figure 4.15a, previously shown in Chapter 2, again highlights the initiation of bone formation within the macroporous spaces primarily within concavities of the substratum, later expanding within the macroporous spaces in tight contact with capillary sprouting and invasion. Bone remodels with the formation of lamellar osteonic-like bone (Fig. 4.15c,d) with osteoclastic activity and remodelling of the newly formed bone together with prominent vascular invasion (Fig. 4.15d *white* arrows). Undecalcified sections show the induction of collagenous condensations which are then mineralized with the induction of bone differentiation and osteoid synthesis (Fig. 4.15e,f).

By day 180, newly formed bone in the heterotopically implanted bioreactors is often extensive with remodelling (Figs. 4.16a,b) to lamellar osteonic bone (Fig. 4.16d) tightly attached to the hydroxyapatite substratum. By day 270, bone further remodelled within the macroporous spaces after filling the bioreactors' macro porosities with osteonic remodelled bone (Figs. 4.17c,d,e,f). One year after heterotopic implantation, newly formed bone further remodels into osteonic bone tightly bound to the coral-derived hydroxyapatite construct (Figs. 4.17g,h).

It is noteworthy that the induction of bone formation by coral-derived macroporous constructs proceeds *via* collagenous condensations against the biomimetic matrices as intramembranous bone formation. The induction of chondrogenesis or the formation of islands of cartilage within the implanted bioreactors has never been observed. To end this chapter on the induction of bone formation by coral-derived macroporous constructs, my laboratories and I would like to present again the unique images of chondrogenesis and the induction of endochondral bone as an island of cartilage induction against the posterior fascia of the *rectus abdominis* muscle (Fig. 4.18). Chondrogenesis and the induction of endochondral bone formed by diffusion gradients set into motion by the recombinant transforming growth factor-β_3 (hTGF-β_3) super activating coral-derived constructs implanted for a morphological and molecular data time study (Bone Research Unit, unpublished data 2014). A cartilaginous island formed in the *rectus abdominis* muscle just over the fascia and above the peritoneum (Fig. 4.18a). In spite of the heterotopic implantation, the cartilaginous matrix still retains the phylogenetically ancient genetic memory of the induction of columnar chondrocytes of the mammalian growth plate (Fig. 4.18b *magenta* arrows). Vascular invasion and chondrolysis are then followed (*white* arrows) by the induction of trabecular bone across the primitive induction of the cartilage anlage (Fig. 4.18b *blue* arrows).

4.4   ANGIOGENESIS, CAPILLARY SPROUTING AND THE INDUCTION OF BONE FORMATION

Since our first experiments in the Chacma baboon *Papio ursinus* in the late 1980s, we have been challenged by the significant induction of angiogenesis and capillary sprouting within the heterotopically implanted macroporous coral-derived bioreactors.

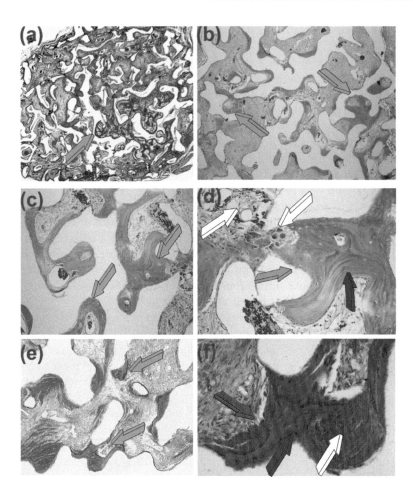

FIGURE 4.15 Set of digital images showing the induction of bone formation across the macroporous spaces of coral-derived bioreactors harvested from the *rectus abdominis* muscle of *Papio ursinus* on day 90 after intramuscular implantation. (a) Low-power view showing newly formed trabeculations by the induced bone (*light blue* arrow) within the macroporous spaces of the heterotopically implanted bioreactor. The digital image, already shown in Chapter 1, Fig. 1.1, highlights the spontaneous and/or intrinsic induction of bone formation by the coral-derived bioreactors (Ripamonti 1991). (b) Induction of bone formation across the macroporous spaces of the coral-derived construct (*light blue* arrows) on day 90 after *rectus abdominis* implantation. (c,d) Details of (b) highlighting the induction of bone formation (*light blue* arrows) with remodelling and with generation of osteonic bone (*dark blue* arrow in d) with osteoclastic activity and angiogenesis (*white* arrows in d). (e,f) Undecalcified K-Plast embedded sections showing the induction of bone formation (*light blue* arrows) with mineralization (*dark blue* arrow in f) of collagenous condensations (*white* arrow in f) with newly formed bone covered by osteoid seams (*red* arrow in f).

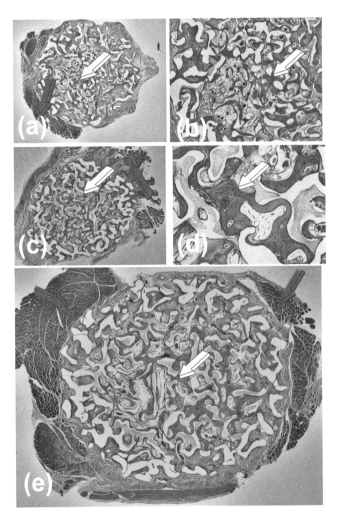

FIGURE 4.16 Substantial induction of bone formation across the macroporous space of the coral-derived bioreactors harvested on day 180 from the *rectus abdominis* muscle of *Papio ursinus*. (a) Newly formed remodelled bone across the macroporous spaces (*white* arrow) surrounded by the *rectus abdominis* muscle (*magenta* arrow) enveloping the heterotopically implanted construct. (b) High-power view showing newly formed bone (*white* arrow) across the macroporous bioreactor with capillary invasion across the macroporous spaces. (c) Low-power view of a macroporous calcium phosphate construct showing bone induction (*white* arrow) across the macroporous spaces on day 180 after heterotopic intramuscular implantation (Ripamonti 1991). (d) High-power view of (c) showing remodelling and osteonic bone patterning (*white* arrow) 180 days after heterotopic *rectus abdominis* implantation. (e) Whole mount histological section of a macroporous construct showing the spontaneous induction of bone formation (*white* arrow) across the macroporous spaces of a coral-derived construct 180 days after intramuscular *rectus abdominis* implantation. Note the tight relationship of the enveloping *rectus abdominis* muscle (*magenta* arrows) surrounding the macroporous bioreactor. Decalcified sections cut at 6 μm from paraffin-embedded blocks stained with toluidine blue in 30% ethanol.

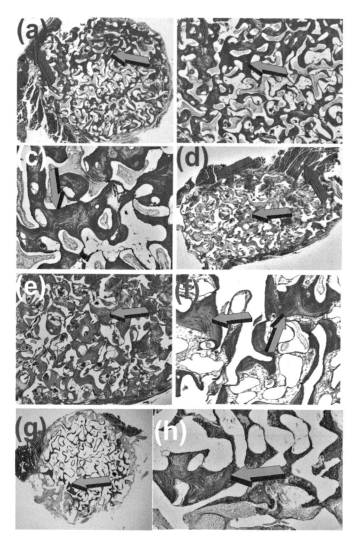

FIGURE 4.17 Morphological analyses of the spontaneous induction of bone formation by days 270 and 365, one year after heterotopic intramuscular implantation in the *rectus abdominis* muscle of adult non-human primates Chacma baboons *Papio ursinus* (Ripamonti 1991; Ripamonti et al. 2009; Ripamonti et al. 2010). (a) Induction of bone formation across the macroporous bioreactor on day 270 after *rectus abdominis* implantation. (b) Remodelled osteonic bone filling the macroporous spaces, with remodelling (c) and the induction of osteonic bone (*light blue* arrow c). (d,e) Complete bone formation by induction in a macroporous bioreactor harvested on day 270 after *rectus abdominis* implantation (*light blue* arrows e,f). Bone forms and remodels across the entirety of the macroporous spaces, constructing a living bioreactor of newly formed lamellar bone within the containing and inductive macroporous bioreactor. (g,h) Bone induction and remodelling by day 360 after heterotopic implantation. Thinning and remodelling of the osteonic bone still filling the macroporous spaces one year after heterotopic implantation (*light blue* arrows) (Ripamonti et al. 2010).

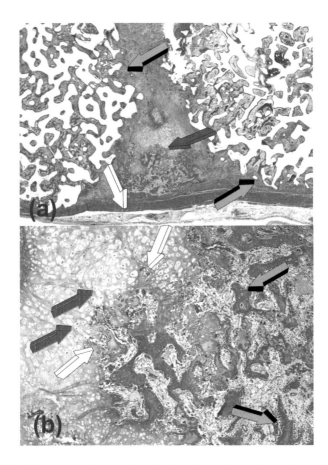

FIGURE 4.18 Chondrogenesis and the induction of endochondral bone formation (*magenta* arrow) just above the *rectus abdominis* fascia (*white* arrow) by diffusion gradients set into motion by doses of recombinant transforming growth factor-β_3 (hTGF-β_3) super activating coral-derived constructs (*blue* arrows) implanted for a morphological and molecular data time study. (a) A cartilaginous island formed in the *rectus abdominis* muscle just over the fascia. (b) In spite of the heterotopic implantation, the cartilaginous matrix still retains the phylogenetically ancient genetic memory of the induction of columnar chondrocytes of the mammalian growth plate (*magenta* arrows). Vascular invasion and chondrolysis are then followed (*white* arrows) by the induction of trabecular bone across the primitive induction of the cartilage anlage (*blue* arrows).

The substantial capillary induction and angiogenesis within the macroporous spaces were pronounced as early as 15 days after heterotopic intramuscular implantation as shown in Figure 4.4 (Figs. 4.4a,b,c). Figure 4.4 epitomizes the morphological scenario of the induction of bone formation sustained by capillary sprouting and the differentiation of the "osteogenetic vessels" of Trueta's definition (Trueta 1963). It further crystallizes the induction of mesenchymal collagenous condensations with capillary sprouting and invasion. *De novo* capillary sprouting within the macroporous spaces provides not only nutrients but soluble morphogens for the induction of bone formation, binding morphogenetic signals, both angiogenic and bone morphogenetic proteins, to the capillary basement membrane components, in particular type IV collagen (Vlodavsky et al. 1987; Folkman et al. 1988; Paralkar et al. 1990; Paralkar et al. 1991). Additionally, the "osteogenetic vessels" provide a continuous flow of progenitor cells migrating along the capillaries (Fig. 4.4e). Angiogenesis within the macroporous spaces is molecularly highlighted by over-expression of *type IV collagen* on day 15 when compared to days 60 and 90 (Klar et al. 2013).

The multifaceted plasticity of the endothelium has been recently highlighted in a series of molecular and morphological studies that have shown that capillaries walls harbour a reserve of progenitor cells that are the "elusive" mesenchymal stem cells of all parenchymatous organs in mammals. Such cells have been identified and renamed as pericytes (Crisan et al. 2008). Perivascular progenitor cells and pericytes surrounding blood vessels and particularly capillaries have been studied and reviewed by Kovacic and Boehm (2009), reporting that resident perivascular cells and pericytes are implicated in vascular remodelling as well as providing progenitors for vascular transformation and angiogenesis, also providing cellular elements for vascular and perivascular stem cells (Kovacic and Boehm 2009). Pericytes' coverage of blood vessels is regulated by PDGF-B and VEGF during angiogenesis (Benjamin et al. 1988), further supporting the accrued knowledge of the pleotropic inductive and differentiating capacities of the pericytes (Crisan et al. 2008; Benjamin et al. 1988).

Further studies by Medici et al. (2011) have shown the conversion of vascular endothelial cells into multipotent stem-like cells capable of the induction of the osteogenic phenotype (Medici et al. 2011). A grand study, in the opinion of the writer of this CRC Press volume, on the spontaneous induction of bone formation has been the experimentation by Kusumbe et al. (2014) that reported a specific vessel sub-type in bone coupling angiogenesis and osteogenesis. The experiments identified a specific bone capillary sub-type, termed type H endothelial cells, proposing that such vessels provide niche signals for perivascular osteoprogenitors with pro-osteogenic capacity (Kusumbe et al. 2014). It was reported that the newly identified capillaries sub-type H mediate angiogenesis in bone, "generate distinct metabolic and molecular microenvironments, maintain perivascular osteoprogenitors and couple angiogenesis to osteogenesis" (Kusumbe et al. 2014).

The study and reported analyses indicate that such vessels sub-types in bone are the "osteogenetic vessels" of Trueta's definition (Trueta 1963), whereby capillary invasion and sprouting within newly formed matrices with cellular niches of progenitors and stem cells together with morphogens bound to basement membrane components (Paralkar et al. 1990; Paralkar et al. 1991; Wlodavsky et al. 1987; Folkman et al. 1988) initiate the ripple-like cascade of "Bone: formation by autoinduction" (Urist 1965).

It is highly possible that the Ca^{++}-rich microenvironment of the calcium phosphate-based macroporous spaces does induce these specific vessels sub-types during the early morphogenetic events of the spontaneous induction of bone formation. Further studies by the same authors showed that endothelial Notch activity promotes angiogenesis and osteogenesis of the bone matrix microenvironment (Ramasamy et al. 2014).

The pleiotropy and vast plasticity of the endothelium are further shown by *in vitro* studies by the addition of highly purified naturally derived osteogenic fractions to aortic endothelial (E8) cells when reaching the typical cobblestone morphology (Heliotis and Ripamonti 1994). A profound alteration of the typical cobblestone morphology was observed after the addition of protein fractions at concentrations of 3, 6 and 10 µg over 24 and 48 hours (Heliotis and Ripamonti 1994). There was an alteration of cell morphology from a cobblestone to a spindle-shaped phenotype after 24 and 48 hours. Importantly, reacquisition of the typical cobblestone appearance could be achieved when cultures were not fed osteogenic fractions for more than 48 hours, provided that osteogenic fractions did not exceed a concentration of 6 µg of proteins/300 µl of medium (Heliotis and Ripamonti 1994). It was noteworthy that if the protein concentrate exceeded 6 µg of proteins per 300 µl of medium and if E8 cells were exposed for more than 24 hours to this concentration, cells would not regain the typical cobblestone appearance (Heliotis and Ripamonti 1994). These data are of interest and indicate that purified bone morphogenetic proteins profoundly alter cell morphology of E8 endothelial cells to a spindle-like shaped phenotype with eventual rounding up and detachment of cells by 72 hours regardless of the protein concentration used (Heliotis and Ripamonti 1994). These differentiation and de-differentiation events may pave the way for *in vivo* rapid phenotypic changes to osteoblast-like cells for the induction of bone formation (Heliotis and Ripamonti 1994; Ripamonti et al. 2009; Ripamonti et al. 2010; Klar et al. 2013).

It is now clear that endothelial cells are more than cells simply covering basement membrane components of vessels of the mammalian body. Recent work has shown that endothelial cell-derived factors or signals regulate *in vivo* morphogenesis (Ramasamy et al. 2015). Importantly, preliminary data have shown that the plasticity of endothelial cells adapts in regeneration and that tissue specific endothelial cells "maintain organ homeostasis and instruct regeneration" (Gomez-Salinero and Rafii 2018).

ACKNOWLEDGEMENTS

The intrinsic and/or spontaneous induction of bone formation by macroporous calcium phosphate-based constructs has been the source of a continuously fascinating quest of tissue induction, morphogenesis, geometry, differentiation and de-differentiation and has culminated, after the discovery of the concavity, the "shape of life", in our understanding of the intrinsic induction of bone formation by a variety of calcium phosphate-based bioreactors epitomizing the "geometric induction of bone formation". I would like to thank the expertise and dedication of the molecular team of the School of Clinical Medicine headed by Raquel Duarte who, together with Roland Klar, mechanistically resolved the "geometric induction of bone formation". and further recognize Caroline Dickens and Therese Dix-Peek for several joint experiments, many discussion hours and for finally resolving the spontaneous induction of bone formation with gene expression analyses regulating the heterotopic induction of bone formation by coral-derived macroporous biore-actors when implanted in the *rectus abdominis* muscle of *Papio ursinus*, where there is no bone.

REFERENCES

Barradas, A.M.C.; Yuan, H.; van Blitterswijk, C.A.; Habibovic, P. Osteoinductive Biomaterials: Current Knowledge of Properties, Experimental Models and Biological Mechanisms. *Eur. Cell Mater.* 2011, *21*, 407–29.

Benjamin, L.E.; Hemo, I.; Keshet, E. A Plasticity Window for Blood Vessels Remodelling Is Defined by Pericyte Coverage of the Preformed Endothelial Network and Is Regulated by PDGF-B and VEGF. *Development* 1988, *125*(9), 1591–8.

Christman, K.L. Biomaterials for Tissue Repair. *Science* 2019, *363*(6425), 340–41.

Crisan, M.; Yap, S.; Casteilla, L. et al. A Perivascular Origin for Mesenchymal Stem Cells in Multiple Human Organs. *Cell Stem Cell* 2008, *3*(3), 301–13.

Daluiski, A.; Engstrand, T.; Bahamonde, M.E.; Gamer, L.W.; Agius, E.; Srevenson, S.L.; Cox, K.; Rosen, V.; Lyons, K.M. Bone Morphogenetic Protein-3 Is a Negative Regulator of Bone Density. *Nat. Genet.* 2001, *27*(1), 84–8.

De Robertis, E.M. Evo-Devo: Variations on Ancestral Themes. *Cell* 2008, *132*(2), 185–95.

Folkman, J.; Klagsbrun; Sasse, J.; et al. A Heparin Binding Angiogenic Protein Basic Fibroblast Growth Factor Is Stored within Basement Membranes. *Am. J. Pathol.* 1988, *130*, 393–400.

Heliotis, M.; Ripamonti, U. Phenotypic Modulation of Endothelial Cells by Bone Morphogenetic Protein Fractions In Vitro. *In Vitro Cell. Dev. Biol. Anim.* 1994, *30A*(6), 353–5.

Gomez-Salinero, J.M.; Rafii, S. Endothelial Cell Adaptation in Regeneration. *Science* 2018, *362*(6419), 1116–7.

Kanatani, M.; Sugimoto, T.; Fukase, M.; Fujita, T. Effect of Elevated Extracellular Calcium on the Proliferation of Osteoblastic MC3T3-E1 Cells: Its Direct and Indirect Effects *via* Human Monocytes. *Biochem. Biophys. Res. Commun.* 1991, *181*(3), 1425–30.

Kerszberg, M.; Wolpert, L. Specifying Positional Information in the Embryo: Looking Beyond Morphogens. *Cell* 2007, *130*(2), 205–9.

Klar, R.M.; Duarte, R.; Dix-Peek, T.; Dickens, C.; Ferretti, C.; Ripamonti, U. Calcium Ions and Osteoclastogenesis Initiate the Induction of Bone Formation by Coral-Derived Macroporous Constructs. *J. Cell. Mol. Med.* 2013, *17*(11), 1444–57.

Klar, M.R.; Duarte, R.; Dix-Peek, T.; Ripamonti, U. The Induction of Bone Formation by the Recombinant Human Transforming Growth Factor-β3. *Biomaterials* 2014, *35*(9), 2773–88.

Kovacic, J.C.; Boehm, M. Resident Vascular Progenitor Cells: An Emerging Role for Non-Terminally Differentiated Vessel-Resident Cells in Vascular Biology. *Stem Cell Res.* 2009, *2*(1), 2–15.

Kusumbe, A.P.; Ramasamy, S.K.; Adams, R.H. Coupling of Angiogenesis and Osteogenesis by a Speficic Vessel Subtype in Bone. *Nature* 2014, *507*(7492), 323–8.

Lander, A.D. Morpheus Unbound: Reimagining the Morphogen Gradient. *Cell* 2007, *128*(2), 245–56.

Luyten, F.P.; Cunningham, N.S.; Ma, S.; Muthukumaran, N.; Hammonds, R.G.; Nevins, W.B.; Wood, W.I.; Reddi, A.H. Purification and Partial Amino Acid Sequence of Osteogenin, a Protein Initiating Bone Differentiation. *J. Biol. Chem.* 1999, *264*(23), 13377–80.

Medici, D.; Shore, E.M.; Lounev, V.Y.; Kaplan, F.D.; Kalluri, R.; Olsen, B.R. Conversion of Vascular Endothelial Cells into Multipotent Stem-Like Cells. *Nat. Med.* 2011, *16*(12), 1400–06.

Müller, P.; Buinheim, U.; Diener, A.; Lüthen, F.; Teller, M.; Klinkenberg, E.-D.; Neumann, H.-G.; Nebe, B.; Liebold, A.; Steinhoff, G.; Rychly, J. Calcium Phosphate Surfaces Promote Osteogenic Differentiation of Mesenchymal Stem Cells. *J. Cell. Mol. Med.* 2008, *12*(1), 281–91.

Paralkar, V.M.; Nandedkar, A.K.N.; Pointer, R.H.; Kleinman, H.K.; Reddi, A.H. Interaction of Osteogenin, a Heparin Binding Bone Morphogenetic Protein, with Type IV Collagen. *J. Biol. Chem.* 1990, *265*(28), 17281–4.

Paralkar, V.M.; Vukicevic, S.; Reddi, A.H. Transforming Growth Factor β Type 1 Binds to Collagen Type IV of Basement Membrane Matrix: Implications for Development. *Dev. Biol.* 1991, *143*(2), 303–10.

Ramasamy, S.K.; Kusumbe, A.P.; Wang, L.; Adams, R.H. Endothelial Notch Activity Promotes Angiogenesis and Osteogenesis in Bone. *Nature* 2014, *507*(7492), 376–80.

Ramasamy, S.K.; Kusumbe, A.P.; Adams, R.H. Regulation of Tissue Morphogenesis by Endothelial Cell-Derived Signals. *Trends Cell Biol.* 2015, *25*(3), 148–57.

Reddi, A.H. Huggins, C.B. Biochemical Sequences in the Transformation of Normal Fibroblasts in Adolescent Rats. *Proc. Natl. Acad. Sci.* USA **1972,** *69*(6), 1601–05.

Reddi, A.H. Cell Biology and Biochemistry of Endochondral Bone Development. *Coll. Rel. Res.* 1981, *1*(2), 209–26.

Reddi, A.H.; Kuettner, K.E. Vascular Invasion of Cartilage: Correlation of Morphology with Lysozyme, Glycosaminoglycans, Protease, and Protease-Inhibitory Activity During Endochondral Bone Development. *Dev. Biol.* 1981, *82*(2), 217–23.

Reddi, A.H. Extracellular Matrix and Development. In: K.A. Piez, A.H. Reddi (eds.) *Extracellular Matrix Biochemistry*, Elsevier, New York, 1984, 247–91.

Reddi, A.H. Bone Morphogenesis and Modeling: Soluble Signals Sculpt Osteosomes in the Solid State. *Cell* 1997, *89*(2), 159–61.

Reddi, A.H. Morphogenesis and Tissue Engineering of Bone and Cartilage: Inductive Signals, Stem Cells, and Biomimetic Matrices. *Tissue Eng.* 2000, *6*(4), 351–9.

Ripamonti, U. Inductive Bone Matrix and Porous Hydroxyapatite Composites in Rodents and Nonhuman Primates. In: J. Yamamuro, L. Wilson-Hench, L. Hench (eds.) *Handbook of Bioactive Ceramics, Volume II: Calcium Phosphate and Hydroxylapatite Ceramics*, CRC Press, Boca Raton, FL, 1990, 245–53.

Ripamonti, U. The Morphogenesis of Bone in Replicas of Porous Hydroxyapatite Obtained from Conversion of Calcium Carbonate Exoskeletons of Coral. *J. Bone Joint Surg. [A]* 1991, *73-A*, 692–703.

Ripamonti, U.; Ma, S.; Cunningham, N.; Yeates, L.; Reddi, A.H. Initiation of Bone Regeneration in Adult Baboons by Osteogenin, a Bone Morphogenetic Protein. *Matrix* 1992a, *12*(5), 369–80.

Ripamonti, U.; Ma, S.; van den Heever, B.; Reddi, A.H. Osteogenin, a Bone Morphogenetic Protein, Adsorbed on Porous Hydroxyapatite Substrata, Induces Rapid Bone Differentiation in Calvarial Defects of Adult Primates. *Plast. Reconstr. Surg.* 1992b, *90*(3), 382–93.

Ripamonti, U.; van den Heever, B.; van Wyk, J. Expression of the Osteogenic Phenotype in Porous Hydroxyapatite Implanted Extraskeletally in Baboons. *Matrix* 1993a, *13*(6), 491–502.

Ripamonti, U.; Yeates, L.; van den Heever, B. Initiation of Heterotopic Osteogenesis in Primates After Chromatographic Adsorption of Osteogenin, a Bone Morphogenetic Protein, onto Porous Hydroxyapatite. *Biochem. Biophys. Res. Commun.* 1993b, *193*(2), 509–17.

Ripamonti, U. Osteoinduction in Porous Hydroxyapatite Implanted in Heterotopic Sites of Different Animal Models. *Biomaterials* 1996, *17*(1), 31–5.

Ripamonti, U.; Duneas, N. Tissue Engineering of Bone by Osteoinductive Biomaterials. *MRS Bulletin* November 1996, 36–9.

Ripamonti, U.; Duneas, N. Tissue Morphogenesis and Regeneration by Bone Morphogenetic Proteins. *Plast. Reconstr. Surg.* 1998, *101*(1), 227–39.

Ripamonti, U.; Crooks, J.; Kirkbride, A.N. Sintered Porous Hydroxyapatites with Intrinsic Osteoinductive Activity: Geometric Induction of Bone Formation. *S. Afr. J. Sci.* 1999, *95*, 335–43.

Ripamonti, U. *Smart* Biomaterials with Intrinsic Osteoinductivity: Geometric Control of Bone Differentiation. In: J.D. Davis (ed.) *Bone Engineering*, M2 Corporation, Toronto, 2000, 215–22.

Ripamonti, U.; Ramoshebi, L.N.; Matsaba, T.; Tasker, J.; Crooks, J.; Teare, J. Bone Induction by BMPs/Ops and Related Family Members in Primates. The Critical Role of Delivery Systems. *J. Bone Joint Surg.* 2001, *83-A*(51), 16–27.

Ripamonti, U. Osteogenic Proteins of the Transforming Growth Factor-ß Superfamily. In: H.L. Henry, A.W. Norman (eds.). *Encyclopedia of Hormones*, Academic Press, San Diego, CA 2003, 80–6.

Ripamonti, U. Soluble, Insoluble and Geometric Signals Sculpt the Architecture of Mineralized Tissues. *J. Cell. Mol. Med.* 2004, *8*(2), 169–80.

Ripamonti, U.; Ramoshebi, L.N.; Patton, J.; Matsaba, T.; Teare, J.; Renton, L., Soluble Signals and Insoluble Substrata: Novel Molecular Cues Instructing the Induction of Bone. In: E.J. Massaro, J.M. Rogers (eds.) *The Skeleton*, Humana Press, Totowa, New Jersey 2004, Chapter 15, 217–27.

Ripamonti, U. Soluble Osteogenic Molecular Signals and the Induction of Bone Formation. *Biomaterials* Leading Opinion Paper 2006, *27*(6), 807–22.

Ripamonti, U.; Ferretti, C.; Heliotis, M. Soluble and Insoluble Signals and the Induction of Bone Formation: Molecular Therapeutics Recapitulating Development. *J. Anat.* 2006, *209*(4), 447–68.

Ripamonti, U.; Heliotis, M.; Ferretti, C. Bone Morphogenetic Proteins and the Induction of Bone Formation: From Laboratory to Patients. *Oral Maxillofac. Surg. Clin. North Am.* 2007, *19*(4), 575–89.

Ripamonti, U.; Richter, P.W.; Nilen, R.W.N.; Renton, L. The Induction of Bone Formation by *Smart* Biphasic Hydroxyapatite Tricalcium Phosphate Biomimetic Matrices in the Non-Human Primate *Papio ursinus*. *J. Cell. Mol. Med.* 2008, *12*(6B), 2609–21.

Ripamonti, U. Biomimetism, Biomimetic Matrices and the Induction of Bone Formation. *J. Cell. Mol. Med.* 2009, *13*(9B), 2953–72.

Ripamonti, U.; Crooks, J.; Khoali, L.; Roden, L. The Induction of Bone Formation by Coral-Derived Calcium Carbonate/Hydroxyapatite Constructs. *Biomaterials* 2009, *30*(7), 1428–39.

Ripamonti, U. Soluble and Insoluble Signals Sculpt Osteogenesis in Angiogenesis. *World J. Biol. Chem.* 2010, *26*(5), 109–32.

Ripamonti, U.; Klar, R.M.; Renton, L.F.; Ferretti, C. Synergistic Induction of Bone Formation by hOP-1, hTGF-β3 and Inhibition by Zoledronate in Macroporous Coral-Derived Hydroxyapatites. *Biomaterials* 2010, *31*(25), 6400–10.

Ripamonti, U. The Concavity: The "Shape of Life"and the Control of Bone Differentiation – Feature Paper – *Science in Africa* May 2012.

Ripamonti, U.; Roden, L.; Renton, L.; Klar, R.M.; Petit, J.-C. The Influence of Geometry on Bone: Formation by Autoinduction. *Science in Africa* 2012. http://www.scienceinafric a.co.za/2012/Ripamonti_bone.htm.

Ripamonti, U., Induction of Bone Formation by Calcium Phosphate-Based Biomimetic Macroporous Constructs. In: M. Ramalingam, P. Vallitu, U. Ripamonti, W.-J. Li (eds.) *CRC Press Taylor & Francis, Boca Raton USA; Tissue Engineering and Regenerative Medicine. Nano Approach*, 2013, Chapter 5, 85–103.

Ripamonti, U.; Duarte, R.; Ferretti, C. Re-Evaluating the Induction of Bone Formation in Primates. *Biomaterials* 2014, *35*(35), 9407–22.

Ripamonti, U.; Dix-Peek, T.; Parak, R.; Milner, B.; Duarte, R. Profiling Bone Morphogenetic Proteins and Transforming Growth Factor-Bs by hTGF-β₃ Pre-Treated Coral-Derived Macroporous Constructs: The Power of One. *Biomaterials* 2015, *49*, 90–102.

Ripamonti, U.; Parak, R.; Klar, R.M.; Dickens, C.; Dix-Peek, T.; Duarte, R. The Synergistic Induction of Bone Formation by the Osteogenic Proteins of the TGF-β Supergene Family. *Biomaterials* 2016, *104*, 279–96.

Ripamonti, U. Biomimetic Functionalized Surfaces and the Induction of Bone Formation. *Tissue Eng.* 2017, *23*(21,22), 1197–209.

Ripamonti, U. Functionalized Surface Geometries Induce: "Bone: Formation by Autoinduction". *Front. Physiol.* 2018, *8*, 1084.

Ripamonti, U.; Duarte, R. Inductive Surface Geometries: Beyond Morphogens and Stem Cells. *S. Afr. Dent. J.* 2019, *74*(8), 421–44.

Trueta, J. The Role of the Vessels in Osteogenesis. *J. Bone Joint Surg.* 1963, *45B*, 402–18.

Turing, A.M. The Chemical Basis of Morphogenesis. *Philos. Trans. R. Soc. Lond.* 1952, *237*, 27–41.

Urist, M.R. Bone: Formation by Autoinduction. *Science* 1965, *220*(3698), 893–99.

Urist, M.R.; Hou, Y.K.; Brownell, A.G.; Hohl, W.; Buyske, J.; Lietze, A.; Tempst, P.; Hunkapiller, M.; DeLange, R.J. Purification of Bovine Bone Morphogenetic Protein by Hydroxyapatite Chromatography. *Proc. Natl. Acad. Sci. U.S.A.* 1984, *81*(2), 371–5.

van Eeden, S.; Ripamonti, U. Bone Differentiation in Porous Hydroxyapatite Is Regulated by the Geometry of the Substratum: Implications for Reconstructive Craniofacial Surgery. *Plast. Reconstr. Surg.* 1994, *93*(5), 959–66.

Vlodavsky, I.; Folkman, J.; Sullivan, R.; Fridman, R.; Ishai-Michaeli, R.; Sasse, J.; Klagsbrun, M. Endothelial Cell-Derived Basic Fibroblast Growth Factor: Synthesis and Deposition into Subendothelial Extracellular Matrix. *Proc. Natl. Acad. Sci. U.S.A.* 1987, *84*(8), 2292–6.

Williams, D. Editorial. The Continuing Evolution of Biomaterials. *Biomaterials* 2011, *32*(1), 1–2.

Wolpert, L. Positional Information and the Spatial Pattern of Cellular Differentiation. *J. Theor. Biol.* 1996, *25*(1), 1–47.

Wozney, J.M.; Rosen, V.; Celeste, A.J.; Mitsock, L.M.; Whitters, M.J.; Kriz, R.W.; Hewick, R.M.; Wang, E.A. Novel Regulators of Bone Formation: Molecular Clones and Activities. *Science* 1988, *242*(4885), 1528–34.

Yamasaki, H. Heterotopic Bone Formation around Porous Hydroxyapatite Ceramics in the Subcutis of Dogs. *Jpn. J. Oral Biol.* 1990, *32*(2), 190–2.

Yamasaki, H.; Sakai, H. Osteogenic Response to Porous Hydroxyapatite Ceramics under the Skin of Dogs. *Biomaterials* 1992, *13*(5), 308–12.

Yuan, H.; Fernades, H.; Habibovic, P.; de Boer, J.; Barradas, A.M.C.; de Ruiter, A.; Walsh, W.R.; van Blitterswijk, C.A.; de Bruijn, J.D. Osteoinductive Ceramics as a Synthetic Alternative to Autologous Bone Grafting. *Proc. Natl. Acad. Sci. U.S.A.* 2010, *107*(31), 13614–9.

Zhang, X.; Williams, D. *Definitions of Biomaterials for the Twenty-First Century.* Elsevier, 2019.

Zheng, B.; Cao, B.; Crisan, M.; Sun, B.; Li, G.; Logar, A.; Yap, S.; Pollett, J.B.; Drowley, L.; Cassino, T.; Gharaibeh, B.; Deasy, B.; Huard, J.; Péault, B. Prospective Identification of Myogenic Endothelial Cells in Human Skeletal Muscle. *Nat. Biotechnol.* 2007, *25*(9), 10125–34.

Zimmerman, L.B.; De Jesus-Escobar, J.M.; Harland, R.M. The Spemann Organizer Signal Noggin Binds and Inactivates Bone Morphogenetic Protein 4. *Cell* 1996, *96*(4), 599–606.

5 Concavities of Crystalline Sintered Hydroxyapatite-Based Macroporous Bioreactors Initiate the Spontaneous Induction of Bone Formation

Ugo Ripamonti

5.1 INTRODUCTION

"Bone: formation by autoinduction" (Urist 1965), and more generally the induction of osteogenesis, the development of the vertebrates and the induction of skeletogenesis providing both the craniofacial and appendicular axial skeletons of the nascent vertebrates, the emergence of mammals, pleiotropic speciation with the appearance of the Australopithecines walking up-right, shortly followed by *Homo habilis* thus able to use tools capable of freeing the upper limbs for more industrious *Homo*-like activities, above all however freed for maternal care, tightly holding the growing newborn for feeding and care, walking erect following skeletogenesis and the induction of bone with modified pelvis and femoral articulations and muscle attachments providing ambulatory capacity to the early hominids, speciation of *Homo erectus*, walking out of Africa, further speciating the *Homo* clade across the planet is the most fascinating opera of Nature's marvellous evolutionary pathway from the fruit fly *Drosophila melanogaster* to the primate *Homo sapiens* (De Robertis 2008; Ripamonti 2009; Ripamonti 2017; Ripamonti 2018).

The induction of bone formation requires three components: soluble osteogenetic molecular signals, responding cells and a complementary substratum upon which cells attach and differentiate (Reddi 1997; Reddi 2000; Ripamonti 2004; Ripamonti et al. 2004). We have learned that that the optimal induction of bone formation is dependent on the combined action of the osteogenetic soluble molecular signals and a complementary substratum, or insoluble signal, which controls and delivers the biological activity of the morphogenetic soluble molecular signals of the transforming growth factor-β (TGF-β) supergene family (Ripamonti 2003; Ripamonti and Reddi 1995; Reddi 2000; Ripamonti 2006). The combination of a soluble osteogenetic molecular signal with a complementary substratum does result in the induction of bone formation in heterotopic extraskeletal sites (Sampath and Reddi 1981; Ripamonti and Reddi 1995; Reddi 2000; Ripamonti et al. 2001a). It also defines an osteoinductive biomaterial, that is, a biomimetic matrix, organic or not, that does initiate the cascade of the induction of bone formation (Ripamonti and Duneas 1996).

We previously stated that, when defining an osteoinductive device and/or biomaterial, it is important to differentiate it from an osteoconductive biomaterial (Ripamonti and Duneas 1996). An osteoinductive biomaterial is one bearing osteogenic activity *per se*; its discriminatory bioaction is the induction of bone formation (Ripamonti and Duneas 1996).

Our research studies into a variety of macroporous naturally derived and sintered crystalline hydroxyapatites have shown that macroporous hydroxyapatites are optimal carrier matrices for the biological activity of highly purified naturally derived or recombinantly produced osteogenic proteins of the TGF-β supergene family in both heterotopic and orthotopic sites of rodents and non-human primates (Ripamonti et al. 1992a; Ripamonti et al. 1992b; Ripamonti et al. 1992c; Ripamonti et al. 1993a; Ripamonti et al. 1993b; Ripamonti and Dunaes 1996; Ripamonti et al. 2001a; Ripamonti et al. 2001b). In contrast, an osteoconductive biomaterial is not inherently osteoinductive (Ripamonti and Duneas 1996); its capacity is to guide and direct the growth of bone at the interfaces into the macroporous structure and to facilitate osteointegration when implanted in orthotopic sites (Ripamonti and Duneas 1996).

5.2   SELF-INDUCTIVE BIOMATERIALS

The question we have asked since the last century is: can an osteoconductive biomaterial be transformed into an osteoinductive biomaterial without the exogenous application of the soluble osteogenetic molecular signals of the TGF-β supergene family? That is, can we modify the discriminatory bioaction from solely an osteoconductive biomaterial into an osteogenic biomaterial that *per se* sets into motion the ripple-like cascade of bone formation by induction without the exogenous applications of osteogenic proteins?

After a series of experiments in the *rectus abdominis* muscle of the Chacma baboon *Papio ursinus*, we described the hydroxyapatite-induced osteogenesis model (Ripamonti 1990; Ripamonti 1991; Ripamonti et al. 1993a; Ripamonti 1996) whereby coral-derived macroporous bioreactors were found to initiate the morphogenesis of bone within the heterotopically implanted macroporous spaces without the exogenous application of the soluble osteogenic molecular signals of the TGF-β supergene family (Ripamonti 1990; Ripamonti 1991; Ripamonti et al. 1993a; Ripamonti 2003).

The analyses and re-analyses of serial sections further cut from decalcified and undecalcified specimen blocks revealed the critical role of a specific geometric configuration that intrinsically initiates the induction of bone formation in a variety of calcium phosphate-based macroporous constructs (Fig. 5.1).

The concept that the geometry had a profound impact on the induction of bone formation (Reddi and Huggins 1973; Reddi 1974; Sampath and Reddi 1984) prompted us to prepare macroporous bioreactors made of crystalline sintered hydroxyapatites as well as solid sintered monolithic hydroxyapatite discs with a series of concavities prepared on both planar surfaces (Fig. 5.2) (Ripamonti et al. 1999; Ripamonti 2000; Ripamonti 2004).

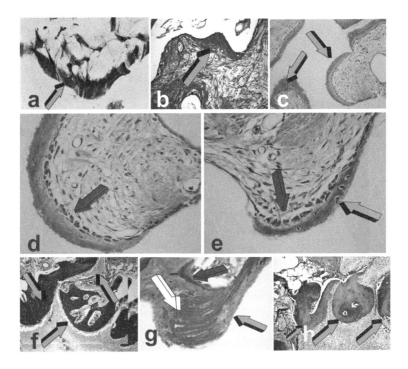

FIGURE 5.1 Series of digital images of coral-derived calcium phosphate-based bioreactors displaying the induction of bone formation within the macroporous spaces of the heterotopically implanted constructs in the *rectus abdominis* muscle of the Chacma baboon *Papio ursinus* harvested on days 60 and 90 after intramuscular implantation. (a) The geometric landscape of the concavity is also displayed following *in vitro* studies that showed that concavities of the *in vitro* bioreactor orient and polarize mouse-derived pre-osteoblast (MC 3T3-E1) cells along concavities of the substratum (Ripamonti et al. 2012a; Ripamonti et al. 2012b; Ripamonti et al. 2013). Note the alignment and orientation of MC 3T3-E1 cells along a concavity of the substratum (*light blue* arrow). (b through h) Morphological conceptualization of the critical role of the concavity initiating the induction of bone formation in macroporous constructs intramuscularly implanted in *Papio ursinus*. (b) Tissue patterning within the macroporous concavities and differentiation and alignment of mesenchymal condensations (*light blue* arrow) against the substratum on day 60 after *rectus abdominis* implantation. (c,d,e) Unequivocal morphological demonstration that the concavity directly initiates the induction of bone formation within the macroporous spaces of the *rectus abdominis* implanted geometric bioreactors (Klar et al. 2013; Ripamonti 2017). (d,e) Newly formed bone is tightly attached to the concavities of the calcium phosphate-based bioreactors (*light blue* arrows). Newly formed bone is populated by contiguous osteoblasts (*red* arrows) in close proximity to substantial angiogenesis and vascular invasion. (f,g,h) The critical role of the concavity, "the shape of life" (Ripamonti 2006; Ripamonti et al. 2006; Ripamonti 2012), in initiating the spontaneous and/or intrinsic induction of bone formation (Ripamonti 1996). Newly formed bone within concavities of the substratum (*light blue* arrows) with the generation of collagenous condensations (*white* arrow in g). In coral-derived calcium phosphate-based constructs, the induction of mesenchymal condensations predates the induction of bone formation with mineralization of the newly formed matrix and induction of bone (*light blue* arrow) with osteoid deposition populated by contiguous osteoblasts (*red* arrow in g).

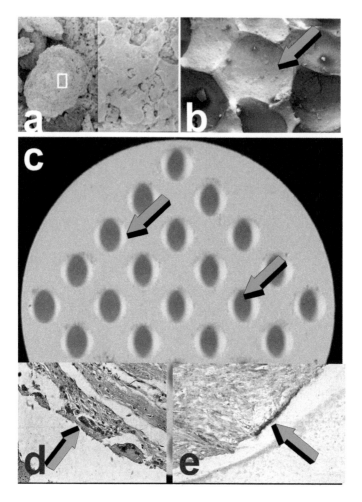

FIGURE 5.2   Self-inducing geometric cues and fabrication of crystalline sintered bioreactors displaying a series of concavities for the spontaneous and/or intrinsic (Ripamonti 1996) induction of bone formation (Ripamonti et al. 1999). (a) Scanning electron photomicrograph of Plasma Biotal No. 3 hydroxyapatite powder for the fabrication of macroporous sintered constructs (Ripamonti et al. 1999). (b) SEM analysis of a sintered macroporous construct that shows the alveolar-like pattern of the macroporous spaces formed by sequences of concavities with defined radii of curvature, and diameters ranging from 400 to 1600 μm (Ripamonti et al. 1999). (c) Sintered crystalline hydroxyapatite construct with a series of concavities on both planar surfaces (*light blue* arrows). To test the hypothesis that bone does initiate within concavities of the substratum only, solid discs of crystalline sintered hydroxyapatites prepared with a series of concavities (*light blue* arrows) were implanted in the *rectus abdominis* muscle of *Papio ursinus* and harvested on day 30 and 90. (d) Immunolocalization of osteogenic protein-1 (OP-1) within mesenchymal stem cells interpreted as osteoblastic-like cells differentiating at the hydroxyapatite interface on day 30 after heterotopic implantation. (e) Immunolocalization of osteogenic protein-1 (OP-1) (*light blue* arrow) now embedded onto the hydroxyapatite sintered biomatrix 30 days after implantation.

In previous and later chapters we describe in detail the rationale of the spontaneous osteoinductivity of hydroxyapatite-based macroporous bioreactors. Examination of several undecalcified and decalcified histological sections cut and re-cut from a multitude of specimen blocks of coral-derived macroporous constructs unequivocally showed that macroporous spaces presenting as concavities provided a unique microenvironment for the spontaneous or intrinsic induction of bone formation (Fig. 5.1) (Ripamonti 2009; Klar et al. 2013; Ripamonti 2017; Ripamonti 2018a; Ripamonti and Duarte 2019). The morphological re-evaluation of several decalcified and undecalcified sections was used to formulate the hypothesis that the induction of bone formation preferentially initiates in specific geometries of the calcium phosphate-based bioreactors architecturally and geometrically conducive and inducive to the induction of bone formation (Fig. 5.1).

The first papers that crystallized the "geometric induction of bone formation" were PCT and WO patents (Ripamonti and Kirkbride 1995; Ripamonti and Kirkbride 1997). The patented findings were later published (Ripamonti et al. 1999), reporting the results of heterotopically implanted macroporous sintered crystalline hydroxyapatites and of solid sintered crystalline hydroxyapatite discs with concavities assembled on both planar surfaces (Fig. 5.2) (Ripamonti et al. 1999). The communication, intended to be published in the *South African Journal of Science* (Ripamonti et al. 1999), was printed with a cover page illustrating the geometric induction of bone formation by sintered calcium phosphate-based macroporous bioreactors harvested from the *rectus abdominis* muscle (Fig. 5.3a) (Ripamonti et al. 1999). The paper was also the first paper to present some mechanistic insights into the spontaneous and/ or intrinsic induction of bone formation by sintered crystalline hydroxyapatite-based bioreactors (Fig. 5.2).

Immunolocalization studies showed osteogenic protein-1 (OP-1, also known as BMP-7) immunolocalized in mesenchymal cellular elements at the concavity interface (Fig. 5.2d) later secreted and embedded onto the hydroxyapatite substratum of the concavity (Fig. 5.2e). The presented images unequivocally showed the critical role of bone morphogenetic proteins as the initiators of the induction of bone formation by calcium phosphate-based bioreactors (Ripamonti et al. 1993a; Ripamonti et al. 1999; Klar et al. 2013; Ripamonti 2017).

5.3 SURFACE GEOMETRY REGULATES BONE DIFFERENTIATION IN HETEROTOPIC SITES

The hypothesis that concavities were conducive and inducive to the spontaneous induction of bone formation without the exogenous applications of the osteogenic soluble molecular signals of the TGF-β supergene family (Ripamonti 2003; Ripamonti 2006) was tested by constructing sintered crystalline bioreactors with macroporous spaces characterized by a series of repetitive concavities (Figs. 5.2a,b) (Ripamonti et al. 1999). To unequivocally determine the critical role of surface geometry and of the concavity *motif* driving the spontaneous induction of bone formation as seen in systematic studies in *Papio ursinus* (Ripamonti 1990; Ripamonti 1991; Ripamonti et al. 1993a; Ripamonti 1996), slurry preparations of Tribo Corr 65–90 μm powder

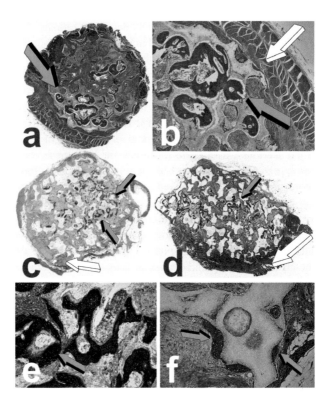

FIGURE 5.3 Low-power microphotographs of self-inducing sintered crystalline macroporous hydroxyapatites implanted in the *rectus abdominis* muscle of *Papio ursinus* and harvested on day 90 after heterotopic intramuscular implantation. (a) Prominent induction of bone formation across the macroporous spaces of a sintered crystalline macroporous bioreactor 90 days after intramuscular implantation. Bone forms within the macroporous spaces (*light blue* arrow) as repetitive concavities within the implanted bioreactor (Ripamonti et al. 1999). (b) High-power view of inducive macroporous spaces (a) highlighting the spontaneous intrinsic induction of bone formation (*light blue* arrow) regulated by the geometry of the substratum. The induction of bone formation initiates in heterotopic intramuscular sites the *rectus abdominis* muscle (*white* arrow) where there is no bone, and without the exogenous application of the osteogenic soluble molecular signals of the TGF-β supergene family (Ripamonti 2003). (c,d) Low-power views of crystalline sintered hydroxyapatites fabricated by impregnating polyurethane foams with slurry preparations of hydroxyapatite powders as previously described (Ripamonti et al. 1999; Ripamonti et al. 2001a). Bioreactors were implanted in the rectus abdominis muscle (*white* arrows) and harvested for morphological analyses on day 90 after implantation (Ripamonti et al. 2001a; Ripamonti et al. 2001b). (e) Higher-power view of a sintered bioreactor generating substantial induction of bone formation (*light blue* arrow) within the macroporous spaces. Undecalcified section cut, ground and polished on the Exakt diamond saw equipment. (f) Sintered macroporous bioreactor inducing bone within concavities of the crystalline substratum. Cross-section images of the concavities generating new bone (*light blue* arrows) were instrumental in designing and constructing hydroxyapatite-coated titanium implants with a series of concavities as seen histologically in f (*light blue* arrow) (Ripamonti et al. 2012a; Ripamonti et al. 2013).

(Tribo Corr, South Africa) were sintered to form solid monolithic hydroxyapatite discs 20 mm in diameter, 4 mm thick, with a series of concavities prepared on both planar surfaces (Fig. 5.2c).

Commercially available polyurethane foams, with continuous and interconnecting pores, were cut into rods 7 mm in diameter and 20 mm in length, and into discs 24 mm in diameter and 4 mm thick for heterotopic intramuscular and orthotopic calvarial implantation, respectively. SEM analyses of sintered artifacts made of different starting powders of hydroxyapatite (Fig. 5.2a) (Plasma Biotal No. 3 and P120 – Plasma Biotal, England; GR 500 P – Cam Implants, the Netherlands; Plasmatex A6020 – Plasma Tecknik, Switzerland; and Tribo-Corr 65–90 μm) (Ripamonti and Kirkbride 1995; Ripamonti and Kirkbride 1997; Ripamonti et al. 1999; Ripamonti and Kirkbride 2000; Ripamonti and Kirkbride 2001). SEM analyses of sintered artefacts showed an alveolar-like pattern of the hydroxyapatite substratum, resulting in porous spaces formed by sequences of concavities with defined radii of curvature and diameters ranging from 400 to 1600 μm (Ripamonti et al. 1999) (Fig. 5.2b).

To crystallize the critical importance of the concavities assembled into calcium phosphate-based constructs in tissue induction and morphogenesis, generated solid monolithic discs 20 mm in diameter, 4 mm thickness, with a series of repetitive concavities prepared on both planar surfaces (Fig. 5.2c), were implanted in heterotopic sites of the *rectus abdominis* muscle of *Papio ursinus* and harvested on days 30 and 90 after intramuscular implantation (Ripamonti et al. 1999).

In a series of systematic experiments in the *rectus abdominis* muscle of several clinically healthy adult non-human primates of the species *Papio ursinus*, crystalline sintered macroporous bioreactors and solid monolithic discs were implanted in 16 animals (Figs. 5.3, 5.4, 5.5, 5.6, 5.7) (Ripamonti et al. 1999). Macroporous sintered crystalline bioreactors were also implanted in orthotopic calvarial sites to study the biology of incorporation of crystalline bioreactors in calvarial membranous sites (Fig. 5.8) (Ripamonti et al. 1999).

Additional experiments in four Chacma baboons *Papio ursinus* studied the biology of incorporation of single-phase hydroxyapatite, biphasic hydroxyapatite/tricalcium phosphate and carbon-impregnated single-phase hydroxyapatite specimens, the latter with fine and coarse porosities (Figs. 5.9, 5.10, 5.11) (Ripamonti et al. 2007a). Single-phase and biphasic hydroxyapatite/tricalcium phosphate (33% hydroxyapatite) were also implanted in orthotopic calvarial sites to study the biology of incorporation of self-inductive biphasic biomimetic matrices in orthotopic sites (Figs. 5.12, 5.13) (Ripamonti et al. 2007a).

Long-term studies in four Chacma baboons *Papio ursinus* were initiated to investigate the induction of bone formation by biphasic hydroxyapatite/β-tricalcium phosphate (HA/βTCP) biomimetic matrices (Figs. 5.14a,b) (Ripamonti et al. 2008). The induction of bone formation was studied after heterotopic *rectus abdominis* and orthotopic calvarial implantation HA/βTCP biomimetic matrices with post-sinter phase content ratios of 19/81 and 4/96 HA/βTCP, respectively (Ripamonti et al. 2008). We reported the striking observation that the induction of resorption lacunae and concavities as carved by osteoclastogenesis set into motion

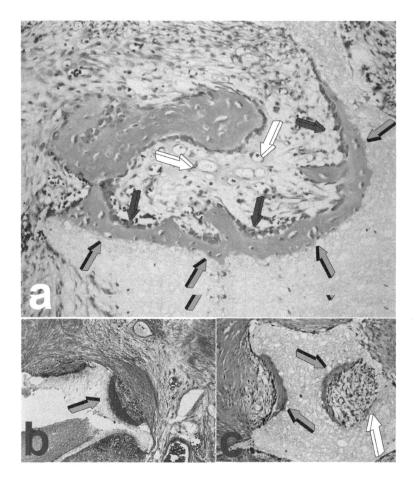

FIGURE 5.4 The geometric landscape of the concavity, "The concavity: the shape of life" (Ripamonti 2006; Ripamonti et al. 2006; Ripamonti 2012), highlighted by a series of digital microphotographs unequivocally demonstrating the "geometric induction of bone formation" as initiated by the concavity sintered in crystalline hydroxyapatite bioreactors (Ripamonti et al. 1999; Ripamonti 2004). (a) Induction of bone formation along the concavity of a sintered crystalline macroporous construct harvested on day 30 after *rectus abdominis* implantation in *Papio ursinus*. Newly formed bone within the concavity is tightly attached to the inducing sintered crystalline substratum (*light blue* arrows). Newly formed bone is populated by contiguous osteoblasts (*red* arrows) supported by prominent angiogenesis (*white* arrows). (b) Induction of bone formation (*light blue* arrow) within a concavity of the sintered crystalline bioreactor. (c) Concavities of the macroporous sintered substratum are initiators of the induction of bone formation (*light blue* arrows). Note (*white* arrow) the very beginning of the induction of bone formation along a concavity of the sintered bioreactor. Harvested macroporous bioreactors, decalcified in a formic-hydrochloric acid mixture, were double-embedded in celloidin and paraffin wax (Ripamonti et al. 1999). Serial sections, 5 μm thick, were stained with Goldner's trichrome (Ripamonti et al. 1999).

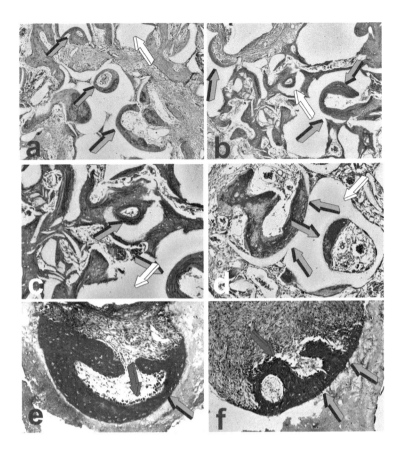

FIGURE 5.5 Progressive and continuous induction of bone formation on day 90 after *rectus abdominis* implantation of crystalline sintered macroporous bioreactors. (a,b,c,d) Digital images showing the induction of bone formation across the macroporous spaces after initiation of the induction of bone formation (*light blue* arrows) by the concavities of the sintered substratum (*white* arrows). (e,f) Remodelling of the newly formed bone within the concavities of the substratum (*light blue* arrows) populated by contiguous osteoblasts (*magenta* arrows).

the intrinsic induction of bone formation, replacing the implanted biomimetic matrices in a *continuum* of resorption/dissolution and induction of bone formation (Ripamonti et al. 2008).

Self-inducing concavities of 19/81 and 4/96 HA/βTCP specimens harvested on days 90 and 365 are shown in Figures 5.14 and 5.15 (Ripamonti et al. 2008). Morphological analyses of calvarial tissue blocks harvested on days 90 and 365 after orthotopic implantation showed significant induction of bone formation with prominent resorption/dissolution of the implanted biomimetic matrices with *restitutio ad integrum* of the treated calvarial defects (Figs. 5.16d, 5.17c,d).

Macroporous sintered hydroxyapatites discs prepared from Plasma Biotal No. 3 (Ripamonti et al. 1999) and implanted in calvarial defects of *Papio ursinus* showed

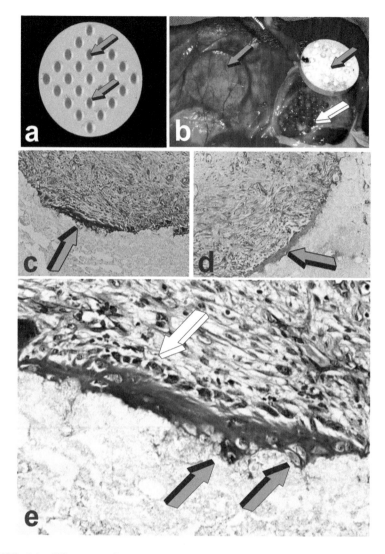

FIGURE 5.6 "The concavity: the shape of life" (Ripamonti 2004; Ripamonti 2006; Ripamonti 2012) controls the induction of bone formation when crystalline sintered bioreactors are implanted in the *rectus abdominis* muscle of the Chacma baboon *Papio ursinus* (Ripamonti et al. 1999; Ripamonti 2004; Ripamonti 2017). (a) Crystalline hydroxyapatite sintered solid disk with concavities (*light blue* arrows) on both planar surfaces (Ripamonti et al. 1999). (b) Tissue harvest of implanted discs (*light blue* arrows) on days 30 and 90 from the *rectus abdominis*; vascularized mesenchymal tissue originating from the *rectus abdominis* has grown into the morphogenetic concavities (*white* arrow). (c,d) Processed tissues cut longitudinally across the concavities show the induction of bone formation (*light blue* arrows) by day 30 after heterotopic implantation of the sintered bioreactors. (e) High-power view of the newly induced bone within the concavity of the substratum tightly embedded onto the crystalline sintered hydroxyapatite with osteocytes attached to the substratum (*light blue* arrows). Newly formed bone matrix is populated by contiguous osteoblasts (*white* arrow).

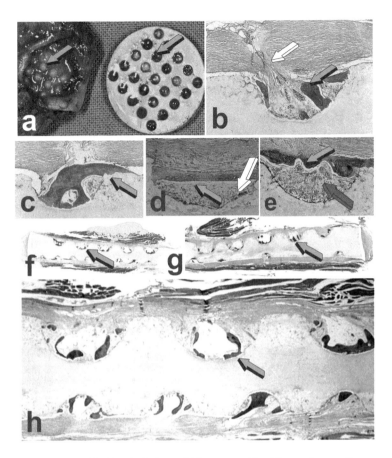

FIGURE 5.7 On day 90, identical sintered bioreactors (a) with morphogenetic concavities (*light blue* arrows in a) initiate (b) the induction of bone formation (*light blue* arrow) tightly controlled by capillary sprouting and vascular invasion (*white* arrow). (c) Newly formed bone engineered by the concavity extends to the planar surfaces of the crystalline construct (*light blue* arrow). (d,e) Morphogenetic drive of the concavity extending newly formed bone from one edge to the opposite edge of the crystalline sintered concavities (*light blue* arrows). (d) Osteoclastic activity along the sintered concavity (*white* arrow), and (e) pronounced vascular support and angiogenesis (*magenta* arrows) subjacent to the newly formed bone bridging the concavities of the crystalline substratum. (f,g,h) Induction of bone formation (*light blue* arrows) with bone marrow generation within concavities of the sintered crystalline substrata on day 90 after intramuscular *rectus abdominis* implantation. Bone formation initiates spontaneously without the exogenous application of osteogenic soluble molecular signals of the TGF-β supergene family (Ripamonti 2003).

extensive and complete penetration of bone within the macroporous spaces by day 90 after implantation (Figs. 5.17c,d). Comparative morphological and histomorphometrical data with later experiments in identical calvarial defects of *Papio ursinus* treated with macroporous Plasma Biotal constructs (Ripamonti et al. 2001b) showed that Plasma Biotal bioreactors *solo* (Ripamonti et al. 1999; Ripamonti et al. 2001a)

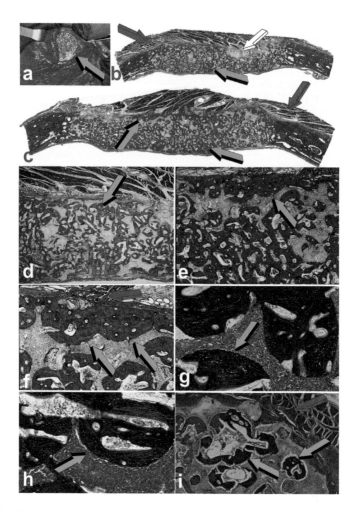

FIGURE 5.8 Morphology of calvarial regeneration and biomimetic matrices incorporation 90 days after implantation of (a) sintered crystalline hydroxyapatites (*light blue* arrow) prepared from Plasma Biotal No. 3 (Ripamonti et al. 1999). The induction of bone formation by the concavity *motif* in macroporous bioreactors heterotopically implanted in *Papio ursinus* is orthotopically translated by assembling repetitive concavities within macroporous crystalline discs for calvarial implantation in adult Chacma baboons *Papio ursinus*. Harvested calvarial tissues were processed for morphological and histological analyses (Ripamonti et al. 1999). (b,c) Complete bone formation by induction across the macroporous spaces extending across the entirety of the implanted macroporous bioreactors (*light blue* arrows) covered by the *temporalis muscle* (*magenta* arrows). In another Plasma Biotal specimen there is lack of bone induction below the temporalis muscle (*white* arrow in b). (d,e) Significant induction of bone formation by the concavity *motif* across the sintered macroporous spaces. Solid blocks of newly formed bone have generated within the macroporous spaces (*light blue* arrows). (f,g,h) Details of the concavity *motif* assembled into sintered crystalline bioreactors controlling the initiation of bone formation without the exogenous applications of the osteogenic proteins of

FIGURE 5.8 (CONTINUED)
the TGF-β supergene family (Ripamonti et al. 1999; Ripamonti 2003). (f,g,h) Newly formed bone is tightly attached to the generating concavity (*light blue* arrows) invading the large macroporous spaces of Plasma Biotal bioreactors (Ripamonti et al. 1999). (i) Identical Plasma Biotal macroporous bioreactors with a cylinder configuration implanted in the *rectus abdominis* muscle of *Papio ursinus* initiate the induction of bone formation without the exogenous applications of the osteogenic proteins of the TGF-β supergene family (Ripamonti et al. 1999; Ripamonti 2003). Figures 5.8i and 5.8c show the pre-clinical translation of the "geometric induction of bone formation" in non-human primate species from heterotopic to orthotopic sites for potential bone induction and regeneration in clinical contexts (Ripamonti et al. 1999; Ripamonti et al. 2001a).

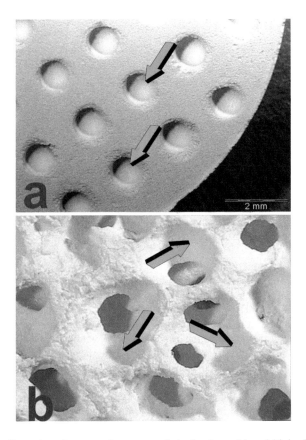

FIGURE 5.9 Scanning electron microscopy of single-phase (a) and biphasic (b) sintered hydroxyapatite biomimetic matrices for heterotopic *rectus abdominis* and orthotopic calvarial implantation in the Chacma baboon *Papio ursinus* (Ripamonti et al. 2007a). (a) Solid discs of single-phase hydroxyapatite with self-inducing geometric cues (Ripamonti et al. 2007a) as imparted by concavities (*light blue* arrows) assembled along the planar surfaces. (b) Macroporous constructs assembled for orthotopic calvarial implantation show macroporous spaces with repetitive sequences of concavities with defined radii of curvatures and diameters ranging from 700 to 1400 μm (*light blue* arrows) (Ripamonti et al. 2007a).

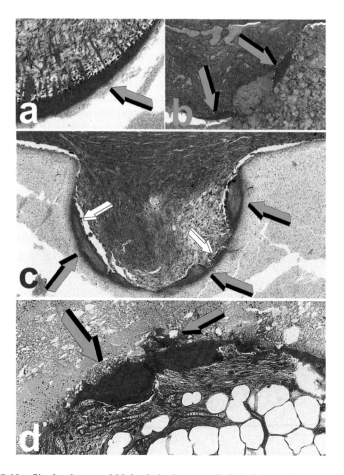

FIGURE 5.10 Single-phase and biphasic hydroxyapatite/tricalcium phosphate biomimetic matrices initiate the induction of heterotopic bone formation within concavities of the substrata when implanted in the *rectus abdominis* muscle of *Papio ursinus* (Ripamonti et al. 2007a). (a) Induction of bone formation by a single-phase hydroxyapatite biomatrix on day 90 after heterotopic implantation. Bone is surfaced by contiguous osteoblasts nesting between palisading fibres generating from the newly formed bone (*light blue* arrow). (b) Induction of bone formation (*light blue* arrows) on day 90 by a biphasic hydroxyapatite/tricalcium phosphate biomimetic matrix in the *rectus abdominis* muscle of *Papio ursinus* (Ripamonti et al. 2007a). (c) Remodelling and maintenance of newly formed bone by concavities prepared in single-phase hydroxyapatite biomatrices (*light blue* arrows) 180 days after intramuscular implantation (Ripamonti et al. 2007a). Note the lamellar remodelled newly formed bone with limited elongated osteocytes throughout the bone matrix still populated by contiguous osteoblasts (*white* arrows). (d) Self-inducing concavity of a biphasic hydroxyapatite/tricalcium phosphate bioreactor with newly formed remodelled bone matrix (*light blue* arrows) with limited osteoblastic activity 180 days after heterotopic intramuscular implantation. Minimal partial resorption of the biphasic biomatrix (*light blue* arrows) pointing to dissolution of the biphasic matrix with concurrent induction of bone formation (*light blue* arrows). Serial sections cut at 5 μm from decalcified specimen blocks were stained with Goldner's trichrome or with 0.1% toluidine blue in 30% ethanol (Ripamonti et al. 2007a).

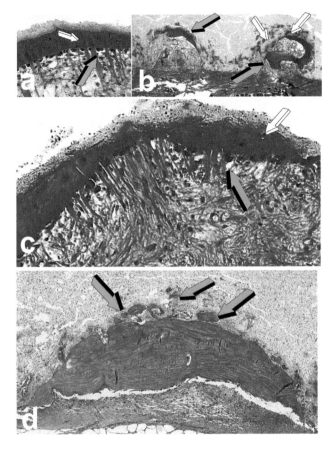

FIGURE 5.11   Self-inducing geometric cues of the concavity initiating tissue patterning and the induction of bone formation by carbon-impregnated single-phase hydroxyapatite biomatrices. (a) Newly formed bone tightly attached and embedded within the concavity of a fine carbon-impregnated single-phase hydroxyapatite on day 90 after intramuscular heterotopic implantation in the *rectus abdominis* of the Chacma baboon *Papio ursinus* (Ripamonti et al. 2007a). Collagenous fibres (*light blue* arrow) originate within the formed bone induced in the concavity (*white* arrow). Extruding collagenous fibres pattern capillary invasion and osteoblastic-like distribution within single protruding fibres (*light blue* arrow). (b) Induction of bone formation (*light blue* arrows) by concavities of coarse carbon-impregnated single-phase hydroxyapatite biomatrices on day 90 after *rectus abdominis* implantation. Anchorage and permeation of bone matrix-like tissue (*white* arrows) within the coarse carbon-impregnated biomatrices (Ripamonti et al. 2007a). (c) Induction of heterotopic bone, remodelling and maintenance of the newly formed bone 180 days after intramuscular implantation of concavities assembled in fine carbon-impregnated single-phase hydroxyapatite biomatrices. Collagenous material protruding from the newly deposited bone matrix (*light blue* arrow) with osteocytes (*white* arrows) scattered along the newly formed and remodelled bone matrix. (d) Maintenance and remodelling of the newly formed bone 180 days after heterotopic intramuscular implantation of carbon-impregnated single-phase hydroxyapatite coarse biomatrices. Partial dissolution of the biphasic matrix with the induction of bone formation (*light blue* arrows) replacing the coarse dissolved matrix (Ripamonti et al. 2007a).

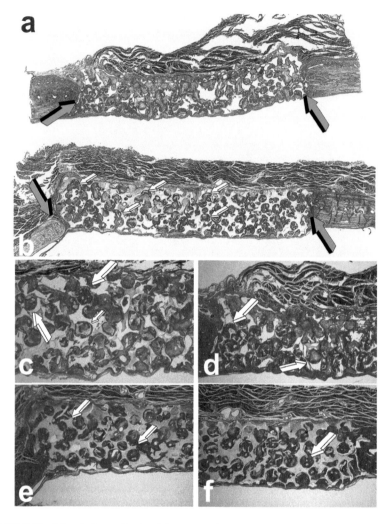

FIGURE 5.12 Morphology of calvarial regeneration and incorporation on days 90 and 180 after implantation of single-phase hydroxyapatite and biphasic hydroxyapatite/tricalcium phosphate biomimetic matrices (Ripamonti et al. 2007a). (a) Low-power view of a biphasic hydroxyapatite/tricalcium phosphate construct inserted into the recipient calvarial defect (*light blue* arrows) with the induction of bone formation within the macroporous spaces on day 90 after orthotopic implantation. (b) Corresponding calvarial specimen on day 90 of a single-phase hydroxyapatite construct with the induction of bone formation across the macroporous spaces. Bone initiates within concavities of the macroporous spaces (*white* arrows) (Ripamonti et al. 2007a). Macro-photographic details of single-phase hydroxyapatite constructs (c,e) vs. biphasic hydroxyapatite/tricalcium phosphate bioreactors (d,f) on days 90 and 180. Arrows highlight the induction of bone formation by self-inducing concavities of both biomimetic matrices. After bilateral carotid perfusion, calvarial specimen blocks were further fixed in 10% neutral-buffered formaldehyde and decalcified in a formic and hydrochloric acid mixture (Ripamonti et al. 2007a). Serial sections cut at 5 μm were stained with 0.1% toluidine blue in 30% ethanol.

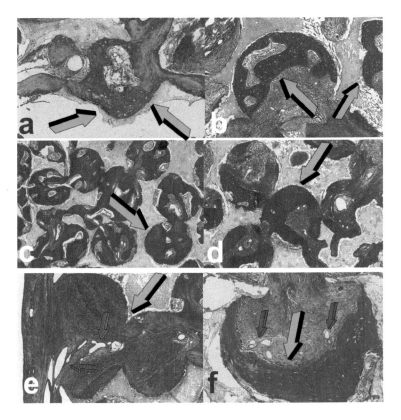

FIGURE 5.13 High-power views of calvarial regeneration and incorporation of biomatrices on days 90 (a,b) and 180 (c,d,e,f) after orthotopic implantation. (a) The induction of bone formation initiates within concavities (*light blue* arrows) of the single-phase hydroxyapatite biomatrix; image on day 90 after orthotopic calvarial implantation. (b) Similarly, the concavity is the geometric cue that initiates the induction of bone formation (*light blue* arrows) by biphasic hydroxyapatite/tricalcium phosphate; image on day 90 after orthotopic implantation. (c,e and d,f) Induction of bone formation controlled by the geometric cue of the concavity (*light blue* arrows) in both single-phase hydroxyapatite (c,e) and biphasic hydroxyapatite/ tricalcium phosphate bioactive matrices (d,f). The induction of bone formation within the concavities of both substrata is always supported by prominent angiogenesis and vascular invasion (*magenta* arrows).

induced greater amounts of bone when compared to Plasma Biotal bioreactors pre-treated with 500 µg recombinant human osteogenic protein-1 (hOP-1, also known as BMP-7) (Ripamonti et al. 2001a).

It is noteworthy that the spontaneous induction of bone formation was greater on both time periods, i.e. 30 and 90 days after implantation, respectively (Ripamonti et al. 2001a; Ripamonti et al. 2001b). The findings that Plasma Biotal macroporous discs *solo* induced greater amounts of bone when compared to hOP-1-treated identical macroporous bioreactors are highly remarkable. Our results were later confirmed by Yuan et al. (2010) who showed that osteoinductive ceramics were equally efficient

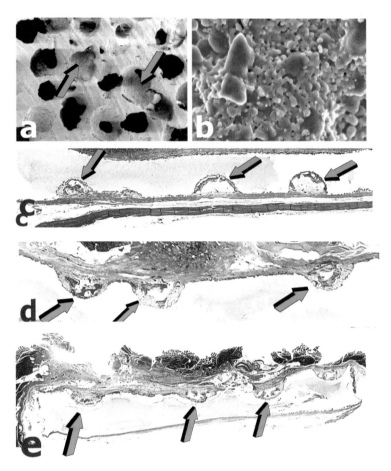

FIGURE 5.14 Preparation and sintering of single-phase hydroxyapatite biomatrices were as described (Ripamonti et al. 2008). Hydroxyapatite and β-tricalcium phosphate (HA/βTCP) were combined to form batches with pre-sinter HA/βTCP content ratios (wt%) of 40/60 and 20/80, respectively. The post-sinter phase content ratios were 19/81 and 4/96, respectively (Ripamonti et al. 2008). (a,b) Scanning electron microphotographs (SEM) of a specimen of hydroxyapatite/β-tricalcium phosphate 19/81 (HA/βTCP pre-sinter wt% 40/60). (a) Blue arrows point to repetitive sequences of concavities with defined radii of curvature and diameters in macroporous constructs for orthotopic calvarial implantation. (b) SEM analysis of the surface microstructure of a post-sinter wt% 19/81. βTCP are shown as larger grains mixed with the smaller grains of the hydroxyapatite. (c,d) Self-inducing geometric cues spontaneously initiating the induction of bone formation by HA/βTCP pre-sinter wt% 20/80 on day 90 after intramuscular *rectus abdominis* implantation. (e) Induction of bone by pre-sinter wt% 40/60 HA/βTCP 90 days after calvarial implantation. Concavities initiate the induction of bone formation (*light blue* arrows) without the exogenous application of osteogenic soluble molecular signals of the TGF-β supergene family (Ripamonti 2003).

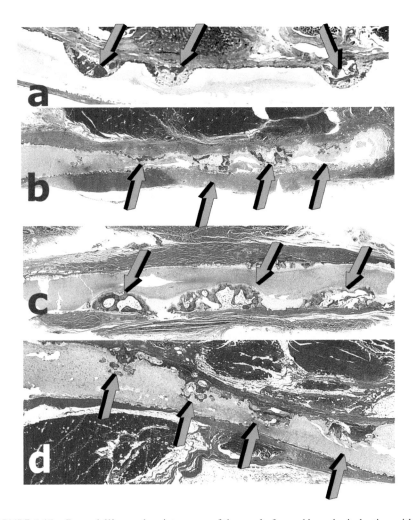

FIGURE 5.15   Remodelling and maintenance of the newly formed bone by induction within concavities exposed for 365 days to the cellular and molecular microenvironments of the *rectus abdominis* muscle of the Chacma baboon *Papio ursinus*. Digital images show the substantial induction of bone formation (*light blue* arrows) in concavities prepared in both HA/βTCP pre-sinter wt% 20/80 (a,c) and 40/60 (b,d). The newly formed bone is maintained for 365 days after implantation, with remodelling and induction of bone marrow formation in spite of the lack of mechanical forces to the macroporous constructs with the exclusion of *rectus abdominis* movements and ventral contractions. (c) Induction of substantial osteogenesis with marrow generation within the newly developed and remodelled ossicles within the concavities of the intramuscularly implanted substrata (*light blue* arrows).

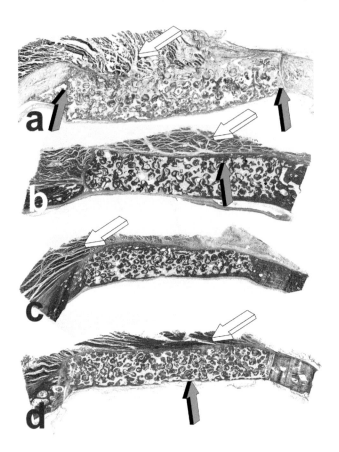

FIGURE 5.16 Calvarial incorporation, induction of bone formation and remodelling after orthotopic implantation of hydroxyapatite/β-tricalcium phosphate (HA/βTCP) pre-sinter wt% 40/60. Induction of bone formation across the macroporous spaces, remodelling and maintenance of generated bone up to one year after orthotopic calvarial implantation. (a) Tissue induction, regeneration and incorporation of the post-sinter wt% ratio 19/81 macroporous biomatrices on day 90 after calvarial implantation. (b,c,d) Induction of tissue formation and bone penetration (*light blue* arrows) across the macroporous spaces of post-sinter wt% ratio 19/81 macroporous construct 365 days after orthotopic implantation.

in initiating new bone formation within the macroporous spaces when compared to autogenous bone grafts and hBMP-2-treated defects of the ovine iliac crest (Yuan et al. 2010).

Whilst *per se* this equivalence of the induction of bone formation by osteoinductive calcium phosphate-based bioreactors *solo* vs. relatively high doses of recombinant hOP-1 (Ripamonti et al. 1999; Ripamonti et al. 2001a; Ripamonti et al. 2001b) or hBMP-2 and autogenous bone grafts (Yuan et al. 2010) is striking, it does indicate that tissue-regenerative mechanisms are controlled by as yet unknown regenerative mechanisms across different mammalian species (Lismaa et al. 2018). Alternatively,

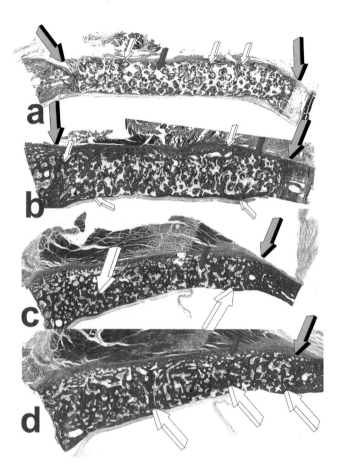

FIGURE 5.17 Induction of bone formation across the macroporous spaces of hydroxyapatite/β-tricalcium phosphate (HA/βTCP) pre-sinter wt% 20/80 ratio after orthotopic calvarial implantation (*light blue* arrows). Implanted biomatrices (*magenta* arrows) after histological processing and decalcification. (a) Bone induction (*white* arrows) initiates throughout the macroporous spaces of the HA/βTCP construct harvested on day 90 after calvarial implantation (Ripamonti et al. 2008). (b) Substantial induction of bone formation across the macroporous spaces of a HA/βTCP pre-sinter wt% 20/80 ratio on day 365 after calvarial implantation. Solid blocks of newly formed bone (*white* arrows) encasing the macroporous bioreactor with less matrix (*magenta* arrow) as compared to day 90 (a). (c,d) Complete bone regeneration across the calvarial defects implanted with macroporous bioreactors HA/βTCP pre-sinter wt% 20/80 ratio on day 365 after calvarial implantation. Solid blocks of compact remodelled bone one year after implantation with resorption/dissolution of the implanted matrix with *restitutio ad integrum* of the non-healing calvarial defects. There is induction of bone formation progressively increasing across time with reduction of the implanted matrix after osteoclastic resorption enhancing the induction of bone formation. After bilateral carotid perfusion, calvarial specimen blocks were decalcified electrolytically in a Sakura TDE™30 decalcifying unit (Sakura, Fintek, USA) and processed for paraffin wax embedding. Serial sections, cut at 4 µm, were stained with a modified Goldner's trichrome stain (Ripamonti et al. 2001; Ripamonti et al. 2008).

recombinant hBMPs are not as yet synthesized with a molecular efficiency to bind to various substrata and delivery systems to optimally initiate the induction of bone formation. Expression, synthesis and embedding of the secreted gene products into the concavities of the substratum implanted *solo* in orthotopic defects of animal models represent an endogenous pathway expressing BMPs' gene products (Ripamonti et al. 2001a; Ripamonti et al. 2001b; Yuan et al. 2010) without the potential synthesis and expression of BMPs' inhibitors as seen when delivering high doses of recombinant hBMPs (Ripamonti et al. 2015).

The heterotopic intramuscular and orthotopic calvarial models for tissue induction and morphogenesis by osteoinductive biomimetic matrices have been described in detail in several experiments in the Chacma baboon *Papio ursinus* since the early 1990s at the University of the Witwatersrand, Johannesburg (Ripamonti 1991; Ripamonti 1992; Ripamonti 1996; Ripamonti et al. 1993a; Ripamonti et al. 1993b; Ripamonti et al. 1999; Ripamonti et al. 2001a; Ripamonti et al. 2001b; Ripamonti et al. 2004; Ripamonti 2005; Ripamonti 2006; Ripamonti et al. 2007a; Ripamonti et al. 2008; Ripamonti et al. 2009; Ripamonti et al. 2010; Ripamonti et al. 2012c; Klar et al. 2013; Ripamonti et al. 2015; Ripamonti 2017; Ripamonti 2018).

In excess of 200 specimen blocks were harvested on days 30, 90, 180 and 365 after heterotopic and orthotopic implantation (Ripamonti et al. 2001a; Ripamonti et al. 2001b; Ripamonti 2017; Ripamonti 2018; Ripamonti and Duarte 2019 for reviews). Tissues were fixed and processed for undecalcified and decalcified sectioning as described (Ripamonti 1991; Ripamonti 1992; Ripamonti 1996; Ripamonti et al. 1993; Ripamonti et al. 1999; Ripamonti et al. 2001; Ripamonti et al. 2004; Ripamonti 2005; Ripamonti 2006; Ripamonti et al. 2007a; Ripamonti et al. 2007b; Ripamonti et al. 2008; Ripamonti et al. 2009; Ripamonti et al. 2010; Ripamonti et al. 2012c; Klar et al. 2013; Ripamonti et al. 2015; Ripamonti 2017; Ripamonti 2018a).

Bioreactor preparation, assembly, implantation in the *rectus abdominis* muscle of *Papio ursinus*, tissue harvest, histological analyses and quantitation methods to study the percentage of bone induction within the concavities of the sintered crystalline hydroxyapatites are detailed in published research output including US, EP and PCT patents (Ripamonti 1994; Ripamonti and Kirkbride 1995; Ripamonti and Kirkbride 1997; Ripamonti et al. 1999; Ripamonti et al. 2001a; Ripamonti et al. 2001b; Ripamonti et al. 2004; Ripamonti 2005; Ripamonti 2006; Ripamonti et al. 2007; Ripamonti et al. 2008; Ripamonti et al. 2012c; Klar et al. 2013; Ripamonti et al. 2015; Ripamonti 2017; Ripamonti 2018a; Ripamonti 2019). The following sections report a variety of heterotopically implanted sintered macroporous bioreactors in the *rectus abdominis* muscle of *Papio ursinus* together with a series of morphological digital images defining the "geometric induction of bone formation" (Ripamonti et al. 1999).

5.4   THE GEOMETRIC DESIGN OF THE CONCAVITY REGULATES THE INDUCTION OF BONE FORMATION

To determine the intrinsic osteoinductivity imparted by surface geometry, sintered macroporous crystalline hydroxyapatite bioreactors with a series of repetitive concavities along the macroporous spaces (Fig. 5.2b) were implanted in heterotopic sites of the *rectus abdominis* muscle of the Chacma baboon *Papio ursinus* (Ripamonti et al. 1999). Harvested specimens on days 30 and 90 showed the spontaneous and intrinsic induction of bone formation across the macroporous spaces characterized by concave surface topographies (Fig. 5.3). In some specimens on day 90, the spontaneous induction of bone formation was remarkable, expanding across the macroporous spaces (Figs. 5.3b,e) (Ripamonti et al. 1999).

High-power views on day 30 (Fig. 5.4) show that the induction of bone formation exclusively initiates within the concavities of the heterotopically implanted sintered macroporous bioreactors (Fig. 5.4). On day 90, digital microscopy shows that the induction of bone formation as initiated within the concavities of the crystalline substrata expands within the macroporous spaces (Fig. 5.5). High-power views unequivocally show how the concavities of the crystalline calcium-phosphate bioreactors establish ideal inducive and conducive macro- and micro-environments, initiating the spontaneous and/or intrinsic induction of bone formation (Figs. 5.5e,f). The induction of bone initiates without the exogenous application of osteogenic proteins when sintered crystalline constructs are implanted in heterotopic intramuscular sites, where there is no bone.

To directly determine the intrinsic osteoinductivity of the concavity signal in heterotopic extraskeletal sites of the *rectus abdominis* muscle of *Papio ursinus*, monolithic discs of sintered crystalline hydroxyapatite with concavities of 800 and 1600 μm prepared on both planar surfaces (Fig. 5.6a) were implanted in heterotopic intramuscular sites of the *rectus abdominis* muscle (Ripamonti et al. 1999). Tissue specimens were harvested on day 30 (Fig. 5.6b) and 90 (Fig. 5.7a), and processed for histological and immuno-histological analyses (Figs. 5.2d,e, 5.6, 5.7) (Ripamonti et al. 1999).

Figures 5.6 and 5.7 show that the geometric configuration of the concavity of the sintered crystalline bioreactor initiates the spontaneous and/or intrinsic induction of bone formation, firstly, by generating vascularized tissue growths replicating the concavities on the intramuscularly implanted bioreactors (Fig. 5.6b *white* arrow) with later the induction of bone formation as early as day 30 (Figs. 5.6c,d,e). Tissue constructs replicating the concavities of the crystalline substrata are also shown on day 90 (Fig. 5.7a *light blue* arrow). Morphological analyses of tissue specimens harvested on day 90 show the induction of bone formation exclusively within concavities of the crystalline substrata (Figs. 5.7f,g,h). It is noteworthy that the formation and later remodelling of the newly formed bone result in the induction of bone marrow within the newly formed generated ossicles that formed within the sintered concavities (Fig. 5.7e).

As we will discuss in Chapter 7, reporting the induction of bone formation by hydroxyapatite-coated titanium implants, tissue specimens, harvested from the *rectus abdominis* muscle on day 5 after implantation, showed the induction of myoblastic, pericytic and possibly endothelial cell trafficking along collagenous condensations formed by the geometric configurations of the concavities (Figs. 7.10, 7.11). The digital SEM images reported in Chapter 7 show the establishment of tractional forces across the margins of the exposed concavities. This resulted in tissue patterning and collagenous bridging across the concavities on day 5 after heterotopic intramuscular implantation. Later, the collagenous bridging across concavities resulted in the induction of bone formation with newly formed mineralized bone bridging the edges of the exposed concavities 90 days after exposure to the *rectus abdominis* multi-faceted pleiotropic microenvironments with selected angiogenic stem cell niches (Figs. 5.7d,e).

The induction of bone is always characterized by the presence of continuous vascularization and capillary sprouting within the concavity macroenvironment (Figs. 5.7b,d,e). Capillary sprouting is prominent below the induction of bone formation (Fig. 5.7e) together with osteoclastic activity still on day 90 critical for bone induction to occur (Figs. 5.7d,e *white* arrows) (Klar et al. 2013; Ripamonti et al. 2014; Ripamonti et al. 2015). The plasticity of the endothelial cell regulating tissue morphogenesis and regeneration has been recently highlighted (Gomez-Salinero 2018; Ramasamy et al. 2014).

Following the demonstration of substantial induction of bone formation in orthotopic calvarial sites of *Papio ursinus* using Plasma Biotal No. 3 (Fig. 5.8) (Ripamonti et al. 1999), a series of experiments were later set up to study the induction of bone formation by concavities prepared in single-phase sintered hydroxyapatite, biphasic hydroxyapatite/tricalcium phosphate (Fig. 5.9), and single-phase hydroxyapatite constructs with enhanced micro porosity following carbon impregnation (Ripamonti et al. 2007a). Comparative morphological analyses showed that carbon-impregnated single-phase hydroxyapatite matrices generated greater amounts of bone within concavities of biomimetic bioreactors heterotopically implanted in the *rectus abdominis* muscle (Figs. 5.10, 5.11) Ripamonti et al. 2007a). Though carbon-impregnated single-phase hydroxyapatite matrices were not implanted in calvarial defects, analyses showed that single-phase hydroxyapatite biomimetic matrices formed greater amounts of bone when compared to biphasic HA/TCP bioreactors implanted in orthotopic calvarial sites (Figs. 5.12, 5.13) (Ripamonti et al. 2007a).

High-power views of orthotopically implanted single-phase (HA single-phase) and biphasic hydroxyapatite/tricalcium phosphate (biphasic HA/TCP) biomimetic matrices showed the induction of bone formation within the macroporous spaces tightly attached to the concavities of the biomimetic matrices (Fig. 5.13). Digital images once again show the preferential induction of bone formation across concavities of the biomimetic substrata (Fig. 5.13 *blue* arrows).

Intramuscular implantation of hydroxyapatite/β-tricalcium phosphate (HA/βTCP) discs post-sinter 4/96 by ratios HA vs. βTCP (Figs. 5.14a,b) (Ripamonti et al. 2008) showed the induction of bone formation within the concavities of the substratum on day 90 in both 4/96 (Figs. 5.14c,d) and 19/81 post-sinter HA/βTCP (Fig. 5.14e). Substantial amounts of bone formed in concavities of post-sinter 4/96 and 19/81 HA/βTCP specimens on day 365 (Figs. 5.15c,d) (Ripamonti et al. 2008).

Orthotopic calvarial implantation of post-sinter 19/81 (Fig. 5.16) and 4/96 HA *vs.* βTCP (Fig. 5.17) macroporous bioreactors showed the induction of bone formation across the macroporous spaces 90 and 365 days after calvarial implantation. Greater amounts of bone formed in 4/96 HA vs. βTCP specimens (Figs. 5.17c,d) both on days 90 and 365 after calvarial orthotopic implantation (Ripamonti et al. 2008).

Orthotopic constructs post-sinter 4/96 ratio HA vs. βTCP showed *restitutio ad integrum* of the regenerated calvarial defects on day 365 with prominent induction of bone formation across the macroporous spaces of the implanted biomimetic matrices (Figs. 5.17c,d).

5.5   THE CONCAVITY: THE SHAPE OF LIFE – PERSPECTIVE AND LIMITATIONS

Regenerative medicine is the grand multidisciplinary challenge of evolutionary biology requiring the integration of molecular, cellular and development biology to explore how to trigger *de novo* induction of tissue and organs in man.

The central question in development biology, and thus regenerative medicine, is the molecular basis of pattern formation (Reddi 1984; Kerszberg and Wolpert 2007; Wolpert 1996). The mechanisms involved in the induction of pattern formation and tissue form and function, or morphogenesis, have been great challenges over the 20th and 21st centuries (Reddi 1984; Kerszberg and Wolpert 2007; Wolpert 1996). These multifaceted pleiotropic research endeavours *par force* led to the hypothesis, and later to the discovery, of entirely novel morphogenetic soluble signals or morphogens that set into motion pattern formation and the induction of tissue morphogenesis, controlling tissues' and organs' development from embryogenesis to post-natal tissue induction (Ripamonti 2006). Morphogens, defined by Turing as "forms generating substances" (Turing 1952), are now at the very crux of tissue induction and morphogenesis (Gilbert and Saxén 1993; Levander 1945; Urist 1965; Reddi 2000; Reddi 2005; Ripamonti et al. 2001; Ripamonti 2004; Ripamonti et al. 2004; Ripamonti 2006).

The question that now arises is: what next beyond morphogens and stem cells (Kerszberg and Wolpert 2007; Ripamonti and Duarte 2019)? The fundamental tenet of tissue engineering, as we have learned from the fundamental studies at the Bone Cell Biology Section of the National Institutes of Health, Bethesda, Maryland, DC, commanded then by A.H. Reddi PhD, is to combine and/or reconstitute soluble molecular signals with insoluble signals or substrata to erect scaffolds of biomimetic biomaterials matrices that mimic the supramolecular assembly of the extracellular matrix of bone (Sampath and Reddi 1981; Sampath and Reddi 1983; Ripamonti and Reddi 1995; Reddi 2000; Ripamonti 2006; Ripamonti 2017). The mechanical reconstitution initiates the induction of bone formation when recombined soluble and insoluble signals are implanted heterotopically in the rodent subcutaneous bioassay (Sampath and Reddi 1981; Sampath and Reddi 1983; Khouri et al. 1991; Ripamonti and Reddi 1995).

Lander lucidly writes

> Whether morphogen gradients cross threshold values at which genes are turned on or off, or influence each other triggering the spontaneous emergence of stable, long-range patterns of morphogen activity, only Morpheus unbound can trigger the cascade of pattern formation resulting in tissue induction and morphogenesis.

(Lander 2007)

There is no bone formation by induction without the osteogenic proteins of the TGF-β supergene family (Ripamonti 2003). As defined by Lander (2007), "Morpheus unbound" needs to bind to extracellular matrices or other carrier matrices to trigger the ripple-like cascade of the induction of bone formation. Synthesis and expression of *bone morphogenetic proteins* genes and secreted gene products must follow the heterotopic implantation of calcium phosphate-based bioreactors, later initiating the spontaneous induction of bone formation as a secondary response (Ripamonti et al. 1993a; Ripamonti 2017; Ripamonti 2018a; Ripamonti and Duarte 2019).

The "intrinsic and/or spontaneous induction of bone formation" of the hydroxy-apatite osteogenesis model (Ripamonti et al. 1993a; Ripamonti 1996) was reported in long-term systematic studies in the Chacma baboon *Papio ursinus* when implanting coral-derived hydroxyapatite macroporous bioreactors in heterotopic sites of the *rectus abdominis* muscle (Ripamonti 1990; Ripamonti 1991). Systematic experiments were then initiated in the late 1980s to study the induction of bone formation by macroporous calcium phosphate-based bioreactors (Ripamonti 1991; Ripamonti et al. 1993a; Ripamonti 1996; Ripamonti et al. 1999; Ripamonti 2009; Ripamonti et al. 2009; Ripamonti et al. 2010).

Independently at Fukuoka University, heterotopic intramuscular implantation of porous and dense hydroxyapatite granules in a canine model showed newly formed bone surrounding the porous hydroxyapatite granules (Yamasaki 1990). In contrast, bone was never observed on day 30 or 90 around dense hydroxyapatite particles (Yamasaki 1990). The lack of bone formation by dense hydroxyapatite particles suggested that "granules possessing micropores may exert an effect on osteogenesis" (Yamasaki 1990). The above findings and interpretations were later published after further experimentation in canines' intramuscular heterotopic sites (Yamasaki and Sakai 1992).

The spontaneous induction of bone formation was further confirmed by systematic long-term studies in *Papio ursinus* (Ripamonti 1991; Ripamonti et al. 1992b; Ripamonti et al. 1993a; Ripamonti 1996; Ripamonti 2009; Ripamonti et al. 1999; Ripamonti et al. 2010).

A collaborative research effort between the University of Leiden, the Netherlands, and Sichuan University, Chengdu, China, reported heterotopic studies in canine models after implantation of porous calcium phosphate-based hydroxyapatite (Klein et al. 1994). It was stated that "the histological observations revealed that porous calcium phosphate ceramics implanted intramuscularly in dogs can induce the formation of osseous substances in the pores" (Klein et al. 1994). The contribution, published in 1994, did not however report the earlier studies of Ripamonti (1990) and Yamasaki (1990), nor the earlier long-term studies in the Chacma baboon *Papio ursinus* (Ripamonti 1991) as well as the morphological and immuno-histochemical studies in *Papio ursinus* later reported in *Matrix* (Ripamonti et al. 1993a).

Since the 1990s, several research articles reported a number of increasing studies on calcium phosphate-based biomaterials in different animal models from different research laboratories across the planet (Yang et al. 1996; Yuan et al. 1999; Barradas et al. 2011; Barradas et al. 2012; Habibovic et al. 2004; Habibovic et al. 2005; Habibovic et al. 2006; Habibovic et al. 2009; Yuan et al. 2006; Le Nihouannen et al. 2008; Yuan et al. 1998; Yuan et al. 1999; Yuan et al. 2006; Yuan et al. 2010; Zhang et al. 2014; Danoux et al. 2016; Sun et al. 2016; Zhang et al. 2014; Zhang et al. 2017; Zhang et al. 2018; Othman et al. 2019; Ripamonti et al. 2009; Ripamonti et al. 2010; Ripamonti and Roden 2010; Klar et al. 2013; Ripamonti et al. 2015; Ripamonti 2017).

The available literature and data, however, only report morphological studies hypothesizing biomaterials' characteristics that may or may not differentiate osteogenic cells leading to the induction of bone formation by calcium phosphate-based biomaterials (Ripamonti et al. 2009 for review). On the other hand, complementary cellular and molecular biology studies to mechanistically resolve the "spontaneous and/or intrinsic hydroxyapatite induced osteogenesis model" (Ripamonti et al. 1993a; Ripamonti 1996) were however delayed in spite of the last century's explosion and understanding of newly developed cellular and molecular biology techniques (Cell Editorial 2014).

Several published papers on calcium phosphate-based biomaterials spontaneously initiating the induction of bone formation when implanted in heterotopic intramuscular sites of a variety of animal models reported multiple possible factors involved in the induction of bone formation (Yuan et al. 1999; Barradas et al. 2011; Barradas et al. 2012; Habibovic et al. 2004; Habivovic et al. 2005; Habibovic et al. 2006; Habibovic et al. 2009; Le Nihouannen et al. 2008; Yuan et al. 2010). Discussed factors included sintering time and temperatures, porosity, surface topography, crystallinity, nanotopographies and sintering phases, however seldom stating that the induction of bone formation ultimately must be initiated *via* the expression and secretion of the soluble osteogenic molecular signals of the TGF-β supergene family (Ripamonti 2003).

Immunolocalization studies of alkaline phosphatase, type I collagen and laminin during the induction of bone formation by coral-derived bioreactors were used to study the expression of the osteogenic phenotype in coral-derived hydroxyapatite constructs implanted intramuscularly in *Papio ursinus* (Ripamonti et al. 1993a). The study concluded that "the hydroxyapatite substratum may act as a solid state matrix for adsorption, storage and controlled release of bone morphogenetic proteins which locally initiate the induction of bone formation" (Ripamonti et al. 1993a).

Following gene expression and molecular biology analyses, the above observations and concluding remarks were indeed confirmed by Northern blot analyses of osteogenic protein-1 and type IV collagen mRNAs on harvested calcium phosphate-based bioreactors heterotopically implanted in *Papio ursinus* (Ripamonti et al. 2007a). Northern blot analyses showed high expression of OP-1 and type IV collagen mRNAs on days 90 and 180 correlating with the induction of bone formation by carbon-impregnated single-phase hydroxyapatite bioreactors (Ripamonti et al. 2007a).

Later studies applied molecular biology techniques to mechanistically resolve the induction of bone formation by coral-derived calcium phosphate-based bioreactors (Klar et al. 2013). To summarize, to evaluate the role of calcium ions (Ca^{++}) and osteoclastogenesis, calcium carbonate/hydroxyapatite macroporous bioreactors with limited conversion to hydroxyapatite (7% HA/CC) were preloaded with either 500 μg of the calcium channel blocker verapamil hydrochloride or 240 μg of the osteoclast inhibitor biphosphonate zoledronate and implanted in the *rectus abdominis* muscle of *Papio ursinus* (Ripamonti et al. 2010; Klar et al. 2013; Ripamonti et al. 2015). Generated tissues on days 15, 60 and 90 were processed for histomorphometry and molecular analyses by quantitative real-time polymerase chain reaction (qRT-PCR).

Zoledronate-treated specimens showed delayed and inhibited tissue patterning and morphogenesis with limited induction of bone formation. Osteoclastic inhibition by bisphosphonate zoledronate yielded minimal bone formation. Similarly, macroporous constructs pre-treated with the Ca^{++} channel blocker verapamil hydrochloride (Ripamonti et al. 2010; Klar et al. 2013). Downregulation of *BMP-2* together with upregulation of *Noggin* correlated with limited induction of bone formation. The results showed unequivocally that the induction of bone is initiated by a local peak of Ca^{++} activating stem cell differentiation, angiogenesis and the induction of bone formation into the concavities of the biomimetic matrices after expression of *BMPs'* genes within the secluded microenvironments of the concavities (Ripamonti et al. 2010; Klar et al. 2013; Ripamonti 2017). The observed lack of bone differentiation by zoledronate-treated macroporous constructs with minimal *osteogenic ptotein-1* expression (Ripamonti et al. 2010) *par force* indicates that the spontaneous induction of bone formation is initiated by the BMPs' pathway. Furthermore, it mechanistically requires osteoclastogenesis that induces nanotopographical surface modifications amenable to the induction of the osteogenic phenotype (Ripamonti et al. 2010; Klar et al. 2013; Ripamonti 2017).

Myoblastic, pericytic, perivascular stem cells and endothelial cell precursors, if not the very endothelial cell, sense the substratum upon which cells attach and migrate (Discher et al. 2005; Engler et al. 2006; Disher et al. 2009; Buxboim and Discher 2010). Cells migrate and attach onto the substratum and are able to convert geometrical and mechanical cues into triggering gene expression pathways including the *bone morphogenetic proteins* pathway. The *connubium* of *smart* biomimetic matrices self-inducing specific gene products resulting in tissue morphogenesis regulated by the geometry of functionalized substrata is the true challenge for regenerative medicine and tissue engineering of the 21st century.

The driving force of the intrinsic induction of bone formation by bioactive bioinspired biomimetic matrices is the shape of the implanted scaffold. The language of shape is the language of geometry; the language of geometry is the language of a sequence of repetitive concavities that biomimetize the remodelling cycle of the primate osteonic bone (Ripamonti 2006; Ripamonti 2009; Ripamonti 2017). The concavity *per se* as cut into calcium phosphate macroporous constructs or generated by osteoclastogenesis during the remodelling cycle of the primate osteonic bone is the geometric signal that initiates the induction of bone formation after osteoclastogenesis throughout the cortico-cancellous bone of the primate skeleton (Ripamonti 2017; Ripamonti 2018a; Ripamonti and Duarte 2019).

We have shown that osteoclasts prime the macroporous spaces of the implanted calcium phosphate-based bioreactors (Ripamonti et al. 2010). Osteoclastic priming of the macroporous surfaces results in nanotopographical geometric modifications of the macroporous surfaces (Ripamonti et al. 2010). Osteoclastogenesis releases calcium ions; Ca^{++} initiates angiogenesis and capillary sprouting together with chemotactic and differentiating signals for myoblastic/myoendothelial and pericytic/perivascular stem cells. Differentiating cells invade the protected microenvironment of the concavities cut by osteoclastogenesis or prepared in the the biomimetic bioreactors so as to initiate the formative phase of the induction of bone formation.

The innovative studies of Discher's laboratories (Discher et al. 2005; Engler et al. 2006; Discher et al. 2009; Buxboim and Discher 2010; Buxboim et al. 2010) on matrix elasticity that directs the lineage specification of stem cells allow us to review the topographical control of cell behaviour as a "primary mechanism controlling the induction of cell differentiation" (Clark et al. 1987; Clark et al. 1990; Curtis and Wilkinson 1997; Li et al. 2008).

A focused review by McNamara et al. (2010) highlights the "nanotopographical control of stem cell differentiation". The geometric control of cellular differentiation was later studied *in vitro* to show that combined micron-/submicron-scale surface roughness induces osteoblast differentiation for potentially improved osteointegration *in vivo* (Gittens et al. 2011).

A critical contribution that extended the effect of geometry on the induction of cell differentiation showed how subcellular cell geometry on micropillars regulates stem cell differentiation (Liu et al. 2016). Nuclear deformation on micropillar assays showed enhanced osteogenesis markers *in vitro*, suggesting that the geometry of cell nuclei is a new cue to regulate the lineage commitment of stem cells at the subcellular level (Liu et al. 2016).

The first and possibly determinant geometric molecular cue for cell differentiation and the regulation of cellular function is the extracellular matrix topography (Reddi 1984; Kim et al. 2012). The matrix provides an optimal substratum for the induction of endochondral bone differentiation (Reddi and Huggins 1972; Reddi 1984) regulating the induction of bone formation (Reddi and Huggins 1973; Reddi 1974; Reddi and Kuettner 1981; Reddi 1981; Sampath and Reddi 1984). Matrix topography directs cell function and differentiation by providing spatial and molecular cues for migration, attachment and differentiation. (Kim et al. 2012)

Nanotopographic cues regulate human mesenchymal stem cells (hMSCs) by controlling internal cytoskeletal networks guiding cell fate decisions (Ahn et al. 2014). It is worth noting that recent experiments showed that a mechano-regulation scheme of endothelial cells is regulated by the geometry of capillary architecture *via* cell–matrix mechanical interactions. (Sun et al. 2014)

To summarize the effect of the nanotopographical surface structure on cell differentiation towards the osteogenic phenotype, recent studies provided evidence that surface roughness induces osteoclast effects on osteoblastic cell differentiation *in vitro* (Zhang et al. 2018). The paper demonstrated the existence of a combinatorial effect of surface roughness, osteoclastogenesis and osteogenesis (Zhang et al. 2018). Of note, the conditioned media from osteoclasts cultured on smooth surfaces, which showed larger cells with active rings and more nuclei per osteoclast compared to osteoclasts cultured on rougher surfaces, significantly promoted the osteogenic differentiation of osteoblastic cells compared to conventional osteogenic media (Zhang et al. 2018). These results have once again indicated that surface roughness and nanotopography are critical factors regulating the induction of gene expression markers essential for the induction of bone formation to occur when initiated by calcium phosphate-based biomimetic matrices. (Klar et al. 2013; Zhang et al. 2018)

Following all the morphological, molecular and gene expression profiled data gathered so far, how does the induction of bone formation proceed after the heterotopic intramuscular implantation of calcium phosphate-based constructs? Which are the mechanisms whereby the initiation of bone formation does occur even without the exogenous application of the soluble osteogenic molecular signals of the TGF-β supergene family (Ripamonti 2003)?

Macroporous constructs were preloaded with 125 or 150 µg recombinant hNoggin, a BMP antagonist (Klar et al. 2014; Ripamonti et al. 2015). In addition, bioreactors were also preloaded with 500 µg of the L-type voltage-gated calcium channel blocker verapamil hydrochloride (Klar et al. 2013). Harvested specimens showed limited if any induction of bone formation. Verapamil hydrochloride strongly inhibited the induction of bone formation (Klar et al. 2013; Ripamonti et al. 2014).

Perhaps however, the more elegant data that mechanistically underpins that the spontaneous and/or intrinsic induction of bone formation is initiated by locally produced and embedded secreted bone morphogenetic proteins gene products onto the substratum is that doses of recombinant human Noggin (125 or 150 µg hNoggin), a BMP antagonist, block the induction of bone formation within the macroporous spaces (Klar et al. 2014; Ripamonti et al. 2015; Ripamonti 2017).

The concavity as cut into biomimetic matrices biomimetizes the ancestral repetitive multi-million-year tested designs and topographies of Nature; the concavity is thus the geometric signal that initiates the induction of bone formation. Concavities are endowed with *smart* functional shape memory geometric cues in which soluble signals induce morphogenesis, and physical forces, imparted by the geometric topography of the substratum, dictate biological patterns, constructing the induction of bone formation and regulating the expression of gene products as a function of the structure.

More recently, studies presented gene expression data using qRT-PCR (Zhang et al. 2014) showing that *Osteocalcin* and *Ostepontin* had significantly higher expression rates when compared to tricalcium phosphate with higher micropore size (Zhang et al. 2014). *RUNX* and *ALP* gene expression was very similar on the tested days, i.e. 4, 7 and 14, with *BMP-2* expression, however, higher in tricalcium phosphate samples characterized by higher micropore size when compared to tricalcium phosphate with lower micropore size (Zhang et al. 2014). The report concluded that adjusting microstructure size could render CaP ceramics osteoinductive (Zhang et al. 2014).

The published data failed, however, to mechanistically explain the observed overexpression of the *BMP-2* gene by tricalcium phosphate samples with higher microstructure (TCP-B) with lack of bone formation in heterotopic intramuscular studies (Zhang et al. 2014). Smaller micropore samples (TCP-S) showed higher expression of *Osteocalcin* with, however, no *RUNX* expression differences compared to tricalcium phosphate samples with higher micropore size (TCP-B) (Zhang et al. 2014).

Results are reported under sub-heading 3.5, Bone-Related Gene Expression, where it is stated that TCP-S (lower surface microstructure/micropore size) enhanced *Osteocalcin* and *Ostepontin* gene expression in human bone marrow stromal cells. On the other hand, TCP-B (greater surface microstructure/micropore size) enhanced *BMP-2* expression, together with synthesis of the BMP-2 gene product (Zhang et al. 2014). *In vivo* experiments in canines after heterotopic paraspinal intramuscular implantation of TCP-S and TCP-B ceramics showed that substantial bone formed by induction in TCP-S ceramics (Zhang et al. 2014). In contrast, and in spite of the over-expression of the *BMP-2* gene, no bone formed in TCP-B implants (Zheng et al. 2014).

Further contradictions with misplaced quotations are evident under *Discussion* where it is reported that "BMPs (e.g. BMP-2 and BMP-7) were detected in cells involved in bone morphogenesis initiated by osteoinductive CaP ceramics [40,41]". Reference [40], as quoted by Zhang et al. (2014), refers to the induction of *BMPs'* gene expression during endochondral bone formation as initiated by naturally derived porcine transforming growth factor-β_1 (pTGF-β_1) delivered by allogeneic insoluble collagenous bone matrix and not by CaP constructs (Duneas et al. 1998). Reference [41], as listed by Zhang et al. (2014), reports the induction of cementogenesis by recombinant human osteogenic protein-1 (hOP-1) when delivered by insoluble collagenous matrices and not CaP ceramics to furcation defects of the Chacma baboon *Papio ursinus* (Ripamonti et al. 1996).

In a more recent paper by Zhang and co-authors (Zhang et al. 2017), it is stated that the induction of "osteogenesis by surface structure is dependent on the geometry and size of the surface structure, its spatial organization and the dynamical changes of the surface properties over time", quoting the critical work of Engler's laboratories on surface topography influencing cell morphology (Kiang et al. 2013).

The above studies compared two types of tricalcium phosphate (TCP) ceramics with different surface topographies, a TCP with a bigger surface structure dimension, and a TCP with a smaller surface structure dimensions, TCP-B vs. TCP-S, respectively (Zhang et al. 2017). The rationale of the experimentation reported that recent studies showed that "TGF-β signalling also initiates osteogenic differentiation" and that such signalling requires the recruitment of TGF-β receptors (TGF-βR) to the primary cilia of the activated cells (Zhang et al. 2017). Cilia, microtubule-like structures attached to the basement membrane, are set to respond to multiple signalling pathways including mechano-transduction (Eggenschwiler and Anderson 2007; Abou et al. 2009; McMurray et al. 2015).

In *vitro* experimentation using human bone marrow stromal cells (hBMSCs) showed that cells achieve better spreading on TCP-S vs. TCP-B together with enhanced osteogenic differentiation as shown by *alkaline phosphatase* activity and *Osteocalcin* expression (Zhang et al. 2017). Histological examination of TCP specimens harvested on day 84 after intramuscular implantation in canine models showed layers of bone formation by induction on surface topographies of TCP-S samples with, however, a lack of bone differentiation by the larger surface topographies of TCP-B (Zhang et al. 2017). The reported data showed that TCP ceramics with sub-micron-scale surfaces initiated the induction of bone formation *in vivo* (Zhang et al. 2017). The study also showed that surface topography regulates cilia structure and TGFβII receptor localization in the cilium (Zhang et al. 2017). The study further proposes that "activation of the TGFβII receptor will enable TGF-β signalling, which is necessary for osteogenic differentiation" (Zhang et al. 2017).

Whilst a twofold activation of the TGFβII receptor is reported in *in vitro* studies using hBMSCs on TCP-S nanotopographies, the data do not correlate the activation of the TGFβII receptor with a specific set of enhanced gene expression pathway *in vitro*, nor the induction of bone formation by TCP constructs *in vivo* by TCP-S samples with smaller surface structure dimensions (Zhang et al. 2017). Which are the genes that set into motion the ripple-like cascade of the induction of bone formation? The reported gene expression data show the expression of the selected osteogenic markers which include collagen type-1, alkaline phosphatase (ALK), Osteocalcin and osteopontin (Zhang et al. 2017).

Various up and down expression patterns are reported by a day-time study and by TCP-B and TCP-S substrata using both basal and osteogenic media (Zhang et al. 2017). The expression patterns of the *Osteocalcin* gene on days 4 and 7 *in vitro* by TCP-S constructs are noteworthy (Zhang et al. 2017). The Osteopontin gene is also shown to be expressed increasingly from day 4 to day 14 during *in vitro* studies on TCP-S substrata (Zhang et al. 2017). It should be noted, however, that Osteopontin gene and gene products are not differentiating factors but merely structural assemblages of few skeletal cells lacking altogether differentiating and morphogenetic capacities.

The study of Zhang et al. (2017) further reports the confocal immunolocalization of p-TGF-βRII in primary cilia of hBMSCs grown on TCP-S substrata as well as a twofold increase of p-TGF-βRII on day 14 in cells grown on TCP-S vs. TCP-B substrata (Zhang et al. 2017). The induction of bone formation after heterotopic intramuscular implantation of TCP-S and TCP-B substrata in the canine (Zhang et al. 2017) shows that TCP-S are spontaneously initiating the induction of bone formation whereas TCP-B failed to induce bone along the bigger structure dimension of the topographically constructed TCP surfaces (Zhang et al. 2017). The reported *in vitro* twofold expression of p-TGF-βRII by TCP-S vs. TCP-B fails however to show which are the downstream activated and expressed genes following the activation of the p-TGF-βRII receptor, and in particular, the expression of genes that will set into motion the ripple-like cascade of "Bone: Formation by autoinduction" (Urist 1965).

Recent studies focused on the role of plasma cell glycoprotein 1 (PC-1) controlling osteoinduction by different calcium phosphate-based ceramics (Othman et al. 2019). PC-1, encoded by the *ectonucleotide pyrophophatase/phosphodiesterase* 1 (*ENPP1*) gene, is involved in the regulation of mineralization (Othman et al. 2019). The knock out of *ENPP1* expression in hMSCs results in increased *BMP-2* expression both at the mRNA and protein levels. This has suggested that *ENPP1* is a negative regulator of BMP-2 signalling (Othman et al. 2019). Whilst the study in its concluding remarks "describes for the first time a role for ENPP1/PC-1 in the osteogenic differentiation of hMSCs" cultured on osteoinductive calcium phosphate-based constructs, it does not however state what it is that biologically initiates the cascade of bone differentiation by the tested calcium phosphate-based ceramics (Othman et al. 2019).

5.6 CHONDROGENESIS IN BIOMIMETIC CALCIUM PHOSPHATE-BASED BIOREACTORS

Finally, the induction of tissue morphogenesis by calcium phosphate-based constructs should also be discussed, highlighting the genetic pathway of the spontaneous and/or intrinsic induction of bone formation by biomimetic self-inductive biomatrices.

It is noteworthy to report that the induction of bone formation by coral-derived or crystalline sintered macroporous calcium phosphate-based bioreactors does not initiate *via* endochondral bone formation, that is *via* the replacement of a cartilage anlage that will induce bone differentiation after vascular invasion, angiogenesis and chondrolysis that set into motion the induction of bone formation (Reddi and Huggins 1972; Reddi 1981; Reddi 2000).

The induction of bone formation by coral-derived constructs initiates *via* the induction and the assemblage of mesenchymal collagenous condensations against the substratum, with later angiogenesis and differentiation of osteoblastic-like cells within the aligned condensations tightly opposed to the substratum (Ripamonti 1990; Ripamonti 1991; Ripamonti et al. 1993a). In contrast, the induction of bone differentiation as initiated by sintered crystalline calcium phosphate-based bioreactors proceeds by differentiating and de-differentiating events of locally recruited mesenchymal cells directly upon the concave macroporous spaces of the sintered crystalline substrata (Ripamonti et al. 1999; Ripamonti 2006; Ripamonti 2017; Ripamonti and Duarte 2019).

Newly formed bone is directly synthesized against the sintered crystalline construct, with osteocytes rapidly embedded within the secreted matrices (Fig. 5.4). Newly deposited bone is tightly attached to the sintered biomatrices, and it is populated by contiguous osteoblasts secreting newly formed, yet to be mineralized matrix or osteoid (Fig. 5.6).

Interestingly, however, the bone morphogenetic events by either coral-derived or sintered crystalline substrata do not show a cartilage induction phase as expected in endochondral osteogenesis (Reddi and Huggins 1972; Reddi 1981; Ripamonti 2017). The spontaneous induction of bone formation by calcium phosphate-based constructs is comparable to a sub-type of intramembranous osteogenesis as seen in the induction of the membranous bone of the craniofacial skeleton that develops *via* intramembranous osteogenesis (Ripamonti 2017).

The lack of chondrogenesis, certainly in *Papio ursinus* as well as in other animal models including canines and ovine models (Yang et al. 1996; Gosain et al. 2002; Gosain et al. 2004), is noteworthy. This may indicate a genetic mechanism regenerating the induction of bone formation by calcium phosphate-based constructs *via* intramembranous osteogenesis only, which may also be controlled by the calcium phosphate-based tri-dimensional macroporous bioreactors.

In unique experimentation in the Carcharhiniform selachian *Carcharhinus obscurus* shark, coral-derived macroporous constructs, 20 mm in length, 11 mm in diameter, were implanted in the dorsal muscles of a number of adolescent *C. obscurus* (Ripamonti 2018b; Ripamonti et al. 2018). Retrieved specimens, often damaged and crushed by the powerful muscular activity of the selachian fishes, were processed

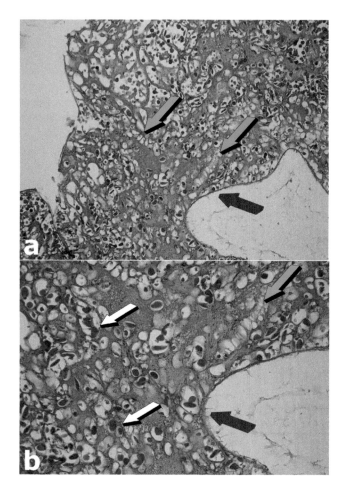

FIGURE 5.18 Chondrogenesis but lack of "bone: formation by autoinduction" (Urist 1965) in coral-derived macroporous constructs intramuscularly implanted in the selachian elasmo-branch *Carcharhinus obscurus* (Ripamonti et al. 2018). (a) Induction and differentiation of chondrocytes with cartilaginous matrix directly against the hydroxyapatite substratum (*dark blue* arrow). Within the chondroid matrix, there are chondrocytes' columnar condensations highly reminiscent of the progressively differentiating chondroblastic patterning of the mammalian growth plate (*light blue* arrows). (b) High-power view highlighting differentiating chondroblasts (*light blue* arrow) and chondroblastic cells within the newly deposited chondroid matrix (*white* arrows).

for histological analyses, and one control specimen was amenable to proper histological processing and sectioning with successful staining of the newly formed tissues within the macroporous spaces (Fig. 5.18). Chondrogenesis developed within the macroporous spaces without any application of soluble morphogenetic signals (Fig. 5.18). Panel (a) shows the induction of cartilage material directly against the calcium phosphate-based bioreactor (*dark blue* arrow) with columns of progressively differentiating chondroblasts (*light blue* arrows), patterning the newly formed cartilage as in mammalian embryonic development (Reddi 2000; Ripamonti 2017; Ripamonti et al. 2018). Our published paper speculated that the micro-inductive micro-environment of the coral-derived macroporous bioreactors engineers cartilaginous columnar condensations as seen in the mammalian growth plate counterpart (Ripamonti et al. 2018). Panel (b) shows the induction of cartilage (Fig. 18b *white* arrow) detailing the columnar arrangement of chondroblasts/chondrocytes as in the mammalian growth plate (Fig. 18b *light blue* arrow).

To the contrary, tissue induction and morphogenesis lacked the induction of bone formation as seen in intramuscular sites of the Chacma baboon *Papio ursinus* (Ripamonti 1991; Ripamonti 2009; Ripamonti 2017). We suggested that in the selachian *Carcharhinus obscurus* shark, the induction programme required for the induction of bone formation is genetically missing due to the evolutionary ablation and lack of genes regulating the induction of bone formation in selachian fishes (Ripamonti 2018b; Ripamonti et al. 2018). Additionally, it is highly likely that the lack of vascular invasion and capillary sprouting that follows chondrolysis, as in the mammalian growth plate, is responsible for the lack of bone formation in sharks (Ripamonti et al. 2018).

In previous communications, we reported that cartilage and perichondrial tissues in Elasmobranchs are avascular after the evolutionary expression and synthesis of powerful inhibitors of angiogenesis (Ripamonti et al. 2018) that block osteogenesis in angiogenesis (Ripamonti 2006; Ripamonti et al. 2006).

The heterotopic implanted topographies of the coral-derived macroporous bioreactors are set to differentiate invading resident mesenchymal cells. The final induction of osteogenesis or chondrogenesis is, however, dependent on the available genetic programmes of the heterotopic implanted animal species, generating the induction of chondrogenesis only if the genetic programme of the induction of bone formation has been ablated as in the cartilaginous fishes, the *Elasmobranchii* (sharks, skates and rays) (Romer 1963; Ripamonti et al. 2018).

Experiments in *Chondrichthyans* including the *Elasmobranchii* (sharks, skates and rays) are *par force* rare, and the reported induction of chondrogenesis within the macroporous spaces of coral-derived calcium phosphate-based bioreactors has indicated that *Carcharhinus obscurus* muscle tissue comprises precursor stem cells for the induction of chondrogenesis in post-natal life (Ripamonti et al. 2018).

More recently, experiments in the cartilaginous skate *Leucoraja erinacea* reported spontaneous cartilage repair after the surgical creation of a cartilage injury defect along the metapterygium of *Leucoraja erinacea* (Fig. 5.19) (Marconi et al. 2020).

The study showed that in *Leucoraja erinacea* transcriptional features of developing cartilage persist into adulthood (Marconi et al. 2020). The surgically created defect along the metapterygium of *Leucoraja erinacea* developed chondrocytes co-expressing genes encoding the transcription factors Sox5, Sox6 and Sox9 (Marconi et al. 2020). Analyses of the reported morphological images showed the induction of well-differentiated chondroblasts and/or chondrocytes attached or in close relationship with the cartilaginous defect of the metapterygium (Figs. 5.19a,c) (Marconi et al. 2020).

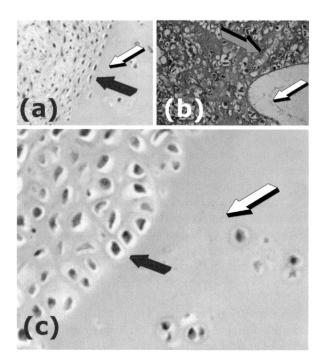

FIGURE 5.19 Repair of hyaline cartilage in the skate *Leucoraja erinacea* (Marconi et al. 2020) (Images courtesy of JA Gillis, University of Cambridge, Cambridge, UK). (a) Sixty days after the induction of a cartilage defect (white arrow) along the metapterygium of *L. erinacea* (Marconi ert al. 2020) there is invasion of fibrous tissue with differentiating chondroblastic cells (dark blue arrow) against the cartilaginous defect along (white arrow). Image courtesy of JA Gillis, University of Cambridge, Cambridge, UK. (b) Compare to tissue induction and differentiation of chondrogenesisi within the macroporous spaces of a coral-derive bioreactor implanted *solo* in the dorsal musculature of *Carcharinus obscurus sharks* where there is no cartilage (Ripamonti et al. 2018). There is the spontaneous induction of chondroblastic cells and the induction of columnar chondrocytic cells reminiscent of the classic mammalian cartilagenous growth plate (light blue arrow b). (c) Ninety days after the induction of the cartilaginous defect in *L. erinacea* (Marconi ert al. 2020) there is differentiation of chondrocytic cells against the cartilaginous defect (white arrow indicating the metapterygium of *L. erinacea*).

The induction of chondrogenesis within the macroporous spaces of the bioreactor implanted intramuscularly in *Carcharhinus obscurus* showed, on the other hand, chondroblastic activity and new matrix deposition with chondroblast differentiation across the observed macroporous spaces of a coral-derived bioreactor implanted in the dorsal musculature where thereis no cartilage (Ripamonti et al. 2018).

The induction of chondrogenesis but lack of osteogenesis within coral-derived constructs when implanted in heterotopic dorsal intramuscular sites of *Carcharhinus obscurus* unequivocally demonstrates the morphogenetic capacity of the calcium phosphate-based bioreactors, and, at the same time, the genetic control of the induction of bone formation. The selachian fishes do not retain the genetic memory of the induction of bone formation.

We have recently reviewed (Ripamonti 2018a; Ripamonti 2018b; Ripamonti et al. 2018; Ripamonti 2017) the grand contribution of Romer who has shown that the cartilaginous endoskeleton of sharks "is not truly a primitive one and that the absence of bone is not an ancestral character, but one due to degeneration from bone-bearing ancestors" (Romer 1963). Romer (1963) stated that

> skeletal evolution in vertebrates is thus "*the reverse of the truth*" that instead beginning with a purely cartilaginous skeleton and later acquiring bone, the early vertebrates first developed bone and that bone is an ancient character rather than a relatively new skeletal material in the history of vertebrates.
>
> **(Romer 1963)**

The DNA of the selachian fishes does not retain "the developmental memory of the osteoinduction programme though it retains the memory of the induction of chondrogenesis as a recapitulation of evolutionary differentiating and de-differentiating events, ultimately lacking angiogenesis and capillary sprouting required for setting into motion chondrolysis" (Ripamonti et al. 2018) and, thus, the induction of bone formation. Perhaps, however, whilst on one hand the chondrogenic images presented in Figure 5.18 are simply fascinating, once again illustrating the morphogenetic capacities of the coral-derived bioreactors when implanted in heterotopic sites of a variety of animal models including sharks, the most convincing images of evo-devo, to "elucidate the contribution of gene loss and developmental constraints to the evolution of animal body plans" (De Robertis 2008), are the detailed images in both panels of Figure 5.18 that show columns of progressively differentiating chondroblasts patterning the newly induced cartilage as in mammalian embryonic development and in the developing growth plate.

Ripamonti et al. (2018) suggested that the calcium phosphate-based bioreactor engineers cartilaginous columnar condensations as seen in the mammalian counterpart (Ripamonti et al. 2018). Re-evaluating those experiments and histological sections of the late 1980s/early 1990s (Ripamonti 2018b; Ripamonti et al. 2018), it is difficult to understand how selachian cartilages retain the columnar patterning typical of the mammalian growth plate (Reddi and Huggins 1972; Reddi 1980; Ripamonti 2007; Ripamonti 2017) owing to significantly different evolutionary pathways of selachians vs. mammals.

We propose in this volume that the mammalian growth plate patterning was imprinted by soluble molecular signals of the TGF-β supergene family in setting tissue patterning and morphogenesis in the ancestral mesenchymal matrix that firstly developed bone and cartilage (Romer 1963), and later blue printing *ab initio* tissue patterning and morphogenesis of the ancestral matrix with the later induction of the cartilaginous growth plate of the mammalian endochondral bones to develop and speciate the vertebrate animals.

In the 1990s, we stated that fruit flies, frogs and fractures are all masterminded by the TGF-β supergene family of proteins (Ripamonti 1994). We conclude this chapter on the concavity spontaneously initiating the induction of bone formation that TGF-β immunolocalizes in shark tissues, mandibular tesserae and during continuously erupting teeth of the shark *C. obscurus* (Ripamonti 2017; Ripamonti 2018b; Ripamonti et al. 2018; Ripamonti 2019). The TGF-β$_3$ ancestrally retains the blue print of the molecular and morphological induction of tissue patterning and morphogenesis of the mammalian growth plate, a multifunctional pleiotropic bioreactor that independently controls the growth of endochondral bones, the development of the skeleton, ambulation and the unique walk into the spectacular biological and molecular scenario of the *Homo* clade.

It was previously stated that "The powerful evolutionary conservation of selected genes and gene products masterminding tissue induction and morphogenesis represents the overall plan of *Nature*'s parsimony in controlling the genesis of form and function or morphogenesis" (Ripamonti and Duarte 2019). The expression pattern of the

> TGF-β$_3$ gene product during folding of the continuously erupting dental lamina in *C. obscurus* epitomizes Nature's parsimony in controlling specialized functions deploying singly, synergistically and synchronously highly conserved pleiotropic morphogens controlling morphogenesis from elasmobranch fishes, fruit flies, frogs and fractures.
>
> **(Ripamonti and Duarte 2019)**

Similarly, the induction of bone formation by concavities of the sintered calcium phosphate-based substrata is masterminded by osteoclastogenesis, Ca^{++} release, angiogenesis, stem cell differentiation, the induction of the osteogenic phenotype and, finally, "Bone: formation by autoinduction" (Urist 1965) within the activated concavities of the heterotopically implanted bioreactors.

ACKNOWLEDGEMENTS

Several colleagues, scientists, technologists and clinicians, mentors and post-graduate students as well as many other human beings who were asked to listen to several ideas, results, thoughts and hypotheses deserve to be acknowledged at the end of this chapter on the concavities set into sintered crystalline substrata or, for that matter, at the end of this CRC Press volume on the geometric induction of bone formation. Barbara van den Heever and Ruqayya Parak were instrumental in preparing a considerable number of histological decalcified and undecalcified mineralized sections including the preparation of impeccable undecalcified sections cut by the Exakt diamond saw cutting and grinding equipment at the Bone Research Unit of the University. This volume could not have born without the dedicated impeccable and continuous and above standard quality of sections from both Barbara and Ruqayya. The work of the late Michel Thomas and William Richter, CSIR Pretoria, has been also fundamental for the preparation of sintered crystalline biomatrices for implantation in orthotopic calvarial and heterotopic *rectus abdominis* sites of the Chacma baboon *Papio ursinus*. The molecular team headed by Raquel Duarte in the School of Clinical Medicine had the capacity to finally resolve the spontaneous and/or intrinsic induction of bone formation by calcium phosphate-based biomimetic matrices. The Unit and I acknowledge the critical molecular help of Raquel Duarte, Caroline Dickens, Therese Dix-Peek and Manfred Klar. To Manfred, a PhD student and later a post-doctoral fellow in our laboratories, a special thanks for having molecularly cracked the induction of bone formation by the coral-derived calcium phosphate-based constructs. A special thanks to Laura Roden, née Yeates, for the significant insights into the mechanisms and for several contributions to the accrued knowledge of the induction of bone formation by the Bone Research Unit. I thank the University for sustaining the Bone Research Laboratory, later fully fledged Unit, since March 1994, when it was officially opened at the Medical School by Hari A. Reddi, then at Johns Hopkins University, Baltimore, USA. To Hari a special thanks for having learned with him the long crucial and distinguished scientific story of "Bone: Formation by autoinduction" and for discussing with me his classic papers on how the geometry of the inductor significantly alters the induction of bone formation.

BIBLIOGRAPHY

Abou Alaiwi, W.A.; Lo, S.T.; Nauli, S.M. Primary Cilia: Highly Sophisticated Biological Sensors. *Sensors* 2009, *9*(9), 7003–20.

Ahn, E.H.; Kim, Y.; Kshitiz; An, S.S.; Afzal, J.; Lee, S.; Kwak, M.; Suh, K.-Y.; Kim, D.-H.; Levchenko, A. Spatial Control of Adult Stem Cell Fate Using Nanotopographic Cues. *Biomaterials* 2014, *35*(8), 2401–10.

Barradas, A.M.C.; Yuan, H.; van Blitterswijk, C.A.; Habibovic, P. Osteoinductive Biomaterials: Current Knowledge of Properties, Experimental Models and Biological Mechanisms. *Eur. Cell. Mater.* 2011, *21*, 407–29.

Barradas, A.M.C.; Yuan, H.; der Stok, J.; Quang, B.; Fernandes, H.; Chaterjea, A.; Hogenes, M.C.H.; Shultz, K.; Donahue, L.R.; Blitterswijk, C.; Boer, J. The Influence of Genetic Factors on the Osteoinductive Potential of Calcium Phosphate Ceramics in Mice. Biomaterials 2012, 33(22), 5696–705. https://doi.org/10.1016/j.biomaterials.2012.04.021.

Benjamin, L.E.; Hemo, I.; Keshet, E. A Plasticity Window for Blood Vessels Remodelling Is Defined by Pericyte Coverage of the Preformed Endothelial Network and Is Regulated by PDGF-B and VEGF. *Development* 1988, *125*(9), 1591–98.

Bettinger, C.; Langer, R.; Borenstein, J. Engineering Substrate Topography at the Micro- and Nanoscale to Control Cell Function. *Angew. Chem. Int. Ed. Engl.* 2009, *48*(30), 5406–15.

Brunet, T.; Larson, B.T.; Linden, T.A.; Vermeij, M.J.A.; McDonald, K.; King, N. Light-Regulated Collective Contractility in a lMulticellular Choanoflagellate. *Science* 2019, *366*(6463), 326–34.

Brunette, D.M. The Effects of Implant Surface Topography on the Behavior of Cells. *Int. J. Oral Maxillofac. Implants* 1988, *3*(4), 231–46.

Buxboim, A.; Discher, D.E. Stem Cells Feel the Difference. *Nat. Methods* 2010, *7*(5), 695–7.

Buxboim, A.; Ivanovska, I.L.; Discher, D.E. Matrix Elasticity, Cytoskeletal lForces and Physics of the Nucleous: How Deeply Do Cells "Feel" Outside and In? *J. Cell Sci.* 2010, *123*(3), 297–308.

Cell Editorial. Pulling It All Together. *Cell* 2014, *157*(1), 1–2, doi:10.1016/j.cell.2014.03.22.

Chen, C.-W.; Montelatici, E.; Crisan, M.; Corselli, M.; Huard, J.; Lazzari, L.; Péault, B.; Péault Perivascular Multi-Lineage Progenitor Cella in Humans Organs: Regenerative Units, Cytokine Sources or Both? *Cytokine Growth Factor Rev.* 2009, *20*(5–6), 429–34.

Clark, P.; Connolly, P.; Curtis, A.S.G.; Dow, J.A.T.; Wilkinson, D.W. Topographical Control of Cell Behaviour I. Simple Step Cues. *Development* 1987, *108*, 635–44.

Clark, P.; Connolly, P.; Curtis, A.S.G.; Dow, J.A.T.; Wilkinson, D.W. Topographical Control of Cell Behaviour: II. Multiple Grooved Substrata. *Development* 1990, *99*, 493–48.

Costa-Rodriguez, J.; Fernandes, A.; Lopes, M.A.; Fernandes, M.H. Hydroxyapatite Surface Roughness: Complex Modulation of the Osteoclastogenesis of Human Precursor Cells. *Acta Biomater.* 2012, *8*(3), 1137–45.

Crisan, M.; Yap, S.; Casteilla, L. *et al.* A Perivascular Origin for Mesenchymal Stem Cells in Multiple Human Organs. *Cell Stem Cell* 2008, *3*(3), 301–13.

Crivellato, E.; Nico, B.; Ribatti, D. Contribution of Endothelial Cells to Organogenesis: A Modern Reappraisal of an Old Aristotelian Concept. *J. Anat.* 2007, *211*(4), 415–27.

Curran, J.M.; Chen, R.; Hunt, J.A. The Guidance of Human Mesenchymal Stem Cell Differentiation In Vitro by Controlled Modifications to the Cell Substrate. *Biomaterials* 2006, *27*(27), 4783–93.

Curtis, A.C.; Wilkinson, C. Topographical Control of Cells. *Biomaterials* 1997, *18*(24), 1573–83.

Danoux, C.; Pereira, D.; Döbelin, N.; Stähli, C.; Barralet, J.; van Blitterswijk, C.; Habibovic, P. The Effects of Crystal Phase and Particle Morphology of Calcium Phosphates on Proliferation and Differentiation of Human Mesenchymal Stromal Cells. *Adv. Healthc. Mater.* 2016, *5*(14), 1775–85, doi:10.1002/adhm.201600184.

De Robertis, E.M. Evo-Devo: Variations on Ancestral Themes. *Cell* 2008, *132*(2), 185–95, doi:10.1016/j.cell.2008.01.003.

Discher, D.E.; Janmey, P.; Wang, Y.-L. Tissue Cells Feel and Respond to the Stiffness of Their Substrate. *Science* 2005, *310*(5751), 1139–43.

Discher, D.E.; Mooney, D.J.; Zandstra, P.W. Growth Factors, Matrices, and Forces Combine and Control Stem Cells. *Science* 2009, *324*(5935), 1673–77.

Doherty, M.J.; Ashton, B.A.; Walsh, S.; Beresford, J.N.; Grant, M.E.; Canfield, A.E. Vascular Pericytes Express Osteogenic Potential In Vitro and In Vivo. *J. Bone Miner. Res.* 1988, *13*, 128–39.

Dong, L.; Gong, J.; Wang, Y.; He, J.; you, D.; Zhou, Y.; Li, Q.; Liu, Y.; Vheng, K.; Qian, J.; Weng, W.; Wang, H.; Yu, M. Chiral Geometry Regulates Stem Cell Fate and Activity. *Biomaterials* 2019, *222*, 119456.

Duneas, N. Crooks, J. Ripamonti, U.Transforming Growth Factor-β1: Induction of Bone Morphogenetic Protein Gene Expression During Endochondral Bone Formation in the Baboon, and Synergistic Interaction with Osteoegenic Protein-1 (BMP-7). *Growth Factors* 1998, *15*, 259–77.

Eggenschwiler, J.T.; Anderson, K.V. Cilia and Developmental Signaling. *Annu. Rev. Cell Dev. Biol.* 2007, *23*, 345–73.

Engler, A.J.; Sen, S.; Sweeney, H.L.; Discher, D.E. Matrix Elasticity Directs Stem Cell Lineage Specification. *Cell* 2006, *126*(4), 677–89.

Ferretti, C.; Ripamonti, U.; Tsiridis, E.; Kerawala, C.J.; Mantalaris, A.; Heliotis, M. Osteoinduction: Translating Preclinical Promises into Reality. *Br. J. Oral Maxillofac. Surg.* 2010, *48*(7), 536–39.

Fiedler, J.; Özdemir, B.; Bartholomä, J.; Pletti, A.; Brenner, R.E.; Ziemann, P. The Effect of Substrate Surface Nanotopogrphy on the Behavior of Multipotent Mesenchymal Stromal Cells and Osteoblasts. *Biomaterials* 2013, *34*(35), 8851–59.

Folkman, J.; Klagsbrun; Sasse, J.; et al. A Heparin Binding Angiogenic Protein Basic Fibroblast Growth Factor Is Stored within Basement Membranes. *Am. J. Pathol.* 1988, *130*, 393–400.

Fu, J.; Wang, Y.-K.; Yang, M.T.; Desai, R.A.; Yu, X.; Liu, Z.; Chen, C.S. Mechanical Regulation of Cell Function with Geometrically Modulated Elastomeric Substrates. *Nat. Methods* 2010, *7*, 733–36. doi:10.1033/NMETH.1487.

Fujibayashi, S.; Nakamura, T.; Nishiguchi, S.; Tamura, J.; Uchida, M.; Kim, H.-M.; Kobuko, T. Bioactive Titanium: Effect of Sodium Removal and the Bone-Bonding Ability of Bioactive Titanium Prepared by Alkali and Heat Treatment. *J. Biomed. Mater. Res.* 2001, *56*(4), 562–70.

Fujibayashi, S.; Neo, M.; Kim, H.-M.; Kobuko, T.; Nakamura, T. Osteoinduction of Porous Bioactive Titanium Metal. *Biomaterials* 2004, *25*(3), 443–50.

Gage, F.H. Adult Neurogenesis in Mammals. *Science* 2019, *364*(6443), 827–8.

Gilbert, S.F.; Saxén, L. Spemann's Organizer: Model and Molecules. *Mech. Developm.* 1993, *41*(2–3), 77–89.

Gittens, R.A.; McLachlan, T.; Olivares-Navarrete, R.; Cai, Y.; Berner, S.; Tannenbaum, R.; Schwartz,Z.; Sandhage, K.H.; Boyan, B.D. The Effects of Combined Micron-/Submicron-Scale Surface Roughness and Nanoscale Features on Cell Proliferation and Differentiation. *Biomaterials* 2011, *32*(13), 3395–403.

Gjorevski, N.; Nelson, C.M. Bidirectional Extracellular Matrix Signaling During Tissue Morphogenesis. *Cytokine Growth Factor Rev.* 2009, *20*(5–6), 459–65.

Gomez-Salinero, J.M.; Rafii, S. Endothelial Cell Adaptation in Regeneration. *Science* 2018, *362*(6419), 1116–17.

Gosain, A.K.; Song, L.; Riordan, P.; Amarante, M.T.; Nagy, P.G.; Wilson, C.R.; Toth, J.M.; Ricci, J.L.A. 1-Year Study of Osteoinduction in Hydroxyapatite –Derived Biomaterials in an Adult Sheep Model: Part 1. *Plast. Reconstr. Surg.* 2002, *109*(2), 619–30.

Gosain, A.K.; Riordan, P.; Song, L.; Amarante, M.T.; Kalantarian, B.; Nagy, P.G.; Wilson, C.R.; Toth, J.M.; McIntyre, B.L.A. 1-Year Study of Osteoinduction in Hydroxyapatite –Derived Biomaterials in an Adult Sheep Model: Part II. Bioengineering Implants to Optimize Bone Replacement in Reconstruction of Cranial Defects. *Plast. Reconstr. Surg.* 2004, *114*(5), 1155–63.

Gospodarowicz, D.; Greenburg, G.; Birdwell, C.R. Determination of Cellular Shape by the Extracellular Matrix and Its Correlation with the Control of Cell Growth. *Cancer Res.* 1978, *38*(11 Pt 2), 4155–71.

Goss, R. *Principles of Regeneration.* Academic Press, New York, 1969.

Habibovic, P.; van der Valk, C.M.; van Blitterswijk, C.A.; de Groot, K.; Meijer, G. Influence of Octacalcium Phosphate |Coating on Osteoinductive Properties of Biomaterials. *J. Mat. Sci. Mat. Med.* 2004, *15*(4), 373–80.

Habibovic, P.; Yuan, H.; Van der Valk, C.M.; Meijer, G.; van Blitterswijk, C.A.; de Groot, K. 3D Microenvironment as Essential Element for Osteoinduction by Biomaterials. Biomaterials 2005, 26, 3565–75.

Habibovic, P.; Sees, T.M.; van den Doel, M.A.; van Blitterswijk, C.A.; de Groot, K.J. Biomed. Mater. Res. 2006, *A77*, 747–62.

Habibovic, P.; Yuan, H.; Van der Valk, C.M.; Meijer, G.; van Blitterswijk, C.A.; de Groot, K.; Hynes, R.O. The Extracellular Matrix: Not Just Pretty Fibrils. *Science* 2009, *326*(5957), 1216–19.

Hopkinson-Wolley, J.; Hughes, D.; Gordon, S.; Martin, P. Macrophage Recruitment During Limb Development and Wound Healing in the Embryonic and Foetal Mouse. *J. Cell Sci.* 1994, *107*(5), 1159–67.

Jung, H.; Baek, M.; D'Elia, K.; Boisvert, C.; Currie, P.D.; Tay, B.-H.; Venkatesh, B.; Brown, S.T.; Heguy, A.; Schoppik, D.; Dasen, J.S. The Ancient Origins of Neural Substrates for Land Walking. *Cell* 2018, *172*(4), 667–82, doi:10.1016/j.cell.2018.01.013.

Karageorgiou, V.; Kaplan, D. Porosity of 3D Biomaterials Scaffolds and Osteogenesis. *Biomaterials* 2005, *26*(27), 5474–91.

Kawai, T.; Takemoto, M.; Fujibayashi, S.; Akiyama, H.; Tanaka, M.; Yamaguchi, S.; Pattanayak, D.K.; Doi, K.; Matsushita, T.; Nakamura, T.; Kobuko, T.; Matsuda, S. Osteoinduction on Acid and Heat Treated Porous Ti Metal Samples in Canine Muscle. *PLOS ONE* 2014, *9*(2), e88366, doi:10.1371/journal.pone.0088366.

Kerszberg, M.; Wolpert, L. Specifying Positional Information in the Embryo: Looking Beyond Morphogens. *Cell* 2007, *130*(2), 205–9.

Khouri, R.K.; Koudsi, B.; Reddi, H. Tissue Transformation In Vivo. A Potential Practical Application. *JAMA* 1991, *266*(14), 1953–5.

Kiang, J.D.; Wen, J.H.; del Alamo, J.C.; Engler, A.J. Dynamic and Reversible Surface Topography Influences Cell Morphology. *J. Biomed. Mater. Res. A* 2013, *101*(8), 2313–21.

Kilian, K.A.; Bugarija, B.; Lahn, B.T.; Mrksich, M. Geometric Cues for Directing the Differentiation of Mesenchymal Stem Cells. *Proc. Natl. Acad. Sci, U.S.A.* 2010, *107*(11), 4872–877.

Kim, D.-H.; Provenzano, P.P.; Smith, C.L.; Levchenko, A. Matrix Nanotopography as a Regulator of Cell Function. *J. Cell. Biol.* 2012, *197*(3), 351–60.

Klar, R.M.; Duarte, R.; Dix-Peek, T.; Dickens, C.; Ferretti, C.; Ripamonti, U. Calcium Ions and Osteoclastogenesis Initiate the Induction of Bone Formation by Coral-Derived Macroporous Constructs. *J. Cell. Mol. Med.* 2013, *17*(11), 1444–57.

Klar, M.R.; Duarte, R.; Dix-Peek, T.; Ripamonti, U. The Induction of Bone Formation by the Recombinant Human Transforming Growth Factor-β3. *Biomaterials* 2014, *35*(9), 2773–88, doi:10.1016/j.biomaterials.2013.12.062.

Klein, C.; de Groot, K.; Weiqun, C.; Yubao, L.; Xingdong, Z. Osseous Substance Formation Induced in Porous Calcium Phosphate Ceramics in Soft Tissues. *Biomaterials* 1994, *15*(1), 31–4.

Kobuko, T.; Miyaji, F.; Kim, H.-M. Spontaneous Formation of Apatite Layer on Chemically Treated Titanium Metals. *J. Am. Ceram. Soc.* 1996, *79*(4), 1127–29.

Kopp,J.L.; Grompe, M.; Sander, M. Stem Cells versus Plasticity in Liver and Pancreas Regeneration. *Nat. Cell. Biol.* 2016, *18*(3), 238–45.

Kovacic, J.C.; Boehm, M. Resident Vascular Progenitor Cells: An Emerging Role for Non-Terminally Differentiated Vessel-Resident Cells in Vascular Biology. *Stem Cell Res.* 2009, *2*(1), 2–15.

Kusumbe, A.P.; Ramasamy, S.K.; Adams, R.H. Coupling of Angiogenesis and Osteogenesis by a Speficic Vessel Subtype in Bone. *Nature* 2014, *507*(7492), 323–28.

Lamers, E.; van Horssen, R.; te Riet, J.; van Delft, F.C.; Luttge, R.; Walboomers, X.F.; Jansen, J.A. The Influence of Nanoscale Topographical Cues on Initial Osteoblast Morphology and Migration. *Eur. Cell Mater.* 2010, *20*, 329–43.

Lander, A.D. Morpheus Unbound: Reimagining the Morphogen Gradient. *Cell* 2007, *128*(2), 245–56.

Lanza, D.; Vegetti, M. Aristotele, A Cura di Diego Lanza e Mario Vegetti: Opere Biologiche UTET, 1971.

A History of Regeneration Research: Milestones in the Evolution of a Science. In: C.E. Dinsmore (ed.) Cambridge Univerrsity Press, Canada, 1991.

Le Nihouannen, D.; Daculsi, G.; Saffarzadeh, A.; Gauthier, O.; Delplace, S.; Pilet, P.; Layrolle, P. Ectopic Bone Formation by Microporous Calcium Phosphate Ceramic Particles in Sheep Muscles. *Bone* 2005, *36*(6), 1086–93.

Le Nihouannen, D.; Saffarzadeh, A.; Gauthier, O.; Moreau, F.; Pilert, P.; Spaethe, R.; Layrolle, P.; Daculsi, G. Bone Tissue Formation in Sheep Muscles Induced by a Biphasic Calcium Phosphate Ceramic and Fibrin Glue Composite. *J. Mat. Sci. Mater. Med.* 2008, *19*(2), 667–75.

Lensch, M.W.; Daheron, L.; Schlager, T.M. Pluripotent Stem Cells and Their Niches. *Stem Cell Rev.* 2006, *2*(3), 185–202.

Levander, G. Tissue Induction. *Nature* 1945, *155*, 148–49.

Li, X.; van Blitterswijk, C.A.; Feng, Q.; Cui, F.; Watari, F. The Effect of Calcium Phosphate Microstructure on Bone-Related Cells *In Vitro*. *Biomaterials* 2008, *29*(23), 3306–16.

Lismaa, S.E.; Kaidonis, X.; Nicks, M.; Bogush, N.; Kikuchi, K.; Naqvi, N.; Harvery, R.P.; Husain, A.; Grahm, R.M. Comparative Regenerative Mechanisms across Different Mammalian Tissues. *NPJ* Reg. *Med.* 2018, *3*(1), 1–20, doi:10.1038/s41536-018-0044-5.

Liu, X.; Liu, R.; Cao, B.; Ye, K.; Li, S.; Gu, Y.; Pan, Z.; Ding, J. Subcellular Cell Geometry on Micropillars Regulates Stem Cell Differentiation. *Biomaterials* 2016, *111*, 27–39.

Lutolf, M.P.; Hubbell, J.A. Synthetic Biomaterials as Instructive Extracellular Microenvironments for Morphogenesis in Tissue Engineering. *Nat. Biotechnol.* 2005, *23*(1), 47–55.

Luyten, F.P.; Cunningham, N.S.; Ma, S.; Muthukumaran, N.; Hammonds, R.G.; Nevins, W.B.; Wood, W.I.; Reddi, A.H. Purification and Partial Amino Acid Sequence of Osteogenin, a Protein Initiating Bone Differentiation. *J. Biol. Chem.* 1999, *264*(23), 13377–80.

Marconi, A.; Hancock-Ronemus, A.; Gillis, J.A. Adult Chondrogenesis and Spontaneous Cartilage Repair in the Skate, *Leucoraja erinacea*. *eLife* 2020, *9*, e53414, doi:10.7554/eLife.53414.

Martin, P.; Parkhurst, S. Parallels between Tissue Repair and Embryo Morphogenesis. *Development* 2004, *131*(13), 3021–34.

Medici, D.; Shore, E.M.; Lounev, V.Y.; Kaplan, F.D.; Kalluri, R.; Olsen, B.R. Conversion of Vascular Endothelial Cells into Multipotent Stem-Like Cells. *Nat. Med.* 2011, *16*(12), 1400–06.

McMurray, R.J. Dalby, M.J. Tsimbouri, P.M. Using biomaterials to study stem cell mechano-transduction, growth and differentiation. *J.Tissue Eng. Regen. Med.* 2015, *9*, 528–39.

McNamara, L.E.; McMurray, R.J.; Biggs, M.J.P.; Kantawong, F.; Oreffo, M.J.; Dalby, M.J. Nanotopographical Control of Stem Cell Differentiation. *J. Tissue Eng.* 2010, *2010*, 120623. doi:10.4061/2010/120623.

McNamara, L.E.; Sjöström, T.; Burgess, K.E.V.; Kim, J.J.W.; Liu, E.; Gordonov, S.; Moghe, P.V.; Meek, R.M.D.; Oreffo, O.C.; Su, B.; Dalby, M.J. Skeletal Stem Cell Physiology on Functionally Distinct Titania Topographies. *Biomaterials* 2011, *32*, 7403–10.

Metavarayuth, K.; Sitasuwan, P.; Zhao, X.; Lin, Y.; Wang, Q. Influence of Surface Topographical Cues on the Differentiation of Mesenchymal Stem Cells In Vitro. *ACS Biomater. Sci. Eng.* 2016, *2*(2), 142–51.

Miyoshi, H.; Adachi, T. Topography Design Concept of a Tussue Engineering Scaffold for Controlling Cell Function and Fate through Actin Cytoskeletal Modulation. *Tissue Eng.* 2014, *20*(6), 609–27.

Müller, P.; Buinheim, U.; Diener, A.; Lüthen, F.; Teller, M.; Klinkenberg, E.-D.; Neumann, H.-G.; Nebe, B.; Liebold, A.; Steinhoff, G.; Rychly, J. Calcium Phosphate Surfaces Promote Osteogenic Differentiation of Mesenchymal Stem Cells. *J. Cell. Mol. Med.* 2008, *12*(1), 281–91.

Nelson, C.M.; Vanduijn, M.M.; Inman, J.L.; Flertcher, D.A.; Bissell, M.J. Tissue Geometry Determines Sites of Mammary Branching Morphogenesis in Organotypic Cultures. *Science* 2006, *314*(5797), 298–300.

Nishiguchi, S.; Kat, H.; Fujita, H.; Kim, H.-M.; Miyaji, F.; Kobuko, T.; Nakamura, T. Enhancement of Bone-Bonding Strengths of Titanium Alloy Implants by Alakli and Heat Treatments. *J. Biomed. Mater. Res. (Appl. Biomater.)* 1999, *48*(5), 689–99.

Önal, P.; Grün, D.; Adamidi, C.; Rybak, A.; Solana, J.; Mastrobuoni, G.; Wang, Y.; Rahn, H.-P.; Chen, W.; Kempa, S.; Ziebold, U.; Rajewsky, N. Gene Expression of Pluripotency Determinants Is Conserved Between Mammalian and Planarian Stem Cells. *EMBO J.* 2012, *31*(12), 2755–69.

Othman, Z.; Fernandes, H.; Groot, A.J.; Luider, T.M.; Alcinesio, A.; de Melo Pereira, D.; Guttenplan, A.P.M.; Yuan, H.; Habibovic, P. The Role of ENPP1/PC-1 in Osteoinduction by Calcium Phosphate Ceramics. *Biomaterials* 2019, *167*, 191–204.

Paralkar, V.M.; Nandedkar, A.K.N.; Pointer, R.H.; Kleinman, H.K.; Reddi, A.H. Interaction of Osteogenin, a Heparin Binding Bone Morphogenetic Protein, with Type IV Collagen. *J. Biol. Chem.* 1990, *265*(28), 17281–84.

Paralkar, V.M.; Vukicevic, S.; Reddi, A.H. Transforming Growth Factor β Type 1 Binds to Collagen Type IV of Basement Membrane Matrix: Implications for Development. *Dev. Biol.* 1991, *143*(2), 303–10.

Ramasamy, S.K.; Kusumbe, A.P.; Wang, L.; Adams, R.H. Endothelial Notch Activity Promotes Angiogenesis and Osteogenesis in Bone. *Nature* 2014, *507*(7492), 376–80.

Ramasamy, S.K.; Kusumbe, A.P.; Adams, R.H. Regulation of Tissue Morphogenesis by Endothelial Cell-Derived Signals. *Trends Cell Biol.* 2015, *25*(3), 148–57.

Reddi, A.H.; Huggins, C. Biochemical Sequences in the Transformation of Normal Fibroblasts in Adolescent Rats. *Proc. Natl. Acad. Sci. U.S.A.* 1972, *69*(6), 1601–5.

Reddi, A.H.; Huggins, C.B. Influence of Geometry of Transplanted Tooth and Bone on Transformation of Fibroblasts. *Proc. Soc. Exp. Biol. Med.* 1973, *143*(3), 634–37.

Reddi, A.H. Bone Matrix in the Solid State: Geometric Influence on Differentiation of Fibroblasts. *Adv. Biol. Med. Phys.* 1974, *15*(0), 1–18.

Reddi, A.H. Cell Biology and Biochemistry of Endochondral Bone Development. *Coll. Rel. Res.* 1981, *1*(2), 209–26.

Reddi, A.H.; Kuettner, K.E. Vascular Invasion of Cartilage: Correlation of Morphology with Lysozyme, Glycosaminoglycans, Protease, and Protease-Inhibitory Activity During Endochondral Bone Development. *Dev. Biol.* 1981, *82*(2), 217–23.

Reddi, A.H. Extracellular Matrix and Development. In: K.A. Piez, A.H. Reddi (eds.) *Extracellular Matrix Biochemistry*, Elsevier, New York, 1984, 247–91.

Reddi, A.H. Bone Morphogenesis and Modeling: Soluble Signals Sculpt Osteosomes in the Solid State. *Cell* 1997, *89*(2), 159–61.

Reddi, A.H. Morphogenesis and Tissue Engineering of Bone and Cartilage: Inductive Signals, Stem Cells, and Biomimetic Matrices. *Tissue Eng.* 2000, *6*(4), 351–59.

Reichman, O.J. Evolution of Regenerative Capabilities. *Am. Nat.* 1984, *123*(6), 752–63.

Richter, P.W.; Thomas, M.E.; Ripamonti, U. Implant Material with Enhanced Surface Bioactivity. South African patent 2001 0575, January 19, 2001.

Richter, P.W.; Thomas, M.E.; Ripamonti, U. Bone Filler Material. South African patent 0945, February 2, 2001.

Ripamonti, U. Inductive Bone Matrix and Porous Hydroxyapatite Composites in Rodents and Nonhuman Primates. In: J. Yamamuro, L. Wilson-Hench, L. Hench (eds.) *Handbook of Bioactive Ceramics, II: Calcium Phosphate and Hydroxylapatite Ceramics*, CRC Press, Boca Raton, FL, 1990, 245–53.

Ripamonti, U. The Morphogenesis of Bone in Replicas of Porous Hydroxyapatite Obtained from Conversion of Calcium Carbonate Exoskeletons of Coral. *J. Bone Joint Surg. [A]* 1991, 73-A, 692–703.

Ripamonti, U. Calvarial Reconstruction in Baboons with Porous Hydroxyapatite. *J. Craniofac. Surg.* 1992a, *3*(3), 149–59.

Ripamonti, U. Calvarial Regeneration in Primates with Autolyzed Antigen Extracted Allogeneic Bone. *Clin. Orth. Rel. Res.* 1992b, *282*(282), 293–303.

Ripamonti, U.; Ma, S.; van den Heever, B.; Reddi, A.H. Osteogenin, a Bone Morphogenetic Protein, Adsorbed on Porous Hydroxyapatite Substrata, Induces Rapid Bone Differentiation in Calvarial Defects of Adult Primates. *Plast. Reconstr. Surg.* 1992, *90*(3), 382–93.

Ripamonti, U.; van den Heever, B.; van Wyk, J. Expression of the Osteogenic Phenotype in Porous Hydroxyapatite Implanted Extraskeletally in Baboons. *Matrix* 1993a, *13*(6), 491–502.

Ripamonti, U.; Yeates, L.; van den Heever, B. Initiation of Heterotopic Osteogenesis in Primates after Chromatographic Adsorption of Osteogenin, a Bone Morphogenetic Protein, onto Porous Hydroxyapatite. *Biochem. Biophys. Res. Commun.* 1993b, *193*(2), 509–17.

Ripamonti, U. A Method for Screening a Selected Material for Its Osteoconductive and Osteoinductive Potential. South African Patent 92/3982, May 25 1994, US patent 5,355,898, October 18, 1994.

Ripamonti, U. *Fruit Flies, Frogs and Fractures: Do They Share Common Morphogenetic Mechanisms? Zoological Society of South Africa: Contemporary Zoology in Southern Africa.* Pietermaritzburg, July 11–14, 1994.

Ripamonti, U.; Kirkbride, A.N. Biomaterial and Bone Implant for Bone Repair and Replacement. PCT/NL95/00181, WO095/3200, November 30, 1995.

Ripamonti, U.; Reddi, A.H. Bone Morphogenetic Proteins: Applications in Plastic and Reconstructive Surgery. Adv. *Plast. Reconstr. Surg.* 1995, *11*, 47–65.

Ripamonti, U. Osteoinduction in Porous Hydroxyapatite Implanted in Heterotopic Sites of Different Animal Models. *Biomaterials* 1996, *17*(1), 31–5.

Ripamonti, U.; Duneas, N. Tissue Engineering of Bone by Osteoinductive Biomaterials. MRS Bulletin November 1996, 36–9.

Ripamonti, U.; Kirkbride, A.N. A Biomaterial and Bone Implant for Bone Repair and Replacement. EP760687A1, March 12, 1997.

Ripamonti, U.; Crooks, J.; Kirkbride, A.N. Sintered Porous Hydroxyapatites with Intrinsic Osteoinductive Activity: Geometric Induction of Bone Formation. *S. Afr. J. Sci.* 1999, *95*, 335–43.

Ripamonti, U. Smart Biomaterials with Intrinsic Osteoinductivity: Geometric Control of Bone Differentiation. In: J.D. Davis (ed.) *Bone Engineering*, M2 Corporation, Toronto, 2000, 215–22.

Ripamonti, U.; Kirkbride, A.N. Biomaterial and Bone Implant for Bone Repair and Replacement. US Patent 6,117,172, September 12, 2000.

Ripamonti, U.; Kirkbride, A.N. Biomaterial and Bone Implant for Bone Repair and Replacement. US Patent 6,302,913 B1, October 16, 2001.

Ripamonti,U.; Ramoshebi, L.N.; Matsaba, T.; Tasker, J.; Crooks, J.; Teare, J. Bone Induction by BMPs/OPs and Related Family Members. The Critical Role of Delivery Systems. *J. Bone Joint Surg.[A]* 2001a, 83-A(Suppl. 1), S116–127.

Ripamonti, U.; Crooks, J.; Rueger, D.C. Induction of Bone Formation by Recombinant Human Osteogenic Protein-1 (hOP-1) and Sintered Porous Hydroxyapatite in Adult Primates. *Plast. Reconstr. Surg.* 2001b, *107*(4), 977–88.

Ripamonti, U. Osteogenic Proteins of the Transforming Growth Factor-ß Superfamily. In: H.L. Henry, A.W. Norman (eds.) *Encyclopedia of Hormones*, Academic Press, San Diego, CA 2003, 80–6.

Ripamonti, U. Soluble, Insoluble and Geometric Signals Sculpt the Architecture of Mineralized Tissues. *J. Cell. Mol. Med.* 2004, *8*(2), 169–80.

Ripamonti, U.; Ramoshebi, L.N.; Patton, J.; Matsaba, T.; Teare, J.; Renton, L. Soluble Signals and Insoluble Substrata: Novel Molecular Cues Instructing the Induction of Bone. In: E.J. Massaro, J.M. Rogers (eds.) *The Skeleton*, Humana Press, Totowa, New Jersey 2004, Chapter 15, 217–27.

Ripamonti, U. Bone Induction by Recombinant Human Osteogenic Protein-1 (hOP-1, BMP-7) in the Primate *Papio ursinus* with Expression of mRNA of Gene Products of the TGF-β Superfamily. *J. Cell. Mol. Med.* 2005, *9*(4), 911–28.

Ripamonti, U. Soluble Osteogenic Molecular Signals and the Induction of Bone Formation. Biomaterials Leading Opinion Paper 2006, *27*(6), 807–22.

Ripamonti, U.; Ferretti, C.; Heliotis, M. Soluble and Insoluble Signals and the Induction of Bone Formation: Molecular Therapeutics Recapitulating Development. *J. Anat.* 2006, *209*(4), 447–68.

Ripamonti, U. Recapitulating Development: A Template for Periodontal Tissue Engineering. *Tissue Eng.* 2007, *13*(1), 51–71.

Ripamonti, U.; Richter, P.W.; Thomas, M.E.; Shape, Self-Inducing Memory Geomteric Cues Embedded within Smart Hydroxyapatite-Nased Biomimetic Matrices. *Plast. Reconstr. Surg.* 2007a, *120*, 1796–07.

Ripamonti, U.; Heliotis, M.; Ferretti, C. Bone Morphogenetic Proteins and the Induction of Bone Formation: From Laboratory to Patients. *Oral Maxillofac. Surg. Clin. North Am.* 2007b, *19*(4), 575–89.

Ripamonti, U.; Richter, P.W.; Nilen, R.W.N.; Renton, L. The Induction of Bone Formation by *Smart* Biphasic Hydroxyapatite Tricalcium Phosphate Biomimetic Matrices in the Non-Human Primate *Papio ursinus. J. Cell. Mol. Med.* 2008, *12*(6B), 2609–21.

Ripamonti, U.; Crooks, J.; Khoali, L.; Roden, L. The Induction of Bone Formation by Coral-Derived Calcium Carbonate/Hydroxyapatite Constructs. *Biomaterials* 2009, *30*(7), 1428–39.

Ripamonti, U. Biomimetism, Biomimetic Matrices and the Induction of Bone Formation. *J. Cell. Mol. Med.* 2009, *13*(9B), 2953–72.

Ripamonti, U. Soluble and Insoluble Signals Sculpt Osteogenesis in Angiogenesis. *World J. Biol. Chem.* 2010, *26*(5), 109–32.

Ripamonti, U.; Roden, L. Biomimetics for the Induction of Bone Formation. *Exp. Rev. Med. Devices* 2010, *74*(4), 469–79.

Ripamonti, U.; Klar, R.M.; Renton, L.F.; Ferretti, C. Synergistic Induction of Bone Formation by hOP- 1, hTGF-β_3 and Inhibition by Zoledronate in Macroporous Coral-Derived Hydroxyapatites. *Biomaterials* 2010, *31*(25), 6400–410.

Ripamonti, U. The Concavity: The "Shape of Life"and the Control of Bone Differentiation – Feature Paper – *Science in Africa* May 2012.

Ripamonti, U.; Roden, L.C.; Renton, L.F. Osteoinductive Hydroxyapatite-Coated Titanium Implants. *Biomaterials* 2012a, *33*(15), 3813–23.

Ripamonti, U.; Roden, L.; Renton, L.; Klar, R.M.; Petit, J.-C. The Influence of Geometry on Bone: Formation by Autoinduction *Science in Africa* 2012b. http://www.scienceinafric a.co.za/2012/Ripamonti_bone.htm.

Ripamonti, U.; Renton, L.; Petit, J.-C. Bioinspired Titanium Implants: The Concavity - The Shape of Life. In: M. Ramalingam, P. Vallitu, U. Ripamonti, W.-J. Li (eds.) *CRC Press Taylor & Francis, Boca Raton, FL 33487-2742 USA;* Tissue Engineering and Regenerative Medicine. A Nano Approach, 2013, Chapter 6, 105–23.

Ripamonti, U.; Duarte, R.; Ferretti, C. Re-Evaluating the Induction of Bone Formation in Primates. *Biomaterials* 2014, *35*(35), 9407–22.

Ripamonti, U.; Dix-Peek, T.; Parak, R.; Milner, B.; Duarte, R. Profiling Bone Morphogenetic Proteins and Transforming Growth Factor-Bs by hTGF-β_3 Pre-Treated Coral-Derived Macroporous Constructs: The Power of One. *Biomaterials* 2015, *49*, 90–102.

Ripamonti, U. Biomimetic Functionalized Surfaces and the Induction of Bone Formation. *Tissue Eng.* 2017, *23*(21,22), 1197–209.

Ripamonti, U. Functionalized Surface Geometries Induce: "Bone: Formation by Autoinduction". *Front. Physiol.* 2018a, *8*, 1084, doi:10.3389/fphys.2017.01084.

Ripamonti, U. Introductory Remarks on Cartilages, Bones and on *"Bone: Formation by Autoinduction". S. Afr. Dent. J.* 2018b, *73*(1), 6–10.

Ripamonti, U.; Roden, L.; van den Heever, B. Sharks, Sharks Cartilages and Shark Teeth: A Collaborative Africa-USA Study to Attempt to Induce *"Bone: Formation by Autoinduction"* in Cartilaginous Fishes. *S. Afr. Dent. J.* 2018, *73*(1), 11–21.

Ripamonti, U. Developmental Patterns of Periodontal Tissue Regeneration. Developmental Diversities of Tooth Morphogenesis Do Also Map Capacity of Periodontal Tissue Regeneration? *J. Periodont. Res.* 2019, *54*(1), 10–26, doi:10.1111/jre.12596.

Ripamonti, U.; Duarte, R. Inductive Surface Geometries: Beyond Morphogens and Stem Cells. *S. Afr. Dent. J.* 2019, *74*(8), 421–44.

Romer, A.S. The "Ancient History" of Bone. *Ann. N Y Acad. Sci.* 1963, *109*, 168–76.

Sammons, R.L.; Lumbikanoda, N.; Gross, M.; Cantzler, P. Comparison of Osteoblast Spreading on Microstructured Dental Implant Surfaces and Cell Behaviour in an Explant Model of Osteointegration. A Scanning Electron Microscopre Study. *Clin. Oral Impl. Res.* 2005, 2005, *16*(6), 657–66, doi:10.1111/j.1600-0501.2005.01168.x.

Sampath, T.K.; Reddi, A.H. Dissociative Extraction and Reconstitution of Extracellular Matrix Components Involved in Local Bone Differentiation. *Proc. Natl. Acad. Sci. U.S.A.* 1981, *78*(12), 7599–603.

Sampath, T.K.; Reddi, A.H. Homology of Bone-Inductive Proteins from Human, Monkey, Bovine, and Rat Extracellular Matrix. *Proc. Natl. Acad. Sci. U.S.A.* 1983, *80*(21), 6591–95.

Sampath, T.K.; Reddi, A.H. Importance of Geometry of the Extracellular Matrix in Endochondral Bone Differentiation. *J. Cell Biol.* 1984, *98*(6), 2192–97.

Sampath, T.K.; Rashka, K.E.; Doctor, J.S.; Tucker, R.F.; Hoffmann, F.M. Drosophila Transforming Growth Factor-β Superfamily of Proteins Induce Endochondral Bone Formation in Mammals. *Proc. Natl. Acad. Sci. U.S.A.* 1993, *90*(13), 6004–08.

Scarano, A.; Perrotti, V.; Artese, L.; Degidi, D.; Piattelli, A.; Iezzi, G.; Iezzi, G. Blood Vessels Are Concentrated within the Implant Surface Concavities: A Histologic Study in Rabbit Tibia. *Odontology* 2014, *102*(2), 259–66.

Sun, J.; Jamilpour, N.; Wang, F.-Y.; Wong, P.K. Geometric Control of Capillary Architecture via Cell-Matrix Mechanical Interactions. *Biomaterials* 2014, *35*(10), 3273–80.

Sung, B.H.; Weaver, A. Direct Migration: Cells Navigate by Extracellular Vesicles. *J. Cell Biol.* 2018, *217*(8), 2613–14.

Takemoto, M.; Fujjibayadshi, S.; Neo, M.; Suzuki, J.; Matsushita, T.; Kobuko, T.; Nakamura, T. Osteoinductive Porous Titanium Implants: Effects of Sodium Removal by Dilute HCl Treatment. *Biomaterials* 2006, *27*(13), 2682–91.

Tanaka, E.M.; Reddien, P.W. The Cellular Basis for Animal Regeneration. *Dev. Cell* 2011, *21*(1), 172–85.

Tomancak, P. Evolutionary History of Tissue Bending. *Science* 2019, *366*(6463), 300–01, doi:10.1126/science.aaz1289.

Trueta, J. The Role of the Vessels in Osteogenesis. *J. Bone Joint Surg.* 1963, *45B*, 402–18.

Turing, A.M. The Chemical Basis of Morphogenesis. *Philos. Trans. R. Soc. Lond.* 1952, *237*, 27–41.

Uchida, M.; Kim, H.-M.; Kobuko, T.; Fujibayashi, S.; Nakamura, T. Effect of Water Treatment on the Apatite-Forming Ability of NaOH-Treated Titanium Metal. *J. Biomed. Mater. Res. (Appl. Biomater.)* 2002, *63*(5), 522–30.

Uchida, M.; Kim, H.-M.; Kobuko, T.; Fujibayashi, S.; Nakamura, T. Structural Dependence of Apatite Formation on Titania Gels in a Simulated Body Fluid. *J. Biomed. Mater. Res.* 2003, *64A*, 164–70.

Urist, M.R. Bone: Formation by Autoinduction. *Science* 1965, *220*(3698), 893–99, doi:10.1126/science.150.3698.893.

Urist, M.R.; Silverman, F.; Büring, K.; Dubuc, F.L.; Rosenberg, J.M. The Bone Induction Principle. *Clin. Orthop. Rel. Res.* 1967, *53*, 243–83.

Urist, M.R.; Hou, Y.K.; Brownell, A.G.; Hohl, W.; Buyske, J.; Lietze, A.; Tempst, P.; Hunkapiller, M.; DeLange, R.J. Purification of Bovine Bone Morphogenetic Protein by Hydroxyapatite Chromatography. *Proc. Natl. Acad. Sci. U.S.A.* 1984, *81*(2), 371–5.

van Eeden, S.; Ripamonti, U. Bone Differentiation in Porous Hydroxyapatite Is Regulated by the Geometry of the Substratum: Implications for Reconstructive Craniofacial Surgery. *Plast. Reconstr. Surg.* 1994, *93*(5), 959–66.

Vlacic-Zischke, J.; Hamlet, S.M.; Friis, et al The Influence of Surface Microroughness and Hydrophilicity of Titanium on the Up-Regulation of TGFβ/BMP Signalling in Osteoblasts. *Biomaterials* 2011, *32*, 7403–10.

Vlodavsky, I.; Folkman, J.; Sullivan, R.; Fridman, R.; Ishai-Michaeli, R.; Sasse, J.; Klagsbrun, M. Endothelial Cell-Derived Basic Fibroblast Growth Factor: Synthesis and Deposition into Subendothelial Extracellular Matrix. *Proc. Natl. Acad. Sci. U.S.A.* 1987, *84*(8), 2292–96.

Watari, S.; Hayashi, K.; Wood, J.A.; Russell, P.; Nealey, P.F.; Murphy, C.J.; Genetos, D.C. Modulation of Osteogenic Differentiation in hMSCs Cells by Submicron Topographically-Patterned Ridges and Grooves. *Biomaterials* 2012, *33*(1), 128–36.

Wilkinson, A.; Hewitt, R.N.; McNamara, L.E.; McCloy, D.; Meek, D.R.M.; Dalby, M.J. Biomimetic Microtopography to Enhance Osteogenesis *In Vitro. Acta Biomater.* 2011, *7*(7), 2919–25.

Wolpert, L. Positional Information and the Spatial Pattern of Cellular Differentiation. *J. Theor. Biol.* 1996, *25*(1), 1–47.

Yamasaki, H Heterotopic Bone Formation around Porous Hydroxyapatite Ceramics in the Subcutis of Dogs. *Jpn. J. Oral Biol.* 1990, *32*(2), 190–2.

Yamasaki, H.; Sakai, H. Osteogenic Response to Porous Hydroxyapatite Ceramics under the Skin of Dogs. *Biomaterials* 1992, *13*(5), 308–12.

Yang, Z.; Yuan, H.; Tong, W.; Zou, P.; Chen, W.; Zang, X. Osteogenesis in Extraskeletally Implanted Porous Calcium Phosphate Ceramics: Variability among Different Kinds of Animals. *Biomaterials* 1996, *17*(22), 2131–7.

Yang, Y.; Wang, K.; Gu, X.; Leong, K.W. Biophysical Regulation of Cell Behaviour-Cross Talk between Substrate Stiffness and Nanotopography. *Engineering* 2017, *3*(1), 36–54.

Yoon, J.-L.; Kim, H.N.; Bhang, S.H.; Shin, J.-Y.; Han, J.; La, W.-G.; Jeong, G.-J.; Kang, S.; Lee, J.-R.; Oh, J.; Kim, M.S.; Jeon, N.L.; Kim, B.-S. Enhanced Bone Repair by Guided Osteoblast Recruitment Using Topographically Defined Implant. *Tissue Eng.* 2016, *22*(7), 654.

Yuan, H.; Kurashina, K.; de Bruijn, J.D.; De Groot, K.; De Bruijn, J.D.; De Groot, K. Osteoinduction by Calcium Phosphate Biomaterials. *J. Mater. Sci. Mater. Med.* 1998, *9*(12), 723–6.

Yuan, H.; Kurashina, K.; de Bruijn, J.D.; Li, Y.; de Groot, K.; Zhang, X. A Preliminary Study on Osteoinduction of Two Kinds of Calcium Phosphate Ceramics. *Biomaterials* 1999, *20*(19), 1799–806.

Yuan, H.; van Blitterswijk, C.A.; de Groot, K.; de Bruijn, J.D. A Comparison of Bone Formation in Biphasic Calcium Phospahte (BCP) and Hydroxyapatite (HA) Implanted in Muscle and Bone of Digs at Different Time Periods. *J. Biomed. Mater. Res.* 2006, *A78*, 139–47.

Yuan, H.; Fernandes, H.; Habibovic, P.; de Boer, P.; Barradas, A.M.; de Ruiter, A.; Walsh, W.R.; van Blitterswijk, C.A.; de Bruijn, J.D. Osteoinductive Ceramics as Synthetic Alternative to Autologous Bone Grafting. *Proc. Natl. Acad. Sci. U.S.A.* 2010, *107*(31), 13614–19.

Zhang, J.; Luo, X.; Barbieri, D.; Barradas, A.M.C.; de Bruijn, J.; van Blitterswijk, A.; Yuan, H. The Size of Surface Microstructure as an Osteogenic Factor in Calcium Phosphate Ceramics. *Acta Biomater.* 2014, *10*(7), 3254–63.

Zhang, J.; Dalbay, M.T.; Luo, X.; Vrij, E.; Barbieri, D.; Moroni, L.; de Bruijn, J.D.; van Blitterswijk, C.A.; Chapple, J.P.; Knight, M.M.; Yuan, H. Topography of Calcium Phosphate Ceramics Regulates Primary Cilia Length and TGF Receptor Recruitment Associated with Osteogenesis. *Acta Biomater.* 2017, *57*, 487–97.

Zhang, J.; Sun, L.; Luo, X.; Barbieri, D.; de Bruijn, J.D.; van Blitterswijk, C.A.; Moroni, L.; Yuan, Y. Cells Responding to Surface Structure of Calcium Phosphate Ceramics for Bone Regeneration. J. Tissue Eng. Reg. Med. 2017, *11*(11), 3273–3283. doi:10.1002/term.2236.

Zhang, Y.; Chen, S.E.; Shao, J.; van den Beucken, J.J.J.P. Combinatorial Surface Roughness Effects on Osteoclastogenesis and Osteogenesis. *ACS Appl. Mater. Interfaces* 2018, *10*(43) 36652–36663. doi:10, 36652-663.

Zheng, B.; Cao, B.; Crisan, M.; Sun, B.; Li, G.; Logar, A.; Yap, S.; Pollett, J.B.; Drowley, L.; Cassino, T.; Gharaibeh, B.; Deasy, B.; Huard, J.; Péault, B. Prospective Identification of Myogenic Endothelial Cells in Human Skeletal Muscle. *Nat. Biotechnol.* 2007, *25*(9), 10125–34.

6 Molecular Pathways Regulating the Geometric Induction of Bone Formation

Synthetizing and Embedding Osteogenic Proteins into Nanotopographic Geometries

Raquel Duarte and Ugo Ripamonti

6.1 INTRODUCTION

Calcium carbonate/hydroxyapatite macroporous bioreactors are replicas of the coral exoskeleton of the genus *Gonipora* (Ripamonti 1991; Shors, 1999). Bioreactors were prepared by hydrothermal exchange with phosphate, followed by conversion to hydroxyapatite, giving rise to calcium carbonate constructs with 7% hydroxyapatite (7% HA/CC, Biomet, Indiana, USA) (Ripamonti et al. 2009; Shors, 1999). Endowed with a permissive surface geometry, these coral-derived macroporous bioreactors intrinsically initiate the spontaneous induction of bone formation when implanted in intramuscular heterotopic sites of the Chacma baboon *Papio ursinus*, where there is no bone (Ripamonti 1991; Ripamonti et al. 2009).

A fundamental question guiding the research of the Bone Research Laboratory of the University of the Witwatersrand, Johannesburg, is: what are the molecular signals that morphogenize and propagate the bone induction-signalling cascade by macroporous calcium phosphate-based coral-derived bioreactors when implanted in heterotopic intramuscular *rectus abdominis* sites? To understand mechanistically what underpins the intrinsic and/or spontaneous induction of bone formation by such calcium phosphate-based coral-derived macroporous constructs, the Bone Research Laboratory initiated studies whereby a series of treated and untreated coral-derived 7% HA/CC macroporous devices were implanted in the *rectus abdominus* muscle of the Chacma baboon. The devices were then analysed morphologically for the induction of bone formation, and at the gene expression level using real-time reverse transcription PCR (RT-PCR) (Klar et al. 2013; Klar et al. 2014: Ripamonti et al. 2015).

In the first of a series of time studies to unravel the temporal and spatial nature of molecular-level changes, the 7% HA/CC macroporous constructs were implanted into the *rectus abdominis* muscle of adult Chacma baboons for a period of 15, 60 and 90 days. To obtain insights into the mechanism underlying the intrinsic and/or spontaneous induction of bone formation by the calcium phosphate-based coral-derived constructs, experimental groups of bioreactors were treated with biological agents, for the same period of time. Zoledronate (Zometa®, Novartis), a bisphosphonate, is an inhibitor of osteoclast activity and was loaded at a concentration of 240 µg to test the hypothesis that osteoclastic modifications of the surface of the 7% HA/CC macroporous devices were a critical event in the spontaneous induction events leading to bone formation (Ripamonti et al. 2010). The second agent was 500 µg verapamil hydrochloride (Isoptin®, Knoll Pharmaceuticals) which acts as a blocker of L-type voltage calcium channels, thereby inhibiting Ca^{++} release. The use of this inhibitor was a result of observations that calcium ion release by osteoclast activity during the bone resorption process was important for the regulation of cellular differentiation events and the induction of angiogenesis (Klar et al. 2013; Kohn et al. 1995; Komori 2020; Munaron 2006; Ripamonti et al. 2015; Zayzafoon 2006).

In parallel experiments, coral-derived hydroxyapatite macroporous constructs were treated with either 125 µg recombinant hTGF-β_3 (Novartis AG, Basel, Switzerland) or 125 ug of the BMP inhibitor, human Noggin (hNoggin, Regeneron Pharmaceuticals, USA), or a combination of both (Klar eta l. 2014; Ripamonti et al. 2015). Several members of the TGF-β superfamily have been shown to induce bone formation in the Chacma baboon (Ripamonti 2005; Ripamonti 2006; Ripamonti et al. 2008; Ripamonti et al. 2015. Thus, the inclusion of this isoform in the experiment was to observe and compare morphological and molecular events occurring in the TGF-β_3 induced bone induction with the untreated, and other treated, bioreactors, to uncover more details on the processes involved. Noggin inhibits several members of the BMP pathway including BMP-2, BMP-6 and BMP-7 (reviewed in Krause et al. 2011). Thus, blocking the BMP signalling cascade would reveal important information regarding the events that lead to morphological patterning of the 7% HA/CC coral-derived macroporous constructs.

To facilitate understanding of the molecular events underpinning the induction of bone formation, gene expression profiling was performed on RNA extracted from the processed macroporous constructs using real-time RT-PCR. Genes representative of the TGF-β superfamily, together with regulators and markers of osteoblastogenesis, were assayed. Morphological and molecular data were collected 15, 60 and 90 days after heterotopic implantation. Analyses deduced changes in the temporal pattern of gene expression accompanying morphological changes as evaluated on decalcified and undecalcified sections of the harvested and processed heterotopic specimens at various time points (Klar et al. 2013; Klar et al. 2014; Ripamonti et al. 2015). Important observations from these experiments are detailed herein and provide a narrative for the events that morphogenize tissue induction in the macroporous spaces of the 7% calcium phosphate-based bioreactors, leading to the activation of inductive signalling pathways ultimately setting into motion the spontaneous and/or intrinsic induction of bone formation.

6.2   TISSUE PATTERNING OF GEOMETRIC CONSTRUCTS: MORPHOLOGICAL AND MOLECULAR OBSERVATIONS

Iconic images captured from 7% HA/CC untreated and treated devices harvested at 15 days after transplantation in the *rectus abdominis* muscle of the Chacma baboon together with the analogous gene expression data are presented.

In Figure 6.1 specimens harvested on day 15 are shown. The untreated control specimens (Fig. 6.1a) reveal that macroporous spaces at the periphery of the devices had been invaded by a highly vascular connective tissue matrix. It is geometric cues provided by the intrinsically self-inducing biomimetic construct, with its specialised surface topography, that initiate angiogenesis. Angiogenesis is marked by an abundance of sprouting capillaries that invade the matrix, leading to highly organised collagenous condensations along the surfaces of concavities of the macroporous bioreactors.

In zoledronate-treated 7% HA/CC macroporous constructs (Fig. 6.1b) the most peripheral macroporous spaces of the devices were invaded by a less organized and very loose fibrovascular tissue with disorganized collagenous condensations. This indicated that the collagenous tissue had failed patterning and assembly. Verapamil-treated macroporous constructs (Fig. 6.1c) exhibited a lack of mesenchymal condensations along the surfaces of the implanted devices. There was an invasion of the most peripheral spaces of the macroporous constructs by fibrovascular tissue with capillary sprouting.

The corresponding gene expression profiles for *BMP-2* and *Noggin* in treated and untreated devices are represented together with representative histological sections for comparative purposes. In untreated control specimens (Fig. 6.1a) and Zometa® bisphosphonate treated–specimens (Fig. 6.1b), there is minimal *BMP-2* expression. In Zometa®-treated samples there is upregulation of the *Noggin* gene (Fig. 6.1b) at this time point. In the untreated control samples, which undergo the normal intrinsic induction of bone formation, there is never any upregulation of *Noggin* and this gene remains downregulated throughout the time course (Fig. 6.1a). In verapamil-treated samples there was an inverse relationship between *Noggin* and *BMP-2* expression with pronounced downregulation of *BMP-2* and upregulation of *Noggin* (Fig. 6.1c).

As observed morphologically on day 15 in untreated control devices, the pronounced angiogenesis was associated with a marked upregulation of the *type IV collagen* gene (Fig. 6.2). Interestingly, *type IV collagen* was similarly upregulated on day 15 in all the constructs, regardless of treatment modality, and thus appeared to be associated with the bursts of vascular invasion and capillary formation following the implantation of the bioreactors in the *rectus abdominis* muscle tissue. This pattern in *collagen type IV* expression was replicated in the hTGF-β_3- and hNoggin-treated samples (Klar et al. 2014).

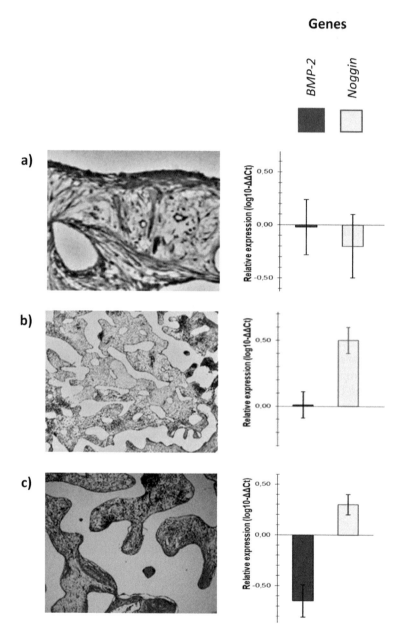

FIGURE 6.1 Morphological and gene expression evaluations on day 15 in (a) untreated, (b) Zometa®-treated and (c) verapamil-HCl-treated 7% HA/CC macroporous bioreactors after implantation in the *rectus abdominis* muscle of *Papio ursinus*. (a) Induction of angiogenesis with extensive capillary sprouting and mesenchymal fibrovascular tissue invasion of the macroporous space. There is pronounced cellular trafficking throughout the macroporous space accompanied by patterning of collagenous condensations tightly lining the macroporous

FIGURE 6.1 (CONTINUED)

surface. Profiling of *BMP-2* and *Noggin* genes reveals that the pronounced vascular invasion and morphogenesis of collagen patterning are associated with a downregulation of *Noggin*, without substantial change in *BMP-2* expression. (b) Zoledronate-treated 7% macroporous constructs show spaces invaded by a very loose and disorganised fibrovascular matrix. There are minimal if any collagenous condensations within the macroporous spaces of the hetero-topic implanted bioreactors. At the molecular level, this disorganized very loose fibrovas-cular invasion with limited if any tissue patterning is associated with upregulated *Noggin* gene with a lack of *BMP-2* expression. (c) Morphologically, verapamil-HCl Ca⁺⁺ channel blocker bioreactors show fibrovascular tissue invasion of the internal spaces of the macropo-rous constructs. There is a complete lack of collagenous condensations and alignment with pronounced *BMP-2* downregulation together with upregulation of the *Noggin* gene.

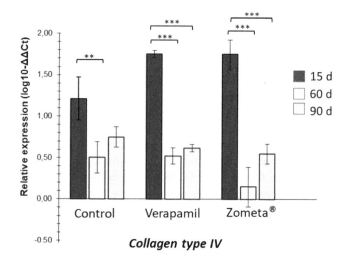

FIGURE 6.2 Relative change in the expression of *collagen type IV* on days 15 (15 d), 60 (60 d) and 90 (90 d) after the implantation of untreated control, Zometa®-treated and vera-pamil-HCl-treated 7% HA/CC macroporous bioreactors. On day 15 after heterotopic implan-tation in the *rectus abdominis* muscle of the Chacma baboon *Papio ursinus*, the expression of *collagen type IV* was upregulated in all bioreactors, regardless of the treatment modality (**p < 0.05; ***p > 0.05) (Klar et al. 2013).

For all treatment modalities by day 90, *collagen type IV* gene expression was still observed but was significantly reduced over the heightened upregulation shown on day 15. This is indicative of the ongoing angiogenesis and sustained capillary sprout-ing for continuously supporting the induction of bone formation. *Type Iv collagen* expression is the vital initial burst in angiogenesis that predates the bone differentia-tion cascade. Type IV collagen is known to be critical for the genesis of the basement membranes of the sprouting capillaries (Reddi 2000) and is thus a good marker for capillary formation. By examining the events at 15 days these experiments detailed the important early patterning events predating the initiation of differentiation path-ways controlling the induction of bone formation in angiogenesis in angiogenesis (Ripamonti 2006).

Morphological and gene expression data for day 60 control and treated specimens are shown in Figure 6.3. The control bioreactors exhibited auto-induced bone formation throughout many of the concavities of the macroporous constructs (Fig. 6.3a). An important feature described by Klar et al. (2013) in the control device is the presence of osteoblasts closely aligned to the basement membranes of the capillaries that had appeared within the macroporous spaces of the control untreated macroporous bioreactors. In Zometa®-treated 7% HA/CC macroporous devices (Fig. 6.3b) collagenous condensations were not properly organised and were facing a very loose vascular tissue matrix while verapamil-treated bioreactors showed remodelling of the collagenous condensations and minimal bone formation on the periphery of the constructs only (Fig. 6.3c). At the molecular level, the pattern of *Noggin* expression had changed such that in the Zometa®- and verapamil-treated samples there is a switch to downregulation of *Noggin* expression. This downregulation is most marked in verapamil-treated macroporous constructs that have started to exhibit some patterning and remodelling of collagenous condensations with the induction of bone at the periphery of the implanted bioreactors.

The morphological and corresponding gene expression data at day 90 for control, Zometa®- and verapamil-treated 7% HA/CC devices are shown in Figure 6.4. There was a significant auto-induced bone in the untreated macroporous constructs, characterised by blocks of remodelled lamellar bone observed throughout the macroporous spaces. The newly formed bone contained osteocytes and a rich supply of capillaries. At the gene level, control devices exhibited upregulation of *BMP-2* corresponding to newly formed bone. In Zometa®-treated samples (Fig. 6.4b), while collagenous condensations against the macroporous constructs appear more organised, there is still the distinct lack of patterning. Very few areas at the periphery of the device show bone formation while the verapamil-treated bioreactors showed fibrovascular tissue invading the macroporous spaces and collagenous condensations were scattered throughout the macroporous spaces.

For all macroporous constructs (treated and untreated), the expression of *BMP-2* is significantly upregulated, and this marks the stage at which the bone formation and remodelling are taking place and follows on from the early phases of angiogenesis and tissue patterning. While tissue patterning is delayed in the treatment group of bioreactors, once the effects of the inhibitors are diminished, there follows the normal differentiation phase that leads to mature osteoblasts secreting bone matrix. The dynamics of the interchanging *BMP-2* and *Noggin* expression in the treated vs. untreated control macroporous constructs can be observed in Figure 6.4.

Data for hTGF-β_3- and hNoggin-treated bioreactors on day 15 are presented in Figure 6.5. In the hTGF-β_3-treated bioreactors (Fig. 6.5a) there was fibrovascular tissue invasion evident across the concavities of the macroporous bioreactors. Abundant red blood cells were embedded in a rich network of fibrin/fibronectin stretching across the concavities of the macroporous spaces. Strikingly, the network of fibrin was organised into concentric rings extending across and invading the concavities of the macroporous bioreactors, and observations suggest that these rings were orchestrating the tissue induction process by organising the extracellular matrix and providing a tri-dimensional scaffold for the transport of various cellular

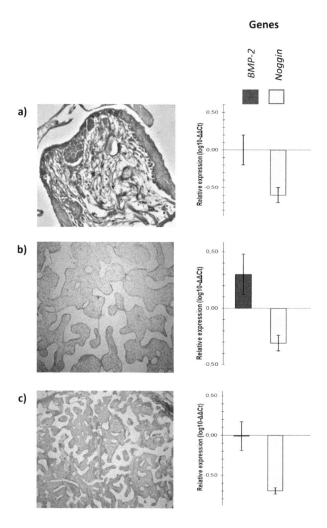

FIGURE 6.3   Morphological and gene expression evaluations in (a) untreated control, (b) Zometa®-treated and (c) verapamil-HCl-treated 7% HA/CC macroporous bioreactors on day 60 after implantation in the *rectus abdominis* muscle of *Papio ursinus*. (a) There is further tissue remodelling and morphogenesis with bone induction tightly formed against the calcium phosphate-based macroporous surfaces. Newly formed bone is surfaced by contiguous osteoblasts facing a highly vascularised mesenchymal tissue. (b) Zometa®-treated bioreactors remodelling and tissue patterning are delayed when compared to control specimens. (c) In the verapamil-HCl-treated bioreactors, there are areas of bone formation at the most peripheral regions of the macroporous specimens. In such areas, there is tissue patterning and morphogenesis of collagenous condensations. In all untreated control and treated samples, *Noggin* expression is downregulated on day 60. *BMP-2* expression in untreated bioreactors is unchanged from the expression level observed on day 15. In Zometa®-treated bioreactors, *BMP-2* expression is now upregulated, while in verapamil-HCl-treated specimens there is recovery in the expression of *BMP-2* from the former downregulation observed on day 15 (Fig. 6.1c).

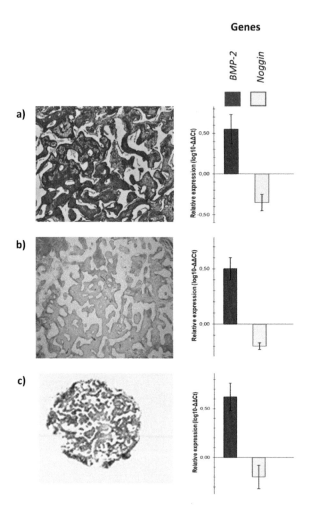

FIGURE 6.4 Morphological and gene expression evaluations on day 90 (a) untreated control, (b) Zometa®-treated and (c) verapamil-HCl-treated 7% HA/CC macroporous constructs after implantation in the *rectus abdominis* muscle of *Papio ursinus*. (a) The macroporous spaces with concavities across the specimens are characterized by bone formation by induction across the macroporous spaces. There are still areas showing ongoing patterning of collagenous condensations. Molecularly, the induction of bone formation as observed in (a) is characterized by *BMP-2* upregulation with *Noggin* downregulation on day 90. (b) Zometa®-treated constructs are undergoing delayed patterning events, marked by collagenous condensations present in loose and disorganised fibrovascular tissue against the surfaces of the concavities. There are areas showing vascular invasion. Molecularly of interest, there is *BMP-2* expression as a possible recovery after inactivity of the preloaded Zometa® as well as limited Zometa® diffusion across the most peripheral regions of the macroporous bioreactors (Klar et al. 2013; Ripamonti et al. 2014; Ripamonti et al. 2015). (c) In bioreactors treated with the Ca [++] channel blocker Verapamil hydrochliride there was a delayed patterning of collagenous condensations with limited bone formation at the periphery of the bioreactors only, with *BMP-2* upregulation and a concomitant downregulation of *Noggin* expression.

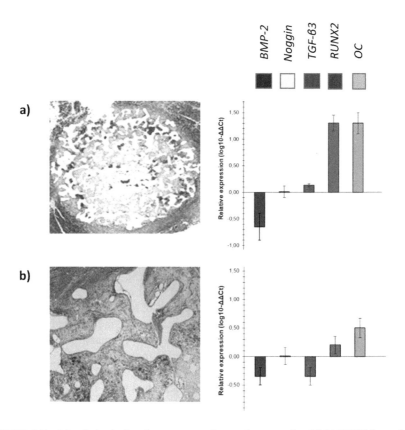

FIGURE 6.5 Morphological and gene expression analyses on day 15 (a) hTGF-β_3- and (b) hNoggin-treated 7% HA/CC macroporous constructs after implantation in the *rectus abdominis* muscle of the Chacma baboon. (a) There is complete fibrovascular tissue invasion across the concavities and macroporous spaces of bioreactor with assembly of a fibrin/fibronectin extracellular matrix network. (b) hNoggin-treated specimens show limited fibrovascular tissue invasion across the concavities of the macroporous constructs. hNoggin treatment results in a lack of tissue patterning with sparse and haphazardly arranged collagenous condensations effectively blocking the induction of bone formation (Ripamonti et al. 2015). At the molecular level, this lack of patterning is associated with a downregulation of both *BMP-2* and *TGF-β_3*. hTGF-β_3-treated bioreactors show marked upregulation of the *RUNX2* and (*OC*) genes compared to hNoggin-treated macroporous constructs.

elements. These rings provide anchorage for nucleated hyperchromatic cells, the osteoblasts, which were being differentiated from invading pericytic precursor cells by the exogenously supplied hTGF-β_3 (see Klar et al. 2014; Ripamonti et al. 2015). In stark contrast, morphological analyses of hNoggin-treated 7% HA/CC macroporous bioreactors showed very little fibrovascular invasion and, when observed, capillary invasion was only evident at the periphery of the devices (Fig. 6.5b).

At the molecular level, this lack of patterning is accompanied by a downregulation of both *BMP-2* and *TGF-β₃* in hNoggin-treated macroporous bioreactors. The hTGF-β₃-treated devices showed a marked upregulation of the transcriptional regulator *RUNX2*, which lies downstream of TGF-β signalling, and the marker of osteoblastogenesis, *Osteocalcin* (*OC*) that is indicative of the differentiation of osteoblasts from osteoblast progenitor cells. This also corresponds with the morphological analysis that shows the extracellular matrix network in the concavities anchoring the newly formed osteoblast cells. The intricacies of signalling events underlying bone formation are illustrated by the downregulation of *BMP-2* expression by hTGF-β₃ treatment. Molecularly in hTGF-β₃-treated 7% HA/CC bioreactors there was an unexpected and significant downregulation of *BMP-2* expression when compared to *BMP-2* expression in the control untreated device, and this downregulation was even more marked than the downregulation of *BMP-2* observed in the hNoggin-treated device (Klar et al. 2014; Ripamonti et al. 2015). *TGF-β₃* expression was also downregulated in the hNoggin-treated devices. This is indicative of the inhibitory function of Noggin which dampens the expression of members of the TGF-β superfamily Rosen 2006. The expression of the *Noggin* gene at this time point was the same in both treated bioreactors and resembled the expression level observed in untreated control bioreactors.

On day 60 substantial bone had formed in the hTGF-β₃-treated macroporous constructs (Fig. 6.6a). The newly formed lamellar bone exhibited extensive vascular invasion and capillary sprouting within the spaces of the devices. Most noteworthy was that the hTGF-β₃-treated constructs exhibited marked induction of bone formation at the periphery of the devices, and despite this prominent induction of bone at the periphery, there was limited bone formation in the central part of the macroporous constructs (Klar et al. 2014; Ripamonti et al. 2015). This phenomenon has raised some very important and interesting questions for the role that TGF-β plays in shaping the microenvironment in the tissue surrounding the macroporous devices and is the subject of much ongoing investigation.

In the hNoggin-treated 7% HA/CC macroporous bioreactors, fibrovascular bundles were rarely observed and only at the periphery (Fig. 6.6b). There was very limited tissue patterning. The delayed organisation and haphazardly patterned tissue condensations were opposed to the surface of the macroporous construct. Higher-power magnification showed that mesenchymal tissue had invaded the macroporous spaces with limited if any tissue patterning. At the molecular level, there is now a switch from the downregulation of *BMP-2* and *TGF-β₃* genes observed at day 15 to upregulation at day 60. This is indicative of activation of the bone induction-signalling cascade and corresponds to the bone induction observed in the hTGF-β₃ macroporous constructs. In the hNoggin-treated macroporous bioreactors the levels of *RUNX2* expression have also increased with the increased expression of *TGF-β₃* and *BMP-2*.

While expression of both *RUNX2* and *OC* had increased markedly at this time point in the hTGF-β₃-treated device corresponding to the induction of bone formation, the expression of *OC* in the hNoggin-treated device has not increased above the day-15 level. This is indicative of the lack of osteoblastogenesis observed at the

Genes

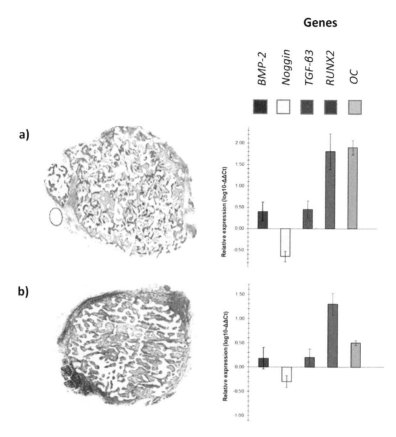

FIGURE 6.6 Morphological and gene expression analyses in (a) hTGF-β_3- and (b) hNog-gin-treated 7% HA/CC macroporous constructs on day 60 after implantation in the *rectus abdominis* muscle of *Papio ursinus*. (a) hTGF-β_3 macroporous bioreactor with bone forma-tion across the macroporous spaces with significant osteogenesis, and extending outside the periphery of the implanted coral-derived construct. There is *BMP-2* and *TGF-β_3* upregulation and particularly RUNX2 and OC expression. At the molecular level, there is now a switch from the downregulation of the *BMP-2* and *TGF-β_3* genes on day 15 to upregulation on day 60. The levels of *RUNX2* expression have also increased. The expression of *OC* had not however increased above the day 15 level. *Noggin* gene expression has been downregulated.

morphological level. *Noggin* gene expression has been also downregulated and, in both devices, there is the expression of the *BMP-2* gene.

The morphological and gene expression analyses of hTGF-β_3- and hNoggin-treated devices at day 90 are presented in Figure 6.7. In the TGF-β_3-treated bioreac-tors, newly formed bone had been remodelled to form solid blocks of lamellar bone and exhibited osteoclastic activity. Vascular invasion was still observed, serving to promote osteoclastogenesis and further remodelling events. The induction of newly formed bone is reflected at the molecular level by the upregulation of *BMP-2*, *RUNX2* and *OC*. The levels of expression of the latter two are significantly upregulated,

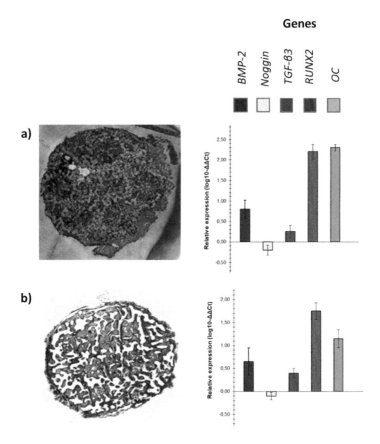

FIGURE 6.7 Morphological and gene expression analyses in (a) hTGF-β_3- and (b) hNog-gin-treated 7% HA/CC macroporous constructs on day 90 after implantation in the *rectus abdominis* muscle of *Papio ursinus*. (a) Solid blocks of newly formed bone have formed across the concavities of the hTGF-β_3-treated bioreactors. The induction of bone formation was associated with the upregulation of *BMP-2* and even more elevated levels of both *RUNX2* and *OC*. (b) There is still a lack of bone differentiation in hNoggin-treated bioreactors with a distinctive lack of collagenous condensations, tissue patterning and alignment. Molecularly, there is an increase in the expression of *BMP-2* and *TGF-β_3*, as well as the *RUNX2* and *OC* genes, possibly indicating the start of the osteogenic signalling in the hNoggin-impeded bio-reactors at the very periphery of the implanted specimens.

indicating the activation of the bone-inducing BMP pathway. In the hNoggin-treated 7% HA/CC macroporous constructs, there was no evidence of bone differentiation. Thus, the hNoggin-treated bioreactors exhibited an extremely delayed and limited patterning of collagenous condensations with the result that there was limited bone formation by day 90. This delay is also reflected at the molecular level that now sees an increase in the expression of the *BMP-2*, *RUNX2* and *OC* genes. However, the expression levels of both *RUNX2* and *OC* are still lagging behind that observed in the hTGF-β_3-treated macroporous bioreactors.

The gene expression profile for hTGF-β_3/hNoggin co-treated macroporous constructs is shown in Figure 6.8a. In hTGF-β_3/hNoggin combined treatment, *BMP-2* expression was downregulated on day 15 and steadily increased over the time course so that by day 90 *BMP-2* was significantly upregulated. On day 90, the levels of *BMP-2* expression were similar to those found in the untreated control and other treated macroporous constructs. That *BMP-2* expression levels on day 90 happen to be similar to all analysed bioreactors suggests the presence of a feedback loop that controls the expression of *BMPs'* genes during the induction of bone formation. On day 15, there was the upregulation of the *Noggin* gene. The *BMP-2* and *Noggin* gene profiles corresponded with the morphological characteristics of the co-treated device that showed very limited fibrovascular invasion of the most peripheral macroporous spaces only.

The fibrovascular bundles merged with haematopoietic materials supported in a fibrin matrix, which lacked the network structure that earmarked the hTGF-β_3-treated construct. On day 60 there was still very limited fibrovascular invasion, delayed tissue patterning and limited alignment of the collagenous condensations. Similar to the hNoggin-treated bioreactors, there was an upregulation of the *RUNX2* gene but the expression of *OC* lagged behind, suggestive of a lack of osteoblast differentiation at this time point. The co-treated devices on day 90 exhibited limited patterning of haphazardly constructed collagenous condensations (Figs. 6.8b,c). There was also little fibrovascular invasion of the central areas of the macroporous constructs. Most often, the collagenous condensations were a single bundle that intersected the macroporous space. These devices also showed very little bone formation which, if present, was only in the most peripheral spaces.

6.3   MECHANISTIC INSIGHTS INTO THE INDUCTION OF BONE FORMATION BY 7% HA/CC MACROPOROUS BIOREACTORS

Deciphering molecular events is key to understanding the underlying mechanisms that drive the regenerative potential of nanotopographically engineered matrices. The studies described here have unravelled some of the molecular complexities that underpin the intrinsic induction of bone formation by 7% HA/CC coral-derived macroporous bioreactors in heterotopic sites of the Chacma baboon *Papio ursinus* (Fig. 6.9).

The combined information from the morphological observations and the gene expression data provides key mechanistic understanding of the process of the spontaneous and/or intrinsic induction of bone formation by the calcium carbonate/hydroxyapatite macroporous constructs. The introduction of treatment modalities has provided a deeper understanding of the intricacies of the temporal nature of bone induction and highlighted early patterning events as being critical in the modelling of newly formed bone by the 7% HA/CC macroporous constructs.

Important observations from these studies were the perturbations in early tissue patterning due to the treatments with Zometa® and verapamil, which impeded osteoclastic activity and Ca^{++} release, respectively. The delay in patterning and consequently later cell differentiation events along the osteoblast lineage can be ascribed

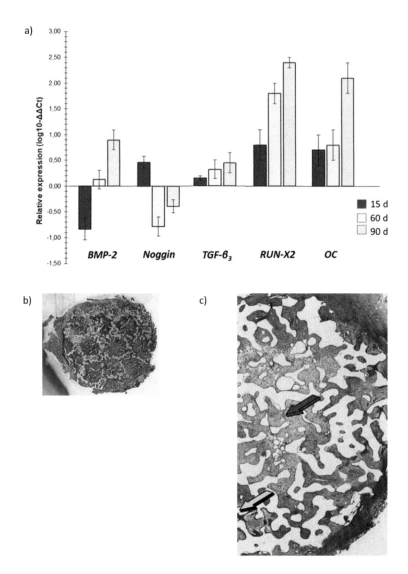

FIGURE 6.8 hTGF-β₃ and hNoggin co-treated macroporous construct analyses. (a) Relative change in expression of *BMP-2, Noggin, TGF-β₃, RUNX2* and *OC* in hTGF-β₃/hNoggin co-treated 7% HA/CC macroporous constructs on days 15, 60 and 90 after heterotopic implantation in the *rectus abdominis* muscle of *Papio ursinus*. (b) Low-power (×2.7) and (c) and high-power (×125) views of hTGF-β₃/hNoggin co-treated macroporous constructs harvested 90 days after implantation in the *rectus abdominis* muscle of *Papio ursinus*. There is limited induction of bone formation at the periphery of the bioreactor (*light blue* arrow c) (Klar et al. 2013). Towards the centre of the macroporous construct (*dark blue* arrow), there is little fibro-vascular invasion with lack of patterning and absence of collagenous condensations.

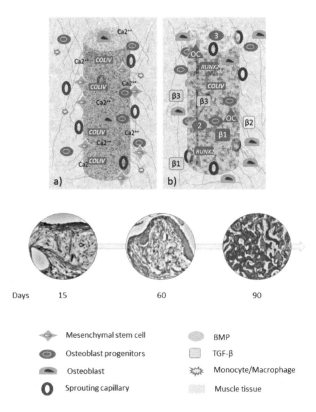

Days   15   60   90

Mesenchymal stem cell   BMP
Osteoblast progenitors   TGF-β
Osteoblast   Monocyte/Macrophage
Sprouting capillary   Muscle tissue

FIGURE 6.9 Proposed mechanisms initiating the intrinsic and/or spontaneous induction of bone formation by 7% HA/CC macroporous bioreactors. (a) Implantation of the 7% HA/CC macroporous constructs results in the recruitment of monocyte/macrophage cells to the site of injury which provides a pool of osteoclast progenitor cells that differentiate into osteoclasts to resorb the calcium phosphate-based surfaces of the macroporous spaces. Following release of Ca^{++} ions from the macroporous surfaces, there is the induction of the signalling pathways to differentiate osteoblasts from mesenchymal stem cells recruited from the surrounding micro-environment to sites of bone induction along the self-inductive nano-patterned surfaces of the primed bioreactors after osteoclastogenesis. This early period is marked by heightened angiogenesis so that by day 15 there is pronounced capillary sprouting as evidenced by upregulated levels of *collagen type IV* gene expression. This early period is coupled with angiogenesis marked by extensive capillary sprouting. There is induction and differentiation against the concavities of the macroporous spaces of mesenchymal collagenous condensations tightly patterned against the biomimetic surfaces. (b) Following on from the critical early events crucial for the proper patterning of collagenous fibres and the induction of osteoblast differentiation pathways, active BMP signalling results in the activation of several members of the TGF-β superfamily. Full commitment to BMP signalling sees further remodelling and morphogenesis throughout the concavities of the macroporous constructs and areas of remodelled bone lining the surfaces of the concavities. The areas of remodelled bone face a highly vascularised invading tissue matrix. By day 90 the concavities are filled with remodelled lamellar bone. Gene expression patterning sees the expression of downstream targets of BMP signalling, such as *RUNX2* and *OC*, that signifies the active bone induction cascade.

to the absence of crucial patterning of the nanotopographical surface structure of the macroporous constructs. This leads to the abrogation of cellular signalling required to induce the differentiation pathways critical for the formation of osteoblasts and bone induction.

In normal skeletal development, collagenous condensation events are critical for skeletal development (Hall and Miyake 2000). The Zometa®-treated samples exhibit a profound delay in the patterning of the collagen fibres, and at the molecular level, this was reflected with the upregulation of the *Noggin* gene. Noggin prevents binding of BMP to its receptor (Zimmerman et al. 1996), thereby inhibiting the bone induction cascade leading to the observed haphazardly organised tissue condensations. Zometa®, a bisphosphonate, acts as an inhibitor of osteoclast proliferation (Coxon et al. 2004) and induces osteoclast apoptosis (Benford et al. 2001). The Zometa®-treated macroporous constructs, due to the inhibition of osteoclastic activity, lack the requisite maturational gradient of collagenous condensations required for normal bone formation.

The role of active free Ca^{++} in physiological processes as well as in normal bone remodelling has been well documented (Clapham 2007; Chai et al. 2012). During bone resorption, Ca^{++} is released from osteoclasts that affect the proliferation and differentiation of osteoblasts. Gradients of extracellular Ca^{++} in cellular microenvironments act as a chemical signal for cell migration and growth and thereby orchestrate the activities of various cells (Orwar et al. 2010; Breitwieser 2008). In bone, high concentrations of Ca^{++} stimulate the migration of osteoblasts to bone resorption sites and facilitate their maturation (Dvorak and Riccardi 2004). In addition, *in vitro* experiments have shown the effect of Ca^{++} on the migration of bone marrow progenitor cells, monocytes, osteoblasts and haematopoietic stem cells (Aguirre et al. 2010; Olszak 2000; Godwin et al. 1997; Adams et al. 2006). Extracellular Ca^{++} has been shown to regulate the migration of osteoblasts through the activation of calcium-sensing receptors (Zayafoon 2006). The influx of Ca^{++} increases the levels of intracellular Ca^{++}, polarising the cell membrane of the osteoblast, with this polarization determining the directional migration pattern of the osteoblast (Ozkucur et al. 2009).

In addition to its migratory influence on osteoblasts, Ca^{++} controls transcriptional programmes *via* calcium/calmodulin signalling, leading to the proliferation and differentiation of osteoblasts (Zayafoon 2006). Intracellular Ca^{++} levels are increased in osteoblasts through calcium-sensing receptors and voltage-gated Ca^{++} channels (Yamauchi et al. 2005; Barradas et al. 2012). Another important observation relevant to findings of the research described in this text is that Ca^{++} increases the expression of *BMP-2* and *BMP-4* genes (Nakade 2001).

Blocking the L-type voltage calcium channel using the inhibitor verapamil HCl resulted in reduced bone formation. The highly disorganised and delayed tissue patterning was not as extreme as that observed in the Zometa®-treated samples. At the molecular level, there was, however, downregulation of *BMP-2* and an analogous upregulation of the BMP inhibitor *Noggin*, suggesting that Ca^{++} release is important in activating signalling pathways controlling the spontaneous induction of bone formation by the 7% HA/CC macroporous constructs. The downregulation of the *BMP-2* in both the verapamil- and Zometa®-treated bioreactors indicated the critical role of BMPs in organising mesenchymal collagenous condensations and tissue patterning.

Thus, the proposed mechanism for the spontaneous induction of bone formation by the 7% HA/CC macroporous constructs involves the critical role of osteoclasts which act upon the surface topography of the biomimetic matrices. Osteoclastogenesis, together with the release of Ca^{++}, triggers angiogenic and cellular differentiation signalling pathways such that newly differentiated osteoblasts can attach to the surface of the macroporous construct to express, secrete and embed BMPs onto the calcium phosphate-based biomatrix. Importantly, these mechanistic insights also show that the spontaneous and/or intrinsic formation of bone is *via* the BMPs' pathway (Klar et al. 2013; Ripamonti et al. 2015).

Cumulative information gathered by experimentation in the Chacma baboon over the years, together with the data associating morphological and molecular data described here, has shown that hTGF-β_3 induces bone *via* the BMP pathway. Further evidence in support of this hypothesis has been revealed in additional gene expression studies showing that TGF-β_3 influences expression of other BMP genes, viz. *BMP-3, BMP-4, BMP-6* and *BMP-7* genes (Ripamonti et al. 2015). This finding in the primate is extended to other members of the family of osteogenic proteins which are also capable of inducing heterotopic bone through the BMP pathway. The induction of heterotopic bone by the other TGF-β isoforms TGF-β_1 and TGF-β_2 has been accompanied by the expression of *BMP-3* and *BMP-7* (Duneas et al. 1998; Ripamonti and Roden 2010).

The data presented in Klar et al. (2014) show that 125 µg hTGF-β_3 is capable of inducing heterotopic ossicles, with the rapid induction of bone taking place outside the border of the macroporous construct. The induction of bone formation by hTGF-β_3 took place with modulation of *BMP-2, TGF-β_3* and *Noggin* expression. At the morphological level, newly formed bone showed prominent osteoid formation and activity of osteoblasts. Indeed upregulated *BMP-2* expression always correlated with bone morphogenesis and was associated with a concomitant downregulation of the *Noggin* gene. The treatment of macroporous constructs with hNoggin or the combination of hNoggin and hTGF-β_3 significantly inhibited the induction of bone formation when compared to the bone that formed in the control macroporous devices.

The treatment of 7% HA/CC macroporous constructs with hNoggin severely inhibited morphogenesis by impeding the early patterning events and collagenous condensation. The hNoggin treatment indicated the importance of these critical events to the induction of bone by the macroporous bioreactors. The use of hNoggin as a BMP inhibitor also points to the fact the intrinsic ability of the 7% HA/CC macroporous constructs to induce bone takes place *via* the BMP pathway, and gives an indication of the important sequence of events during the bone morphogenesis cascade. The proper activation of the BMP pathway is critical for the formation and alignment of collagenous condensations. These requisite tissue-patterning events take place before the differentiation of osteoblasts which later express *BMPs* to drive the deposition of bone.

The mammalian TGF-β isoforms are unique morphogens, and, unlike other growth factors, are secreted in a latent form and deposited into the extracellular matrix (ECM) (Pfeilschifter et al. 1990). Transforming growth factors deliver a bio-logical action once activated, usually in response to any perturbations in the ECM (Annes et al. 2003). The activation of latent TGF-β underlies the spatial and tem-poral regulation of TGF-β during physiological processes (Morikawa et al. 2016). TGF-β acts as a "molecular sensor", responding to perturbations in the environment by releasing an active ligand to trigger downstream signalling effects (Annes et al. 2003). To date, many TGF-β activators have been identified, including metallopro-teinases, lower pH and reactive oxygen species (Yu and Stamenkovic 2000; Jenkins 2008; Adams and Lawler 2004; Munger et al. 1999; Lyons et al. 1990), and once activated, the TGF-β ligands initiate signalling events in responding cells to elicit cellular effects.

Members of the TGF-β superfamily, which includes the BMPs, are responsible for proliferation and differentiation of bone progenitors and the TGF-β isoforms are master regulators of mesenchymal cell fate and are involved in controlling early osteoblast differentiation (reviewed in Grafe et al. 2018). To transduce their signal, TGF-β ligands require two classes of serine/threonine kinase receptors, TGF-β receptor type I (TβRI), or ALK-5, and TGF-β receptor type II (TβRII) (Shi and Massagué 2003). The classical intracellular signalling pathway down-stream of the receptors involves the Smad pathway (Feng and Derynck 2005). The Smads are composed of a group of receptor-regulated Smads (R-Smads), Smad1, 2, 3, 5 and 8, the inhibitory Smads (Smad6 and 7), and the common mediator Smad (co-Smad), Smad 4. Intracellular signalling initiates when TβRI phosphorylates R-Smads and the activated R-Smads heterodimerize with Smad-4, followed by translocation of the entire complex to the nucleus to drive transcrip-tional responses (Ross and Hill 2008).

During osteoblast differentiation, the TGF-β isoforms signal *via* the canonical Smad2/3 pathway and non-classical TAK1-MMK-p38 pathways, whilst the BMPs signal through type I and II BMP and ALK2 receptor. Signal transduction is *via* Smads1/5/8 (Derynck and Zhang 2003). Signalling *via* the TGF-β and BMP pathways converges to activate the expression of the master transcriptional regulator RUNX2 (Grafe et al. 2018). However, there is considerable complexity in the pathways gov-erning bone formation, and the BMP and TGF-β pathways exert their effects by cross-talk with several other pathways (reviewed in Guo and Wang 2009). The BMP pathway cross-reacts with components of the Wnt, Notch and fibroblast growth fac-tor (FGF) pathways, whilst elements of the TGF-B pathway cross-react with the FGF, Wnt and the pituitary hormone (PTH) pathways. A further level of complex-ity is added by the activities of antagonists, such as Noggin, Chordin, Gremlin and Cerberus which control the activities of the growth factors (see Wu et al. 2016). The inclusion of the BMP inhibitor in these crucial experiments detailed in this chapter hints at the complexities of the signalling pathways controlling the induction of bone formation by the 7% HA/CC macroporous constructs in the Chacma baboon.

A hint of the additional complexity in the signalling pathways governing the spontaneous induction of bone formation by 7% HA/CC macroporous bioreactors comes to the fore by observing the gene expression in the devices following the hTGF-β_3 treatment. There is downregulation of *BMP-2* expression at 15 days in the hTGF-β_3-treated bioreactors but by days 60 and 90 *BMP-2* expression levels have recovered to similar levels to those observed in the control untreated bioreactors.

The endpoint of BMP and TGF-β signalling is the regulation of the *RUNX2* gene and the osteoblastic specific differentiation marker OC. The expression of these genes increases steadily and dramatically throughout the bone formation. The importance of activation of these genes in the intrinsic induction of bone by 7% HA/CC macroporous constructs is shown in the hNoggin-treated bioreactors. There is a pronounced downregulation of *RUNX2* in the hNoggin-treated bioreactors that is not observed in the devices treated with hTGF-β_3. This is also evident in hNoggin/hTGF-β_3 co-treated sample where the effect of hTGF-β_3 treatment is negated by the hNoggin treatment.

The role of the microenvironmental niche in the bone induction cascade by the macroporous constructs in *Papio ursinus* is illustrated in the hTGF-β_3-treated bioreactors. The recruitment of stem cells results in the rapid bone induction cascade when the 7% coral-derived macroporous construct is treated with hTGF-β_3 and embedded in the *rectus abdominis* muscle of *Papio ursinus*. Morphologically, bone induction occurs rapidly after 15 days and is characterised by pronounced vascular invasion and capillary sprouting. This distinctive morphology is accompanied at the gene level by upregulation of *collagen type IV*, signifying angiogenesis, and the important *RUNX2* and *OC* genes. The latter two genes drive the osteogenic induction programme by controlling the differentiation of stem cells along the osteoblastic lineage (Grafe et al. 2018).

The corresponding morphological analyses of day-15 bioreactors showed an engineered microenvironment in the surrounding muscle tissue that was rich in progenitor cells and differentiating osteoblasts. The extent of structural organisation that accompanies hTGF-β_3 administration is indicative of the role of the powerful morphogen in reprogramming myoblastic and perivascular cells recruited from the microenvironment into osteoblasts that then secrete the bone matrix onto the newly deposited matrix comprising concentric rings of fibrin/fibronectin. The extracellular matrix rings provide a 3D network for the docking of osteoblast precursor cells that will be re-programmed into differentiated osteoblasts to lay down the newly formed bone in the heterotopic site of the Chacma baboon *Papio ursinus*.

6.4   CONCLUSION

Whilst some mechanistic insights have been garnered from the studies described herein, there remains much to be resolved regarding the induction of bone formation by coral-derived calcium carbonate hydroxyapatite macroporous constructs in the Chacma baboon *Papio ursinus*. Through the timed implantation of a series of treated and untreated 7% HA/CC bioreactors, accompanied by a morphological and molecular assessment, important events underlying the spontaneous induction of bone by these constructs have been uncovered. Early events that pattern the surface of the nanotographically engineered constructs are critical for creating a permissive environment for the differentiation of osteoblasts from progenitor cells that are recruited to the site of implantation.

The newly formed collagenous matrix forms a network onto which the osteoblasts become embedded to secrete newly formed bone matrix. The early patterning events involve osteoclastic activity and Ca^{++} release on the surface of concavities of the macroporous bioreactors. Without these events, there is a delay in the differentiation of BMPs and perturbation of the proper organisational structure of the newly formed bone. Molecular profiling has unequivocally shown that bone formation takes place *via* the BMPs' pathway. This was shown by the effect of the inhibitor Noggin that abrogated the signalling of the TGF-β superfamily of genes. Moreover, how hTGF-β_3 is capable of inducing substantial bone when added to the 7% HA/CC macroporous devices requires further investigation. Studies now need to focus on the global changes that take place at both the gene and protein levels. Only by understanding how the microenvironment and molecular pathways function, can mechanistic aspects underlying the intrinsic induction of bone formation be manipulated for optimal results in pre-clinical and clinical contexts.

REFERENCES

Adams, G.B.; Chabner, K.T.; Alley, I.R.; Olson, D.P.; Szczepiorkowski, Z.M.; Poznansky, M.C.; Kos, C.H.; Pollak, M.R.; Brown, E.M.; Scadden, D.T. Stem Cell Engraftment at the Endosteal Niche Is Specified by the Calcium-Sensing Receptor. *Nature* 2006, 439(7076), 599–603.

Adams, J.C.; Lawler, J. The Thrombospondins. *Int J Biochem Cell Biol* 2004, 36(6), 961–8.

Aguirre, A.; Gonzalez, A.; Planell, J.A.; Engel, E. Extracellular Calcium Modulates In Vitro Bone Marrow-Derived Flk-1+ CD34+ Progenitor Cell Chemotaxis and Differentiation Through a Calcium-Sensing Receptor. *Biochem Biophys Res Commun* 2010, 393(1), 156–61.

Annes, J.P.; Munger, J.S.; Rifkin, D.B. Making Sense of Latent TGFbeta Activation. *J Cell Sci* 2003, 116(2), 217–24.

Barradas, A.M.; Fernandes, H.A.; Groen, N.; Chai, Y.C.; Schrooten, J.; van de Peppel, J.; van Leeuwen, J.P.; van Blitterswijk, C.A.; de Boer, J.A. A Calcium-Induced Signaling Cascade Leading to Osteogenic Differentiation of Human Bone Marrow-Derived Mesenchymal Stromal Cells. *Biomaterials* 2012, 33(11), 3205–15.

Benford, H.L.; McGowan, N.W.; Helfrich, M.H.; Nuttall, M.E.; Rogers, M.J. Visualization of Bisphosphonate-Induced Caspase-3 Activity in Apoptotic Osteoclasts In Vitro. *Bone* 2001, 28(5), 465–73.

Breitwieser, G.E. Extracellular Calcium as an Integrator of Tissue Function. *Int J Biochem Cell Biol* 2008, 40(8), 467–80.

Chai, Y.C.; Carlier, A.; Bolander, J.; Roberts, S.J.; Geris, L.; Schrooten, J.; Van Oosterwyck, H.; Luyten, F.P. Current Views on Calcium Phosphate Osteogenicity and the Translation into Effective Bone Regeneration Strategies. *Acta Biomater* 2012, 8(11), 3876–87.

Clapham, D.E. Calcium Signaling. *Cell* 2007, 131(6), 1047–58.

Coxon, J.P.; Oades, G.M.; Kirby, R.S.; Colston, K.W. Zoledronic Acid Induces Apoptosis and Inhibits Adhesion to Mineralized Matrix in Prostate Cancer Cells via Inhibition of Protein Prenylation. *BJU Int* 2004, 94(1), 164–70.

Derynck, R.; Zhang, Y.E. Smad-Dependent and Smad-Independent Pathways in TGF-Beta Family Signaling. *Nature* 2003, 425(6958), 577–84.

Duneas, N.; Crooks, J.; Ripamonti, U. Transforming Growth Factor-Beta 1: Induction of Bone Morphogenetic Protein Genes Expression During Endochondral Bone Formation in the Baboon, and Synergistic Interaction with Osteogenic protein-1 (BMP-7). *Growth Factors* 1998, 15(4), 259–77.

Dvorak, M.M.; Riccardi, D. Ca2+ as an Extracellular Signal in Bone. *Cell Calcium* 2004, 35(3), 249–55.

Feng, X.H.; Derynck, R. Specificity and Versatility in TGF-Beta Signaling through Smads. *Annu Rev Cell Dev Biol* 2005, 21, 659–93.

Godwin, S.L.; Soltoff, S.P. Extracellular Calcium and Platelet-Derived Growth Factor Promote Receptor-Mediated Chemotaxis in Osteoblasts through Different Signaling Pathways. *J Biol Chem* 1997, 272(17), 11307–12.

Grafe, I.; Alexander, S.; Peterson, J.R.; Snider, T.N.; Levi, B.; Lee, B.; Mishina, Y. TGF-β Family Signaling in Mesenchymal Differentiation. *Cold Spring Harb Perspect Biol* 2018, 10(5), a022202.

Guo, X.; Wang, X.F. Signaling Cross-Talk Between TGF-Beta/BMP and Other Pathways. *Cell Res* 2009, 19(1), 71–88.

Hall, B.K.; Miyake, T. All for One and One for All: Condensations and the Initiation of Skeletal Development. *BioEssays* 2000, 22(2), 138–47.

Jenkins, G. The Role of Proteases in Transforming Factor-Beta Activation. *Int J Biochem Cell Biol* 2008, 40(6–7), 1068–78.

Klar, R.M.; Duarte, R.; Dix-Peek, T.; Dickens, C.; Ferretti, C.; Ripamonti, U. Calcium Ions and Osteoclastogenesis Initiate the Induction of Bone Formation by Coral-Derived Macroporous Constructs. *J Cell Mol Med* 2013, 17(11), 1444–57.

Klar, R.M.; Duarte, R.; Dix-Peek, T.; Ripamonti, U. The Induction of Bone Formation by the Recombinant Human Transforming Growth Factor-β3. *Biomaterials* 2014, 35(9), 2773–88.

Kohn, E.C.; Alessandro, R.; Spoonster, J.; Wersto, R.P.; Liotta, L.A. Angiogenesis: Role of Calcium-Mediated Signal Transduction. *Proc Natl Acad Sci USA* 1995, 92(5), 1307–11.

Komori, T. Molecular Mechanism of Runx2-Dependent Bone Development. *Mol Cells* 2020, 43(2), 168–75.

Krause, C.; Guzman, A.; Knaus, P. Noggin. *Int J Biochem Cell Biol* 2011, 43(4), 478–81.

Lyons, R.M.; Gentry, L.E.; Purchio, A.F.; Moses, H.L. Mechanism of Activation of Latent Recombinant Transforming Growth Factor Beta 1 by Plasmin. *J Cell Biol* 1990, 110(4), 1361–67.

Morikawa, M.; Derynck, R.; Miyazono, K. TGF-β and the TGF-β Family: Context-Dependent Roles in Cell and Tissue Physiology. *Cold Spring Harb Perspect Biol* 2016, 8(5), a021873.

Munaron, L. Intracellular Calcium, Endothelial Cells and Angiogenesis. *Recent Pat Anticanc Drug Discov* 2006, 1(1), 105–19.

Munger, J.S.; Huang, X.; Kawakatsu, H.; Griffiths, M.J.; Dalton, S.L.; Wu, J.; Pittet, J.F.; Kaminski, N.; Garat, C.; Matthay, M.A.; Rifkin, D.B.; Sheppard, D. The Integrin Alpha v Beta 6 Binds and Activates Latent TGF Beta 1: A Mechanism for Regulating Pulmonary Inflammation and Fibrosis. *Cell* 1999, 96(3), 319–28.

Nakade, O.; Takahashi, K.; Takuma, T.; Aoki, T.; Kaku, T. Effect of Extracellular Calcium on the Gene Expression of Bone Morphogenetic Protein-2 and −4 of Normal Human Bone Cells. *J Bone Miner Metab* 2001, 19(1), 13–19.

Olszak, I.T.; Poznansky, M.C.; Evans, R.H.; Olson, D.; Kos, C.; Pollak, M.R.; Brown, E.M.; Scadden, D.T. Extracellular Calcium Elicits a Chemokinetic Response from Monocytes In Vitro and *In Vivo*. *J Clin Invest* 2000, 105(9), 1299–305.

Orwar, O.; Lobovkina, T.; Gozen, I.; Erkan, Y.; Olofsson, J.; Weber, S.G. Protrusive Growth and Periodic Contractile Motion in Surface-Adhered Vesicles Induced by Ca(2+)-Gradients. *Soft Matter* 2010, 6, 268–72.

Ozkucur, N.; Monsees, T.K.; Perike, S.; Do, H.Q.; Funk, R.H. Local Calcium Elevation and Cell Elongation Initiate Guided Motility in Electrically Stimulated Osteoblast-Like Cells. *PLOS ONE* 2009, 4(7), e6131.

Pfeilschifter, J.; Bonewald, L.; Mundy, G.R. Characterization of the Latent Transforming Growth Factor Beta Complex in Bone. *J Bone Miner Res* 1990, 5(1), 49–58.

Reddi, A.H. Morphogenesis and Tissue Engineering of Bone and Cartilage: Inductive Signals, Stem Cells, and Biomimetic Biomaterials. *Biomater. Tissue Eng* 2000, 6(4), 351–9.

Ripamonti, U. The Morphogenesis of Bone in Replicas of Porous Hydroxyapatite Obtained from Conversion of Calcium Carbonate Exoskeletons of Coral. *J Bone Joint Surg Am.* 1991, 35(5), 692–703.

Ripamonti, U. Bone Induction by Recombinant Human Osteogenic Protein-1 (hOP-1, BMP-7) in the Primate *Papio ursinus* with Expression of mRNA of Gene Products of the TGF-β Superfamily. *J Cell Mol Med* 2005, 9(4), 911–28.

Ripamonti, U. Soluble Osteogenic Molecular Signals and the Induction of Bone Formation. *Biomaterials* 2006, 27(6), 807–22.

Ripamonti, U.; Crooks, J.; Khoali, L.; Roden, L. The Induction of Bone Formation by Coral-Derived Calcium Carbonate/Hydroxyapatite Constructs. *Biomaterials* 2009, 30(7), 1428–39.

Ripamonti, U.; Roden, L. Induction of Bone Formation by Transforming Growth Factor-β_2 in the Non-Human Primate *Papio ursinus* and Its Modulation by Skeletal Muscle Responding Stem Cells. *Cell Prolif* 2010, 43(3), 207–18.

Ripamonti, U.; Klar, R.M.; Renton, L.F.; Ferretti, C. Synergistic Induction of Bone Formation by hOP-1 hTGF-β, and Inhibition by Zoledronate in Macroporous Coral-Derived Hydroxyapatites. *Biomaterials* 2010, 31(25), 6400–10.

Ripamonti, U.; Dix-Peek, T.; Parak, R.; Milner, B.; Duarte, R. Profiling Bone Morphogenetic Proteins and Transforming Growth Factor-Bs by hTGF-β3 Pre-Treated Coral-Derived Macroporous Bioreactors: The Power of One. *Biomaterials* 2015, 49, 90–102.

Rosen, V. BMP and BMP Inhibitors in Bone. *Ann N Y Acad Sci* 2006, 1068, 19–25.

Ross, S.; Hill, C.S. How the Smads Regulate Transcription. *Int J Biochem Cell Biol* 2008, 40(3), 383–408.

Shi, Y.; Massagué, J. Mechanisms of TGF-Beta Signaling from Cell Membrane to the Nucleus. *Cell* 2003, 113(6), 685–700.

Shors, E.C. Coralline Bone Graft Substitutes. *Orthop Clin North Am* 1999, 30(4), 599–613.

Wu, M.; Chen, G.; Li, Y. TGF-β and BMP Signaling in Osteoblast, Skeletal Development, and Bone Formation, Homeostasis and Disease. *Bone Res* 2016, 4, 16009.

Yamauchi, M.; Yamaguchi, T.; Kaji, H.; Sugimoto, T.; Chihara, K. Involvement of Calcium-Ssensing Receptor in Osteoblastic Differentiation of Mouse MC3T3-E1 Cells. *Am J Physiol Endocrinol Metab* 2005, 288(3), E608–16.

Yu, Q.; Stamenkovic, I. Cell Surface-Localized Matrix Metalloproteinase-9 Proteolytically Activates TGF-Beta and Promotes Tumor Invasion and Angiogenesis. *Genes Dev* 2000, 14(2), 63–176.

Zayzafoon, M. Calcium/Calmodulin Signalling Controls Osteoblast Growth and Differentiation. *J Cell Biochem* 2006, 97(1), 56–70.

Zimmerman, L.B.; De Jesús-Escobar, J.M.; Harland, R.M. The Spemann Organizer Signal Noggin Binds and Inactivates Bone Morphogenetic protein 4. *Cell* 1996, 86(4), 599–606.

7 The Spontaneous Induction of Bone Formation by Intrinsically Osteoinductive Bioreactors for Human Patients
Osteoinductive Hydroxyapatite-Coated Titanium Implants

Ugo Ripamonti

7.1 GEOMETRY AND THE RATIONALE FOR FABRICATING SELF-INDUCING OSTEOINDUCTIVE HYDROXYAPATITE-COATED TITANIUM IMPLANTS

The remarkable discovery of the spontaneous and/or intrinsic osteoinductive activity (Ripamonti 1996) of coral-derived bioreactors (Ripamonti 1990; Ripamonti 1991) has stressed the *connubium* of biomaterial sciences with molecular and cellular biology techniques. This has mechanistically resolved the spontaneous induction of bone formation by macroporous calcium phosphate-based bioreactors regulating gene expression and the induction of bone formation (Fig. 7.1) (Ripamonti et al. 1993; Ripamonti 2006; Ripamonti et al. 2009; Klar et al. 2013; Ripamonti 2017).

Previous experimentation has shown differentiating osteogenic cells directly on the hydroxyapatite substratum (Ripamonti et al. 1993) (Fig. 7.1a). Cell differentiation and proliferation revealed that the substratum acts as a solid-state matrix for cell attachment, orientation, spreading and osteoblast cell differentiation (Fig. 7.1a). Osteoblast differentiation is followed by the expression, synthesis and controlled release of endogenously produced bone morphogenetic proteins (BMPs) (Ripamonti

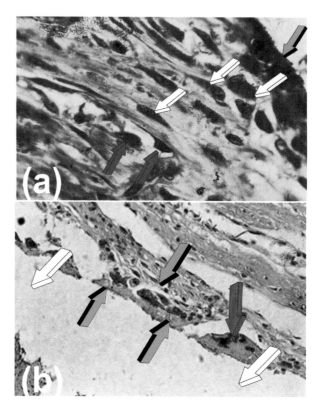

FIGURE 7.1 Induction of osteoblast-like cells directly onto a coral-derived bioreactor implanted in the *rectus abdominis* of a chacma baboon *Papio ursinus* and harvested on day 60 after intramuscular implantation. (a) Transformation and induction of osteoblastic cells (*light blue* arrow) directly at the calcium phosphate surface. *White* arrows point to several perivascular pericytic cells from the vascular compartment migrating to the osteogenic compartment against the coral-derived hydroxyapatite bioreactor. Note the hyperchromatic large nuclei of the migrating cells, particularly when in contact with the morphogenetic gradient of the osteogenetic compartment. Note also the hyperchromatic large nuclei of endothelial cells in the sprouting capillaries (*magenta* arrows) close to the osteogenetic and morphogenetic microenvironments of the coral-derived hydroxyapatite surface. Fig. 7.1a is at the crux of the spontaneous and/or intrinsic induction of bone formation by calcium phosphate-based bioreactors (Ripamonti et al. 1993). (b) Induction of the osteogenic phenotype in responding mesenchymal perivascular, pericytic and/or myoblastic cells invading the concavity of the calcium phosphate bioreactor. *Light blue* arrows point to the expression of osteogenic protein-1 (OP-1, also known as BMP-7) within resident differentiated mesenchymal cells at the surface of a concavity within a solid disc of crystalline sintered hydroxyapatite (Ripamonti et al. 1999; Ripamonti 2004). Osteoclasts (*magenta* arrow) are also seen resting in the concavity of the crystalline substratum (*white* arrows). (a) Undecalcified section prepared from tissue blocks harvested on day 60 after heterotopic intramuscular implantation. Specimens were fixed in 70% ethanol and embedded, undecalcified, in Historesin (LKB Bromma, Sweden).

1990; Ripamonti 1991; Ripamonti et al. 1993; Ripamonti et al. 2009; Klar et al. 2013; Ripamonti 2017; Ripamonti 2018). Expressed (Fig. 7.1b) and secreted proteins embedded onto the hydroxyapatite substratum initiate the hydroxyapatite-induced osteogenesis model (Ripamonti et al. 1993; Ripamonti 1996). The induction of bone formation, defined as intrinsic osteoinductive activity (Ripamonti 1996), has been shown in systematic studies in the non-human primate species *Papio ursinus* (Ripamonti 1991; Ripamonti et al. 1993; Ripamonti 1996; Ripamonti 2006; Ripamonti et al. 2009; Ripamonti et al. 2001a; Ripamonti et al. 2010; Klar et al. 2013; Ripamonti 2017; Ripamonti 2018).

Whilst gene expression studies (Klar et al. 2013; Ripamonti 2017) have molecularly resolved the spontaneous and/or intrinsic induction of bone formation (Ripamonti et al. 1993; Ripamonti 1996), the paradigmatic shift of the geometric induction of bone formation spontaneously initiating "Bone: Formation by induction" (Urist 1965) has not yet been routinely incorporated in the clinical surgical *armamentarium*.

Do the geometric theme and the biological/surgical concept of the "geometric induction of bone formation" (Ripamonti et al. 1999) bear the molecular and surgical keys for the induction of bone formation in clinical contexts? Our systematic studies in the Chacma baboon *Papio ursinus* have shown that the induction of bone formation by coral-derived or sintered macroporous crystalline bioreactors is often substantial. This is shown by the remarkable images of undecalcified and decalcified sections as seen in Chapters 4 and 5 (Ripamonti 1991; Ripamonti et al. 1993; Ripamonti et al. 1999; Klar et al. 2013; Ripamonti 2004; Ripamonti et al. 2004; Ripamonti 2006; Ripamonti 2009; Ripamonti et al. 2009; Ripamonti et al. 2010; Ripamonti 2017; Ripamonti 2018).

In non-human primates there are limited studies that compare the hydroxyapatite-induced osteogenesis model (Ripamonti et al. 1993; Ripamonti 1996) to autogenous bone grafts, naturally derived or recombinant human bone morphogenetic proteins reconstituted with different calcium phosphate-based macroporous biomatrices for the induction of bone formation in clinical contexts (Ripamonti 1992a; Ripamonti 1992b; Ripamonti et al. 1992; Ripamonti et al. 2001b). Almost all studies reporting the spontaneous and/or intrinsic induction of bone formation (Ripamonti 1996) conclude that results obtained in whatever animal model would be translatable in clinical contexts. Nothing is so far away from the biological truth. Regenerative potential in *Homo sapiens* is altogether different from other animals, including non-human primate genera and species (Lismaa et al. 2018). Several studies have highlighted the diversity of factors that constrain regeneration (Lismaa et al. 2018). Factors include immunological responses, inflammation and cellularity of inflammatory responses, angiogenic and neurogenic capacity, stem cell pools and niches and extracellular matrix composition (Lenhoff and Lenhoff 1991; Goss 1969; Martin and Parkhurst 2004; Reichman 1984; Kopp et al. 2016; Tanaka and Reddien 2011; Hynes 2009; Hopkinson-Woolley et al. 1994; Lismaa et al. 2018).

Regeneration of tissues has been always a controversial topic. This was often because regeneration attempts with morphological signs of regeneration were boasted, with claims of substantial regeneration, particularly in clinical contexts

(Ripamonti et al. 2014). In context, claims that neurons could be generated in post-natal mammalian brains were often controversial, though a recent report in *Science* once again raises the concept that mammalian regeneration, including neurogenesis, is regulated by several behavioural factors, and that running powerfully "induces neurogenesis, promoting the proliferation of neuronal progenitor cells" (Gage 2019).

The paper of Lismaa et al. (2018) reports comparative regenerative mechanisms across animal phylogeny and taxa, "showing wide variations in regenerative competence" (Lismaa et al. 2018). The paper of De Robertis, "Evo-Devo: Variations on ancestral themes", shows "that conserved gene networks already present in the archetypal ancestor were modified to generate the wonderful diversity of animal life on Earth today", and that the Hox genes are conserved between *Drosophila* and vertebrates (De Robertis 2008). There are thus phylogenetically ancient pathways between animal phyla with high degrees of conservation, including genes responsible for tissue regeneration (De Robertis 2008; Lismaa et al. 2018). Results obtained in non-human primate species may only indicate that similar repair mechanisms may operate in *Homo sapiens*, though translation in clinical contexts is altogether rather speculative (Ferretti et al. 2010; Ripamonti et al. 2014).

Limited studies compared osteoinductive macroporous ceramics or bioreactors to autogenous bone grafts (Ripamonti 1992a; Ripamonti 1992b) with or without the addition of naturally derived (Ripamonti et al. 1992) or recombinant hBMPs (Ripamonti et al. 2001a; Ripamonti et al. 2001b). A study reporting osteoinductive ceramics as a synthetic alternative to autogenous bone grafting appeared in the *Proceedings of the National Academy of the Sciences* USA in 2010 (Yuan et al. 2010). The study reported the performance of osteoinductive calcium phosphate ceramics with and without hBMP-2, *vs.* autologous bone grafting in critical-size defects in the ilium of ovine models (Yuan et al. 2010).

Remarkably, osteoinductive synthetic ceramics were found to be equally effective in inducing bone when compared to autogenous bone grafts (Yuan et al. 2010). Of interest, calcium phosphate ceramic *solo* induced greater amounts of bone when compared to identical ceramics pre-combined with doses of hBMP-2 (Yuan et al. 2010).

Our extensive experimentation in non-human primate species covers a variety of naturally derived and sintered crystalline calcium phosphate-based macroporous bioreactors (Ripamonti 1990; Ripamonti 1991; Ripamonti et al. 1993; Ripamonti et al. 1999; Ripamonti et al. 2001b; Ripamonti 2004; Ripamonti 2006; Ripamonti et al. 2006; Ripamonti et al. 2007a; Ripamonti et al. 2007b; Ripamonti 2009; Ripamonti 2017; Ripamonti 2018).

Our systematic studies showed substantial induction of bone formation when different calcium phosphate-based bioreactors were tested in heterotopic intramuscular sites of *Papio ursinus*. We, however, realized that the intrinsic and/or spontaneous hydroxyapatite osteogenesis model (Ripamonti et al. 1993; Ripamonti 1996) is not the *panacea* for focal skeletal defects including the axial skeleton, where loading and torsional strengths limit severely the use of calcium phosphate-based biomaterials as therapeutic alternatives to autogenous bone grafts. Indeed, such limitations apply to calcium phosphate-based macroporous bioreactors even when preloaded with osteogenic soluble molecular signals of the transforming growth factor-β (TGF-β) supergene family (Ripamonti 2003).

7.2   SOLID TITANIUM BIOREACTORS WITH THE CONCAVITY *MOTIF*

In the late 1980s, our research experimentation shifted from macroporous coral-derived and sintered crystalline hydroxyapatite-based constructs to solid titanium bioreactors with specific geometric configurations coated by crystalline sintered hydroxyapatite (Ripamonti 1994; Ripamonti and Kirkbride 1995; Ripamonti and Kirkbride 1997; Ripamonti and Kirkbride 2001; Ripamonti et al. 1999; Ripamonti et al. 2012a; Ripamonti 2012; Ripamonti et al. 2012b; Ripamonti et al. 2013).

The research proposed a translational shift from macroporous bioreactors endowed with spontaneous osteoinductivity to solid implants incorporating the geometric *motif* of the concavity along the planar surfaces (Ripamonti 2012; Ripamonti et al. 2012a; Ripamonti et al. 2012b; Ripamonti et al. 2013; Ripamonti 2018). The long-term research plan was then to construct hydroxyapatite-coated titanium bioreactors *per se* endowed with the striking capacity to initiate the induction of bone formation in clinical contexts (Ripamonti et al. 2012a; Ripamonti et al. 2012b).

The decision to shift from macroporous calcium phosphate bioreactors to solid titanium-based bioreactors followed the examination and the re-evaluation of a number of undecalcified and decalcified sections prepared from coral-derived as well as sintered macroporous bioreactors (Ripamonti et al. 2012a; Ripamonti 2017; Ripamonti 2018). Such bioreactors are endowed with the striking prerogative of initiating the induction of bone formation when implanted in heterotopic intramuscular sites of *Papio ursinus* without the exogenous application of the osteogenic soluble molecular signals of the TGF-β supergene family (Ripamonti 1990; Ripamonti 1991; Ripamonti et al. 1993; Ripamonti 2003).

Research experiments implanting different biomimetic matrices in the *rectus abdominis* muscle of *Papio ursinus* using coral-derived and sintered crystalline hydroxyapatites revealed the critical role of the concavity in initiating the induction of bone formation. Publications describing the specific geometric configuration initiating the induction of bone formation were, however, delayed (Ripamonti et al. 1999) pending the granting and publication of US, WO, PCT and EP patents (Ripamonti 1994; Ripamonti and Kirkbride 1995; Ripamonti and Kirkbride 1997; Ripamonti and Kirkbride 2001). Several histological sections from both coral-derived and crystalline macroporous hydroxyapatites were reviewed, tissue blocks were further cut and sections stained to assess the role of the concavity in the induction of bone formation (Figs. 7.2, 7.3, 7.4) (Ripamonti et al. 1999; Ripamonti 2004; Ripamonti 2017; Ripamonti 2018).

A set of experiments that confirmed the critical role of the concavity in initiating the induction of bone formation involved the heterotopic *rectus abdominis* implantation of coral-derived constructs designed with different geometric configurations (van Eeden and Ripamonti 1994). Configurations were calcium phosphate-based macroporous cylinders 20 mm in length and 8 mm in diameter vs. particulate

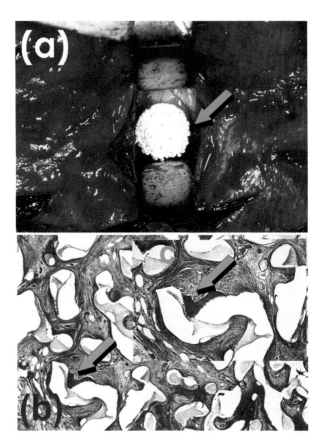

FIGURE 7.2   Composite digital images showing the role of geometry and of the geometric configuration of the inductor to initiate, control and block the induction of bone formation. (a) Granular particulate coral-derived calcium phosphate-based bioreactors were pelletized to form a large pellet (*light blue* arrow) for implantation in the exposed *rectus abdominis* muscle of *Papio ursinus* (van Eeden and Ripamonti, 1994). (b) Low-power view showing lack of bone differentiation by granular particulate constructs on day 90 after heterotopic implantation. In one concavity of the particulate substratum, there is the induction of bone formation (*light blue* arrow in b). Inset c details the spontaneous induction of bone formation within the concavity of the intramuscularly implanted substratum (*light blue* arrow). Standard macro porous cylinders, 20 mm in height, 8 mm in diameter, initiated, on the other hand, the substantial induction of bone morphogenesis within the macro porous spaces (van Eeden and Ripamonti 1994) as reported in other experiments (Ripamonti 1991; Ripamonti et al. 1993; Ripamonti 1996).

coral-derived calcium phosphate-based granular hydroxyapatites, 400 to 620 µm in diameter (van Eeden and Ripamonti 1994).

The standard multi-tested macroporous cylinders, 20 mm in length and 8 mm in diameter, as reported in previous studies (Ripamonti 1990; Ripamonti 1991; Ripamonti et al. 1993), showed the reproducible induction of bone formation across the macroporous spaces on days 60 and 90 after heterotopic intramuscular

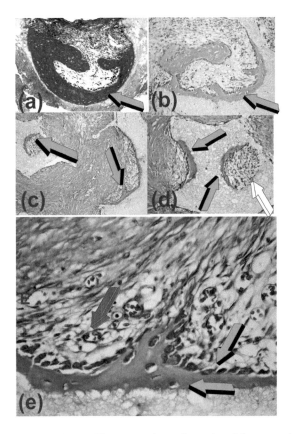

FIGURE 7.3 The effect of a specific geometric configuration, "the concavity: the shape of life" (Ripamonti 2004; Ripamonti 2006; Ripamonti 2012; Ripamonti et al. 2012a; Ripamonti et al. 2012b; Ripamonti et al. 2013), on cellular and morphogenetic events initiating the spontaneous induction of bone formation on days 30 and 90 after heterotopic intramuscular implantation in *the rectus abdominis* of adult chacma baboons *Papio ursinus*, where there is no bone (Ripamonti et al. 1999; Ripamonti 2000; Ripamonti 2004). (a) Classical induction of bone formation within a concavity of the substratum. Multiple osteoblastic cells surfacing the newly formed bone (*light blue* arrow) by day 90 after intramuscular implantation. (b,c,d) Morphogenetic concavities of crystalline hydroxyapatite discs harvested 30 days after intramuscular implantation in *Papio ursinus*. Note how the geometric signal of the concavity initiates the induction of bone formation. The induction of bone initiates without the exogenous application of the osteogenetic soluble molecular signals of the transforming growth factor-β (TGF-β) supergene family. Bone (*light blue* arrows) forms within the concavities of the substratum with palisades of osteoblastic cells. Osteocytes are trapped within the newly secreted matrix. *White* arrow in d points to the very beginning of the induction of bone formation against the concavity of the heterotopically implanted biomatrix. (e) Detail of the spontaneous induction of bone formation within a concavity of the crystalline substratum. Note the tight attachment of the newly formed bone onto the hydroxyapatite surface (*dark blue* arrow) with contiguous osteoblasts secreting bone matrix (*light blue* arrow) and osteocytes within the newly deposited matrix in a morphological and molecular *connubium* with invading sprouting capillaries within the concavity of the crystalline substratum (*magenta* arrows).

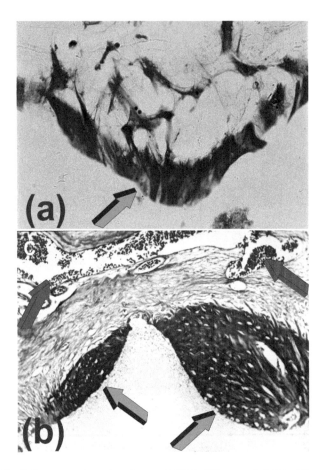

FIGURE 7.4 "The concavity: the shape of life" (Ripamonti 2004; Ripamonti 2006; Ripamonti 2012; Ripamonti et al. 2012a; Ripamonti et al. 2012b; Ripamonti et al. 2013) set into motion the design and engineering of solid titanium-based bioreactors that *per se*, without the osteogenetic soluble molecular signals of the TGF-β supergene family, spontaneously initiate the induction of bone formation, where there is no bone, like in the striated *rectus abdominis* muscle of the non-human primate *Papio ursinus*. (a) *In vitro* experiments using MC 3T3-E1 and MC 3T3-E1 showed cellular palisading and polarized alignment (*light blue* arrow) along concavities of coral-derived constructs. (b) Concavities of crystalline sintered bioreactors spontaneously initiate the induction of bone formation. Figs. 2b, 4a and b were instrumental for the conceptualization of the concavity as a primary signal and microenvironment for the spontaneous and/or intrinsic induction of bone formation (Ripamonti 1991; Ripamonti et al. 1993; Ripamonti 1996; Ripamonti et al. 1999). At the same time, Fig. 4b was critical for the pre-clinical and clinical translation of the concavity *motif*, as the digital image does mimic a cross section of the threads of a titanium implant constructed for oral rehabilitation (Ripamonti al. 2012a; Ripamonti et al. 2012b; Ripamonti et al. 2013).

implantation (van Eeden and Ripamonti 1994). To avoid premature diffusion and possible migration of particulate materials, granular particulate hydroxyapatites were pelletized by adding 1 mg of chondroitin-6-sulphate (Sigma Chemical Co., St. Louis, Mo) and 2 mg of baboon type I collagen to 400 mg of granular hydroxyapatite per implant in individual sterile polypropylene tubes. Granular substrata were washed in two changes of 70% ethanol, followed by a last wash in absolute ethanol and lyophilized for *rectus abdominis* intramuscular implantation (Fig. 7.2a) (van Eeden and Ripamonti 1994).

Generated tissue specimens were harvested 60 and 90 days after intramuscular *rectus abdominis* implantation. Histological analyses on pelletized granular particulate calcium phosphate-based constructs showed the lack of bone differentiation by induction. One specimen of particulate coral-derived hydroxyapatite showed the induction of bone formation within a concavity of a granular construct (Fig. 7.2b *blue* arrow). Inset c (Fig. 7.2c *blue* arrow) details the bone that formed within the concavity of the substratum. This image was instrumental for reviewing several decalcified and undecalcified histological sections to study the effect of a specific geometric configuration, the concavity, controlling the induction of bone formation after heterotopic implantation of calcium phosphate-based bioreactors.

Analyses of several decalcified and undecalcified sections prepared from specimen blocks of both coral-derived and sintered crystalline hydroxyapatite constructs showed beyond doubt that the concavity is a geometric cue that is inducive and conducive to a series of molecular, cellular and morphological cascades, leading to the induction of bone formation (Fig. 7.3) (Ripamonti et al. 1993; Ripamonti et al. 1999). "The geometric induction of bone formation" (Ripamonti et al. 1999) initiates without the exogenous application of the osteogenic soluble molecular signals of the TGF-β supergene family (Ripamonti et al. 1999; Ripamonti 2003). The digital images represented in Figure 7.3 show the inductive and conductive microenvironments of the concavity initiating the induction of bone formation only within geometrically configured calcium phosphate-based bioreactors (Fig. 7.3).

Further morphological and histological images were instrumental for the preparation of solid bioreactors coated by crystalline sintered hydroxyapatites as geometrically prepared titanium bioreactors. Figure 7.4 was instrumental for the preparation of titanium constructs with a series of repetitive concavities prepared along the substratum. Coral-derived macroporous constructs were also used to investigate the effect of the geometric configuration and surface topography *in vitro* (Fig. 7.4a) (Ripamonti 2012; Ripamonti et al. 2012a; Ripamonti et al. 2012b). Mouse-derived fibroblasts (NIH3T3) and pre-osteoblasts (MC3T3-E1) were seeded onto coral-derived macroporous constructs. *Ex-vivo* bioreactors pre-seeded with NIH-3T3 and MC3T3-E1 were harvested, fixed in Bouin's fluid and cut at 2–3 μm and stained by toluidine blue in 30% ethanol (Fig. 7.4a) (Ripamonti et al. 2012a; Ripamonti et al. 2012b). Bioreactors composed of NIH3T3 and MC3T3-E1 seeded onto coral-derived constructs showed that cells grown *in vitro* nested and aligned along concavities of the substratum for further differentiation into the osteogenic phenotype (Fig. 7.4a)

(Ripamonti et al. 2012a; Ripamonti et al. 2012b). Digital images of histological sections prepared from sintered crystalline macroporous bioreactors suggested that the concavity *motif* (Fig. 7.4b) could be re-assembled into geometrically functionalized titanium bioreactors.

These series of studies introduced the concept of the "concavity: the shape of life" (Fig. 7.4b) (Ripamonti 2004; Ripamonti 2006; Ripamonti 2012; Ripamonti et al. 2012a; Ripamonti et al. 2012b). The concavity is a geometric configuration which is inducive and conducive to initiate the induction of bone formation by firstly concentrating Ca^{++} released by osteoclastogenesis which, together with the induction of angiogenesis, de-differentiates local responding somatic cells into osteoblastic-like cells (Ripamonti et al. 1993; Ripamonti 2004; Klar et al. 2013; Ripamonti 2017; Ripamonti 2018). Differentiated osteoblasts express and secrete osteogenic proteins (Fig. 7.1b) later embedded onto the hydroxyapatite substratum (Ripamonti et al. 1999; Ripamonti 2004), initiating the induction of bone formation as a secondary response (Ripamonti et al. 1999; Ripamonti 2004; Klar et al. 2013; Ripamonti 2017; Ripamonti 2018).

We previously stated that perhaps the most fascinating and novel way to initiate the induction of bone formation is to construct biomimetic bioactive matrices that *per se*, in their own right, initiate the morphogenesis of bone (Ripamonti et al. 1993; Ripamonti et al. 1999; Ripamonti 2004; Ripamonti 2006; Ripamonti et al. 2066; Ripamonti et al. 2007a; Ripamonti et al. 2007b; Ripamonti 2009; Ripamonti et al. 2014; Ripamonti 2017; Ripamonti 2018). Nature has had the capacity after several million years of evolution to generate and mastermind highly sophisticated tissue and organs, including the evolution of the skeleton. This fundamental step in vertebrate evolution has provided the emergence of the vertebrates:

> deambulation and body erection, freeing the upper limbs for Homo like activities including the use of tools for hunting, foraging above all, however, for maternal care contributing thus to the speciation of the genus Homo ultimately directing the emergence of *Homo sapiens* and of the Homo clade. (Ripamonti 2009)

Nature has masterminded the evolution of the skeleton, deploying common yet limited mechanisms to direct the emergence not only of the skeleton but of several specialized tissue and organs. The TGF-β and BMP families reflect Nature's parsimony in controlling multiple specialized functions or pleiotropy, deploying several osteogenic molecular signals with minor variation in amino acid *motifs* within highly conserved carboxy terminal domains (Ripamonti 2003; Ripamonti et al. 2004; Ripamonti 2006). The evolutionary conservation of the TGF-β supergene family is superbly demonstrated by the remarkable observation that recombinant decapentaplegic and 60A proteins, gene products of the fruit fly *Drosophila melanogaster*, induce bone formation in mammals (Sampath et al. 1993).

> Nature has thus usurped phylogenetically ancient amino acid motifs and sequences deployed for dorso-ventral patterning in *Drosophila melanogaster* to set the unique vertebrate trait of the induction of bone formation rather than evolving new gene products for the induction of bone and the emergence of the skeleton, and thus of the vertebrates. (Ripamonti 2003; Ripamonti 2004; Ripamonti 2006; Ripamonti 2009)

With the induction of skeletogenesis and body erection, locomotion was *par force* the next evolutionary step characterizing the emergence of the vertebrates and vertebrates' speciation. Bipedalism and deambulation were hallmark steps in hominid's evolution, controlling the emergence of *Homo habilis* and later of *Homo erectus* walking out of Africa, colonizing the planet's landscape. Studies showed conserved gene families, including the remarkably conserved Hox transcription factor-dependent programme in ancient primitive marine vertebrates, the skate *Leucoraja erinacea*, at about 420,000 years before the present (Jung et al. 2018). Such skates share highly conserved neuronal circuitry that is essential for land deambulation (Jung et al. 2018). The communication by Jung et al. (2018) has indicated that circuitry that is essential for walking "evolved through adaptation of a genetic regulatory network shared by all vertebrates with paired appendages" capable thus of locomotion (Jung et al. 2018).

The genesis of bone has further invocted "bone in the solid state" (Reddi 1997), a solid mineralized matter with the supramolecular assembly of structural proteins, collagens and vascular structures permeating the osteonic bone in contact with bone marrow. The precision self-assembly of the bone unit or osteosome (Reddi 1997), with the remodelling processes of the osteonic bone, has been a superior example of Nature's creativity, design and architecture (Reddi 1997). The remodelling of the skeleton, the formation of bone by osteoblasts and the resorption of bone by osteoclasts are a closely integrated homeostatic system (Reddi 1997) and, at the same time, provide the geometric pattern of the remodelling osteonic cycle (Fig. 7.5).

Prominently, activation is followed by osteoclastic activity at any given region of the trabeculae of bone; osteoclasts resorb mineralized bone and form resorption lacunae across the trabeculae; lacunae, pits, excavations along the trabeculae are simply concavities of different diameters and depths and variable radii of curvatures along the trabeculae of bone (*blue* arrows Figs. 7.5a,b). Osteoclastogenesis then ceases, and within the concavities, recruited osteoblastic cells are now laying newly formed osteoid, yet to be mineralized bone matrix (*red* arrows Figs. 7.5b,c).

The multiple concavities formed by osteoclastogenesis across the trabeculae are thus regulators of the induction of bone formation, initiating the formative phase of the remodelling osteonic cycle with newly formed bone deposited within the concavities cut by osteoclastogenesis (Fig. 7.5). The concavity, thus, as carved in calcium phosphate-based biomaterials, spontaneously initiates the induction of bone formation (Fig. 7.5d).

To determine the critical role of the substratum geometry in bone differentiation, tissue induction and morphogenesis, different slurry preparations of hydroxyapatite powder were sintered to form solid monolithic hydroxyapatite discs, 20 mm in diameter, 4 mm thick, with a series of concavities prepared on both planar surfaces (Fig. 7.6a) (Ripamonti et al. 1999).

As reported in Chapter 5, solid discs of sintered crystalline hydroxyapatites (Fig. 7.6a) were implanted in the *rectus abdominis* muscle of *Papio ursinus* and harvested on days 30 (Fig. 7.6b) and 90 (Fig. 7.6e) (Ripamonti et al. 1999). The presented series of digital images shows unequivocally that the spontaneous and/or intrinsic induction of bone formation initiates solely within the concavities of the substratum

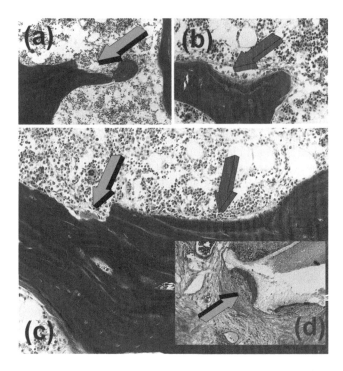

FIGURE 7.5 Why has the "Concavity: the shape of life" (Ripamonti 2004; Ripamonti 2006; Ripamonti 2012; Ripamonti et al. 2012a; Ripamonti et al. 2012b; Ripamonti et al. 2013) been programmed or re-programmed to initiate the induction of bone formation? The concavity biomimetizes the remodelling cycle of the cortico-cancellous bone. The driving force of the spontaneous and/or intrinsic induction of bone formation as initiated by bioactive calcium phosphate-based biomimetic matrices is the shape of the implanted scaffolds. The language of shape is the language of geometry (Ripamonti et al. 1999; Ripamonti 2009; Ripamonti 2017). The language of geometry is the language of a sequence of repetitive concavities that biomimetize the remodelling cycle of the cortico-cancellous bone (Ripamonti 2009; Ripamonti 2010; Ripamonti 2017; Ripamonti 2018). (a) Lacunae, pits and concavities as cut by osteoclastogenesis (*light blue* arrow) initiate the remodelling of the cortico-cancellous bone, and provide resorption lacunae in the form of concavities that initiate the formative phase (b) with osteoid deposition within concavities cut by osteoclastogenesis (*magenta* arrows b,c). Note in c osteoclastogenesis (*light blue* arrow) with formation of a concavity within the trabecular-mineralized bone. A few hundreds of microns away along the surface of the trabecula, there is bone formation with osteoid deposition within a previously cut concavity by osteoclastogenesis (*magenta* arrows). Biomimetic matrices biomimetize the process of cortico-cancellous bone remodelling whereby the calcium phosphate-based matrices sustain osteoclastogenesis, Ca^{++} release, angiogenesis, somatic cell de-differentiation and the induction of bone formation as a secondary response (inset d). The concavity *motif*, either cut by osteoclastogenesis during the remodelling cycle of the cortico-cancellous bone, or prepared as a repetitive microenvironment into calcium phosphate-based macroporous bioreactors, biomimetizes the ancestral repetitive multi-million-years tested designs and topographies of Nature. Differentiating perivascular, myoblastic, myogenic endothelial and pericytic cells sense the substratum upon which cells attach and migrate. Migrating and attaching cells onto the substratum are able to convert geometrical and mechanical cues into triggering gene expression pathways initiating the induction of bone formation.

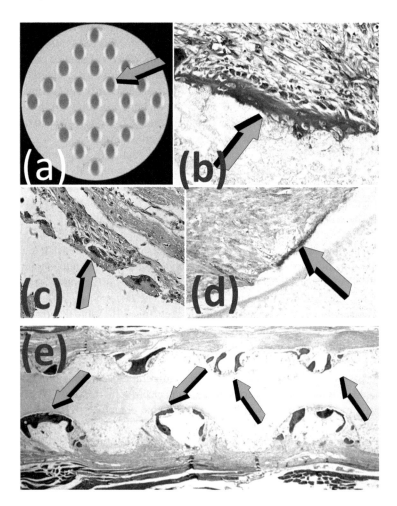

FIGURE 7.6 The induction of bone formation by the concavity *motif* assembled in solid discs of crystalline sintered hydroxyapatite implanted in the *rectus abdominis* of the chacma baboon *Papio ursinus* and harvested on days 30 and 90 after heterotopic implantation (Ripamonti et al. 1999). (a) Sintered discs of crystalline hydroxyapatite were assembled with a series of repetitive concavities on both planar surfaces (Ripamonti et al. 1999). (b) Harvested specimens showed that the concavity *motif* assembled into solid discs of crystalline sintered hydroxyapatite initiated the induction of bone formation (*light blue* arrow) by day 30 after implantation. (c) Within concavities, resting differentiating mesenchymal cells express and secrete osteogenic protein-1 (OP-1) as immuno-localized within the cytoplasm (*light blue* arrow). (d) The protein is then secreted and embedded onto the sintered crystalline bioreactor as shown by immune localization of the expressed and secreted protein within the substratum (d, *light blue* arrow). (Ripamonti et al. 1999). (e) Digital image on day 90 showing unequivocally that the concavity initiates the spontaneous and/or intrinsic induction of bone formation when implanted within the *rectus abdominis* striated muscle of the chacma baboon *Papio ursinus* (Ripamonti et al. 1999; Ripamonti 2017). Bone with bone marrow (*light blue* arrows) forms exclusively within the assembled concavities of the crystalline hydroxyapatite substratum (Ripamonti et al. 1999).

on days 30 and 90 after heterotopic intramuscular implantation (Figs. 7.6b,e). The induction of bone formation follows the expression (Fig. 7.6c) and synthesis of osteogenic proteins along the crystalline solid substratum, which is followed on day 30 by the embedding of the secreted proteins onto the concavities of the sintered crystalline hydroxyapatite (Fig. 7.6d). On day 90, concavities show the induction of bone formation with remodelling and the formation of bone marrow within the newly formed lamellar bone (Fig. 7.6e), even in the absence of osseous load with the exclusion of contracting movements of the *rectus abdominis* muscle.

A solid titanium construct was thus fabricated, replicating the images showed in Figure 7.6 (Fig. 7.6e) (Ripamonti et al. 1999; Ripamonti et al. 2012a). The digital image shows the concavity *motif* assembled into sintered crystalline discs of hydroxyapatite implanted in the *rectus abdominis* of *Papio ursinus* and harvested 90 days after heterotopic implantation (Figs. 7.6e, 7.7a) (Ripamonti et al. 1999).

Figure 7.7 (Fig. 7.7b) (Ripamonti et al. 2012a; Ripamonti et al. 2012b; Ripamonti et al. 2013) shows how the concavity *motif* is re-assembled and translated into geometrically functionalized titanium constructs orthotopically implanted into the exposed tibia of a chacma baboon *Papio ursinus* (Fig. 7.7b) (Ripamonti et al. 2012a; Ripamonti et al. 2012b; Ripamonti et al. 2013). The constructs are functionalized with the concavity *motif* along the planar surfaces (Fig. 7.7b). Note that the concavity (*light blue* arrow in b) preferentially adsorbs plasma and plasma products during the implantation procedures with several still unknown circulating proteins within the blood coagulum in the concavity (Fig. 7.7b *light blue* arrow).

The landscape of the inductive geometric configuration is presented in Figure 7.8. The composite image shows a series of scanning electron microscopy (SEM) images of the geometric *motif* as carved within the titanium constructs. The repeated concavity *motif* plasma-sprayed by crystalline hydroxyapatite-coated implants is highlighted by a series of arrows across the implant's configuration (Fig. 7.8). High-power views (Figs. 7.8e,f) show topographical surface modifications characterized by lacunae, pits and micro-concavities (*light blue* arrows in e,f) consistent with geometrical and topographical modifications as carved and excavated by osteoclastogenesis *in vivo* during the remodelling phases of the osteosome in long-lived vertebrates (Reddi 1997) (Fig. 7.5).

7.3 DOES PURE TITANIUM METAL *PER SE* INITIATE THE SPONTANEOUS INDUCTION OF BONE FORMATION?

Does titanium *per se* initiate *de novo* induction of bone formation? The only experiment that shows that titanium metals implanted in heterotopic intramuscular sites of animals induced heterotopic bone has been reported by Fujibayashi et al. in *Biomaterials* (Fujibayashi et al. 2004). The titanium constructs were porous blocks as well as titanium fibre mesh cylinders (Fujibayashi et al. 2004). Bone was found only in titanium porous blocks after chemical treatment. Firstly, titanium macroporous blocks were in 5 M aqueous NaOH solution at 60°C for 24 h. This was followed by immersion in distilled water at 40°C for 48 h and, lastly, by a thermal treatment to 600°C at a rate of 5°C/min, maintained at 600°C for 1 h. Of interest, bone formation

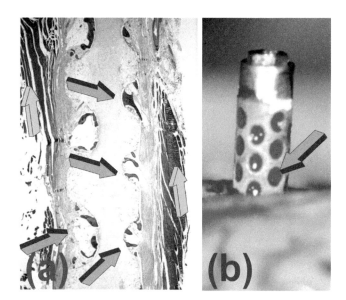

FIGURE 7.7 Translational research of the concavity *motif* from macroporous sintered crystalline bioreactors to hydroxyapatite-coated titanium implants for translation in clinical context. Bioactive self-inducing bioreactors spontaneously initiating the induction of bone formation (Ripamonti et al. 2012a; Ripamonti et al. 2012b; Ripamonti 2013; Ripamonti 2017; Ripamonti 2018). (a) Repetitive sequences of concavities cut into crystalline sintered hydroxyapatite discs conclusively initiate the induction of bone formation with marrow morphogenesis within the concavity of the substratum (*light blue* arrows). (b) Translation of the concavity *motif* into crystalline sintered hydroxyapatite plasma-sprayed titanium implants constructed with an identical geometric *motif*. The implant inserted into the trephined mandibular bone is partially re-exposed for photographic reasons. The digital image (*light blue* arrow) indicates how the concavity preferentially adsorbs plasma and plasma products plus several unidentified plasma proteins within the concavity of the bioreactor. Bioreactors are implanted in the exposed tibiae and edentulous mandibular ridges of the chacma baboon *Papio ursinus*. Samples were also implanted into the *rectus abdominis* muscle of *Papio ursinus* to study the intrinsic osteoinductivity of the coated geometric implants (Ripamonti et al. 2012a; Ripamonti et al. 2012b).

occurred only in chemically and thermally treated blocks (Fujibayashi et al. 2004). Bone was not found in pure untreated titanium blocks nor in titanium fibre mesh cylinders (Fujibayashi et al. 2004).

Fujibayashi et al. (2004) reported that titania endowed with osteoinductive potential needed to be chemically and thermally treated. Macroporous bioactive titanium metals showed superior *in vitro* apatite-forming ability, directly bonding to living bone *in vivo* (Fujibayashi et al. 2004). As stated above, pure titanium macroporous blocks did not induce bone, raising the critical role of the deposition of apatite as a fundamental key for the induction of bone formation to occur. Apatite was formed after a series of thermal and alkali treatments during the preparation of heterotopically implanted macroporous constructs (Kobuko et al. 1996; Nisiguchi et al. 1999;

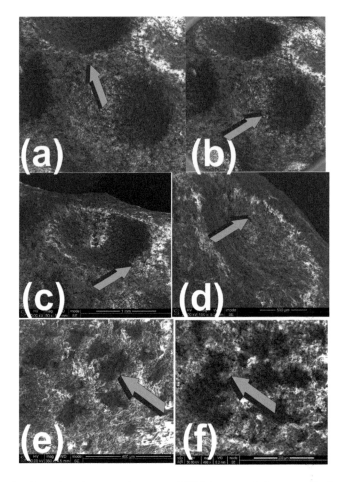

FIGURE 7.8 Scanning electron microscopy (SEM) images of the concavity landscape on titania bioreactors with concavities prepared on the planar surfaces of the bioreactors later coated with crystalline high porosity sintered hydroxyapatite. (a,b,c,d) Images of the concavities as prepared on the titanium planar surface. Blue arrows show the edges and/or margins of the concavities across the implant (*light blue* arrows). (e,f) High-power views showing the surface geometric topography of the crystalline hydroxyapatite with concavities of various dimensions to activate and differentiated mesenchymal cells into the osteoblastic lineage.

Fujibayashi et al. 2001; Uchida et al. 2002; Fujibayashi et al. 2004; Takemoto et al. 2006; Kawai et al. 2014).

A question we have asked is now very pertinent, i.e. how does titanium initiate *de novo* induction of bone formation in heterotopic extraskeletal sites of animal models (Ripamonti et al. 2012a)? Is titanium endowed *per se* with the capacity to initiate osteoblast-like cell differentiation with the induction of bone formation as a secondary response? Careful evaluation of the available literature indicates that the induction of bone formation by macroporous titanium constructs is strictly dependent on the chemical and thermal treatment of macroporous titanium blocks after the deposition of a layer of apatite (Fujibayashi et al. 2004). A review of the literature shows that Kobuko et al. (1996), whilst highlighting the mechanisms for artificial materials to bond to living bone, reported that chemically treated titanium metals showed a dense and uniform deposition of bone-like apatite that formed on titanium substrata after alkali and heat treatments below 600°C (Kokubo et al. 1996). The experiments reported that "the essential requirements for artificial materials to bond to living bone is the formation of bonelike apatite" (Kokubo et al. 1996).

Results showed that a dense and uniform layer of bone-like apatite formed on the surfaces of titania when treated with alkali and heat treatments below 600°C (Kokubo et al. 1996; Uchida et al. 2003). When tensile stresses were applied to treated substrata, fractures often occurred not at the apatite/titanium interface but in the apatite layer (Kokubo et al. 1996). The data indicated that simple chemical surface treatment with alkali and chemically treated titanium metals provides the treated titanium with bioactivity, and that the apatite layer formation tightly enhances the adhesion bond strength to living bone to the point that fractures occur within the deposited apatite and not at the apatite/titanium interface (Kokubo et al. 1996).

The bioactivity of bone-bonding ability after induction of apatite deposition on titanium substrata was further evaluated by Nishiguchi et al. (1999). Smoothed-surface rectangular blocks of titanium with different alkali and heat treatments were inserted transcortically into methaphyses of lagomorphs' tibiae (Nishiguchi et al. 1999). Untreated titanium blocks showed almost no bonding even at 16 weeks after orthotopic insertion. In marked contrast, alkali- and heat-treated titanium blocks showed bonding to bone at several time periods after the induction of apatite layers attached to the titanium substrata (Nishiguchi et al. 1999). Morphological analyses showed that alkali- and heat-treated titania bonded directly to bone at several time points, once again due to apatite deposition on titania alloys (Nishiguchi et al. 1999).

Other studies and experiments conclusively showed that the induction and deposition of apatite along the treated titanium surfaces is a critical biochemical step to engineer the bioactivity of titanium surfaces and metals (Fujibayashi et al. 2001; Uchida et al. 2002; Fujibayashi et al. 2004; Takemoto et al. 2006; Kawai et al. 2014). The deposition of apatite along treated titania surfaces is the biological rationale for the spontaneous and/or intrinsic osteoinductivity of treated titanium macroporous surfaces (Fujibayashi et al. 2001; Uchida et al. 2002; Fujibayashi et al. 2004; Takemoto et al. 2006; Kawai et al. 2014). This is supported by the experimental evidence that untreated pure titanium metals fail to initiate the spontaneous induction of bone formation when implanted in heterotopic intramuscular sites of canine models (Fujibayashi et al. 2004; Kawai et al. 2014).

To summarize, the studies of Fujibayashi's group show the heterotopic induction of bone formation by macroporous titanium blocks (Kobuko et al. 1996; Nishiguchi et al. 1999; Fujibayashi et al. 2001; Uchida et al. 2002; Fujibayashi et al. 2004; Takemoto et al. 2006; Kawai et al. 2014). Such macroporous blocks were, however, acid- and heat-treated to form apatite layers on the treated titania surfaces (Kobuko et al. 1996; Nishiguchi et al. 1999; Fujibayashi et al. 2001; Uchida et al. 2002; Fujibayashi et al. 2004; Takemoto et al. 2006; Kawai et al. 2014).

As per Kawai et al.'s observations, the observed osteoinduction is based on acid- and alkali-treated macroporous titania (Uchida et al. 2002; Kawai et al. 2014). The most convincing data that show that pure untreated titanium is not intrinsically or inherently osteoinductive are that pure untreated titanium constructs are not osteo-inductive when heterotopically implanted in canine heterotopic intramuscular sites (Fujibayashi et al. 2004). The induction of bone formation is strictly dependent on acid- and heat-treated macroporous titanium bioreactors (Uchida et al. 2002; Fujibayashi et al. 2004; Takemoto et al. 2006; Kawai et al. 2014). The experiments of Kawai et al. (2014) reported that titanium specimens that showed the lack of bone differentiation also featured almost zero surface charges and limited if any apatite-forming ability (Kawai et al. 2014).

As reported in our studies, though with crystalline hydroxyapatite titanium coating, bone formed several months after intramuscular implantation (Ripamonti et al. 2012a). Similarly, Fujibayashi et al. (2004) reported the lack of bone differentiation 90 days after heterotopic implantation in canine models. Bone formed 12 months after heterotopic intramuscular implantation (Fujibayashi et al. 2004). The study concluded that the induction of bone formation by titanium-porous metal is strictly related to positive surface charges that facilitate the formation of apatite at the titanium surfaces *in vitro* (Fujibayashi et al. 2004; Kawai et al. 2014).

Reconsidering now those experiments in the late 1980s, we should have also implanted titanium constructs with the above geometric configurations without, however, the crystalline hydroxyapatite coating. The question that we have indeed posed in our paper on the osteoinductive crystalline-hydroxyapatite constructs was: "Can bone be formed by uncoated titanium substrata?" (Ripamonti et al. 2012a).

The above studies raise the scientific question of whether the crystalline hydroxyapatite coating is a necessary requirement to achieve the heterotopic induction of bone formation. Is the presence of hydroxyapatite a *conditio sine qua non* for the induction of bone to occur?

Our previous studies, reporting the remarkable and substantial induction of bone formation by coral-derived macroporous constructs, were later followed by the intrinsic induction of bone formation by sintered crystalline hydroxyapatite bioreactors (Ripamonti et al. 1999). The results showed that the induction of bone formation was by the binding of expressed and locally produced BMPs embedded onto the coral-derived or sintered hydroxyapatite substrata (Figs. 7.6c,d) (Ripamonti 1991; Ripamonti et al. 1993; Ripamonti 1996; Ripamonti et al. 1999; Ripamonti et al. 2001b; Ripamonti 2003; Ripamonti 2004; Klar et al. 2013; Ripamonti 2017; Ripamonti 2018).

It is noteworthy that BMPs' purification from natural animal sources, i.e. bovine and/or baboon demineralized bone matrices after chaotropic extraction by dissociative agents such as 6 M urea or 4 M guanidinium hydrochloride (Gdn. HCl) (Sampath and Reddi 1981; Sampath and Reddi 1983), deploys adsorption chromatography on hydroxyapatite gels (Urist et al. 1984; Luyten et al. 1999; Ripamonti et al. 1992c).

In recent experiments in *Papio ursinus*, large mandibular defects were implanted with 250 μg of hTGF-β_3. The experiment was initiated to study the long-term incorporation of mandibular regenerates engineered by 250 μg hTGF-β_3, and at the same time, to study the biology of integration of titanium bioreactors with and without geometric configurations in the form of concavities along the substratum but without crystalline hydroxyapatite coating at early time points (Bone Research Unit, unpublished data 2019). μCT scans on day 30 showed the lack of bone formation and integration within concavities of the titanium substratum without hydroxyapatite coating when implanted into hTGF-β_3 mandibular regenerates (Fig. 7.9).

We conclude thus that pure titanium metal is not *per se* endowed with the capacity of initiating the induction of bone formation. Rather, the reported induction of bone formation by macroporous titanium constructs is dependent on the synthesis and deposition of apatite layers onto the titanium bioreactors following alkali and temperature treatments (Uchida et al. 2002; Fujibayashi et al. 2004; Takemoto et al. 2006; Kawai et al. 2014).

Expression of BMPs' species and immunolocalization of the expressed and secreted proteins onto the calcium phosphate-based substrata was later accomplished (Ripamonti et al. 1993; Ripamonti et al. 1999) (Figs. 7.1b, 7.6c,d). Such studies indicated that osteogenic proteins were locally expressed, synthesized and secreted onto the calcium phosphate constructs. Embedded proteins would intrinsically initiate the induction of bone formation as a secondary response (Ripamonti et al. 1999; Klar et al. 2013; Ripamonti et al. 2007a; Ripamonti 2017; Ripamonti 2018). Further *in vivo* experiments in the non-human primate *Papio ursinus* showed that mRNAs of osteogenic protein-1 (OP-1, also known as BMP-7) are expressed within the concavities of the substratum. Osteogenic proteins are expressed by resident mesenchymal cells (Fig. 7.1b), later embedding the secreted molecular signals into the concavities of the substratum (Fig. 7.6d). This would initiate the ripple-like cascade of the induction of bone differentiation as a secondary response (Ripamonti et al. 1999; Ripamonti et al. 2007a; Klar et al. 2013; Ripamonti 2017; Ripamonti 2018).

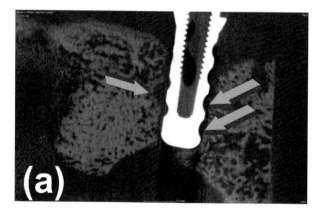

FIGURE 7.9   μCT scan of a geometric implant with concavities along the linear profile of a titanium bioreactor without crystalline hydroxyapatite coating implanted into a hTGF-β_3 mandibular regenerate of a chacma baboon *Papio ursinus* on day 10 after implantation. Note the lack of bone growth and osteointegration within the concavities (*light blue* arrows).

mRNA studies by northern blot analyses on implanted geometric discs and macroporous bioreactors showed that differentiating cells resting within the concavities of the substratum locally express, secrete and embed the secreted proteins directly onto the concavities of the substratum (Ripamonti et al. 2007a; Klar et al. 2013; Ripamonti 2017; Ripamonti 2018). The key finding thus is that responding mesenchymal cells and de-differentiating somatic perivascular/pericytic cells locally differentiate into osteoblastic cells, expressing and secreting osteogenic proteins of the TGF-β supergene family (Ripamonti et al. 2007a; Klar et al. 2013; Ripamonti 2017; Ripamonti 2018). These findings rule out the embedding of osteogenic proteins derived from circulation or surrounding tissues as previously suggested (De Groot 1998; reviewed by Ripamonti et al. 2009). The molecular work provided so far on the spontaneous and/or intrinsic induction of bone formation has revealed the critical role of macrophages and osteoclastogenesis in releasing Ca^{++} to initiate osteogenesis in angiogenesis (Klar et al. 2013; Ripamonti 2017; Ripamonti 2018).

The fundamental questions we have asked were: how does the spontaneous induction of bone formation, or osteogenesis, occur? How do invading responding cells interact with signals released by the invading differentiating cells and the supporting extracellular matrix? How do sequestered bone morphogenetic and angiogenic proteins bound to the extracellular matrix components of the basement membrane of the invading and sprouting capillaries activate specific receptors on responding cells resting with the concavities of the substratum? Further, where do the signals come from or how are signals generated to initiate the spontaneous induction of bone formation without the addition of the osteogenic soluble molecular signals of the TGF-β supergene family? Not least, how do signals spatio-temporally integrate with available responding cells to construct tissue formation and organogenesis, i.e. the spontaneous induction of bone formation with the induction of marrow cavities? From the base of the concavity *vs.* the *rectus abdominis* muscle, or *vice versa* from the enveloping striated muscle with its angiogenic invading component? Finally, how are morphogenetic gradients spatio-temporally organized within the morphogenetic concavities? The functional differentiation into the bone/bone marrow organ is critically dependent on the tri-dimensional architecture of the geometric configuration of the titanium constructs coated by sintered crystalline hydroxyapatite.

Critical data were harnessed by harvesting hydroxyapatite-coated titanium implants inserted into the *rectus abdominis* muscle of *Papio ursinus*. Implanted specimens harvested on day 5 after heterotopic implantation were processed for scanning electron microscopic analyses (SEM) (Ripamonti et al. 2012a). Scans showed a multicellular-driven cellular attachment and differentiation (Fig. 7.10) that was foremostly evident in retrieved geometric implants with a series of concavities along the planar surfaces (Figs. 7.10c,d,e,f).

Carbon-coated samples (coating 5 nm thick) were evaluated on an FEI Nova Nanolab SEM (FEI Company, Oregon, USA), at 30kV (Ripamonti et al. 2012a). There was limited if any cellular attachment on day 5 along the surfaces of the planar control implants (Figs. 7.10a,b). To the contrary, geometric implants with a series of repetitive concavities showed the morphogenesis of cellular condensations and proliferation along and within the concavities of the substratum (Figs. 7.10c,d,e,f).

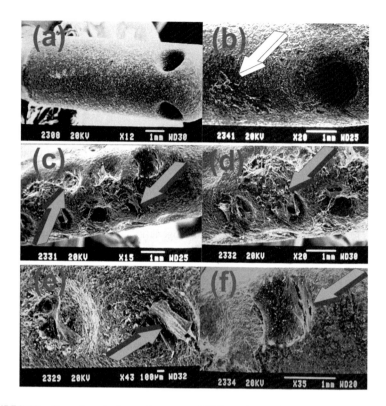

FIGURE 7.10 Scanning electron microscopy (SEM) analyses of hydroxyapatite-coated titanium bioreactors without and with geometric configuration in the form of repetitive concavities along the titanium-coated bioreactors. Devices were harvested from the *rectus abdominis* muscle of *Papio ursinus* 5 days after intramuscular implantation (Ripamonti et al. 2012a). (a,b) Linear hydroxyapatite-coated titanium constructs showing minimal if any cellular attachment on the intramuscularly implanted devices. (c,d and e,f) Pronounced cellularity and invasion within the geometric cue of the concavities with cell adhesion and attachment on the margins of the concavities embedded within the *rectus abdominis* muscle. (e,f) Cell attachment and proliferation within the exposed concavities by the enveloping *rectus abdominis* muscle. There is patterning of the formed newly deposited collagenous condensations across the margins of the concavities. Collagen deposition forms from one side to the other side of the margins of the exposed concavities. There is patterning of the collagenous condensations across the concavities with the induction of a tent-like structure from margin to margin of the concavity that is thus morphogenetic, and on day 90 engineers membranous bone across the margins of the concavities (Figs. 8.11c,d).

SEM images indicated collagen deposition and binding to the edges of the concavities of the geometric implants (Figs. 7.10e,f). SEM images shown in Figures 7.10e and 7.10f show how the concavity constructs the induction of tissue patterning by providing an ideal substratum for mesenchymal fibroblastic cells to move from one edge to the other edge of the concavity suspended within the bioreactor whilst secreting collagen back and forward from the edges of the concavity (Figs. 7.10e,f).

The induction of a collagenic tent above the concavity and within the microenvironment of the bioreactor enveloped by the surrounding *rectus abdominis* muscle is noteworthy, and once again indicates the morphogenetic inductive force of the concavity as "the shape of life" (Ripamonti 2004; Ripamonti 2006) when in contact with the *rectus abdominis* muscle, a striated muscle known to harbour several mesenchymal perivascular/pericytic and myoblastic stem cell niches (Lensh et al. 2006; Zengh et al. 2007; Kovacic and Boehm 2009).

The morphogenetic drive of the concavity is further shown in Figure 7.11 that once again shows the induction of collagenic depositions across the concavity of the implanted hydroxyapatite-coated titanium bioreactor (Figs. 7.11a, inset b). Morphological digital images (Figs. 7.11c,d) show the remarkable induction of bone formation across the concavity bioreactors that bridges the margins of the exposed concavities embedded within the *rectus abdominis* muscle 90 days after implantation (Figs. 7.11c,d). The concavity is thus morphogenetic, and induces heterotopic bone bridging across the margins, developing from earlier collagenic constructs across the concavity bioreactors exposed to the multifaceted *rectus abdominis* microenvironment (Fig. 7.11).

Is the concavity of a titanium substratum coated by crystalline sintered hydroxyapatite endowed with the striking capacity to spontaneously initiate *de novo* induction of bone formation where there is no bone, like in the *rectus abdominis* muscle? Figure 7.12 remarkably shows the spontaneous and/or intrinsic induction of bone formation by geometric titanium bioreactors coated with crystalline sintered hydroxyapatite after implantation in the *rectus abdominis* muscle of the chacma baboon *Papio ursinus*, where there is no bone (Fig. 7.12) (Ripamonti et al. 2012a). These results are the first reported instances of osteoinductive hydroxyapatite-coated titanium constructs developed for implantation in clinical contexts (Ripamonti et al. 2012a).

Instrumental for the self-initiating induction of bone formation by the concavity, "the shape of life" (Ripamonti 2004; Ripamonti 2006), was the demonstration of cell alignment and orientation as dictated by the concavity *in vitro* (Fig. 7.4a) (Ripamonti et al. 2012a; Ripamonti et al. 2012b) and *in vivo* by calcium phosphate-based biomatrices, including sintered crystalline substrata (Figs. 7.3, 7.4b, 7.6) (Ripamonti et al. 2012a; Ripamonti et al. 2012b).

Simultaneously, osteoinductive hydroxyapatite-coated titanium bioreactors were implanted in edentulous ridges of hemi mandibles prepared in chacma baboons *Papio ursinus* (Fig. 7.13). Undecalcified Exakt cut and polished ground sections of Technovit embedded blocks showed the tight osteointegration of the newly formed bone blending directly into the plasma-sprayed crystalline hydroxyapatite layer (Figs. 7.13c,d). Superior osteointegration has been achieved by geometric implants when compared to planar linear implants on day 90 after orthotopic implantation (Ripamonti et al. 2012a; Ripamonti et al. 2012b; Ripamonti et al. 2013). Undecalcified sections show the quality of the osteointegration with the induction of lamellar osteonic bone with bone marrow within the concavities prepared along the titanium substrata (Figs. 7.14c,d,e,f).

It is noteworthy to compare the morphological structural correlation between the concavities prepared in crystalline sintered hydroxyapatites (Figs. 7.14a,b) vs. the

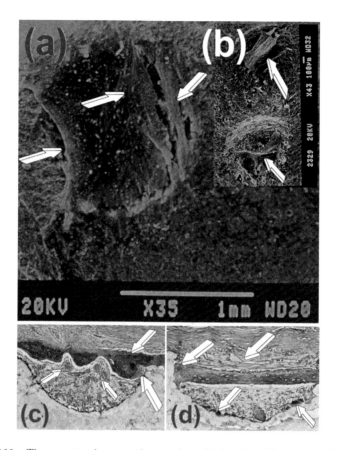

FIGURE 7.11 Tissue patterning, morphogenesis and induction of bone formation by geometric cues of the concavity that engineers the induction of a membranous bone formation across the geometry of the morphogenetic substratum. (a) Collagen deposition across the concavity stretching across the margins of the geometric cue on day 5 after intramuscular *rectus abdominis* implantation. (Inset b) Other concavities of the geometric hydroxyapatite-coated bioreactors engineer and pattern collagenous tractional fields across the margins of the exposed concavities enveloped by the *rectus abdominis* muscle of *Papio ursinus*. (c,d) Collagenous condensations patterned by the crystalline sintered hydroxyapatite coating further remodel and engineer membranous bone formation across the concavities of sintered crystalline hydroxyapatites 90 days after heterotopic intramuscular implantation (*light blue* arrows). Note the pronounced vascular invasion (*magenta* arrows) with capillary sprouting both below and above the newly formed bone stretching across the concavities.

concavities prepared on the biomimetic titanium constructs (Figs. 7.14d,e,f,). The recurrent theme of the concavity sets into motion the induction of bone formation without the exogenous application of the osteogenic soluble molecular signals of the TGF-β supergene family (Ripamonti 2003), additionally resulting in faster and superior osteointegration (Ripamonti et al. 2001b; Ripamonti et al. 2012a; Ripamonti et al. 2013; Ripamonti 2017; Ripamonti 2018).

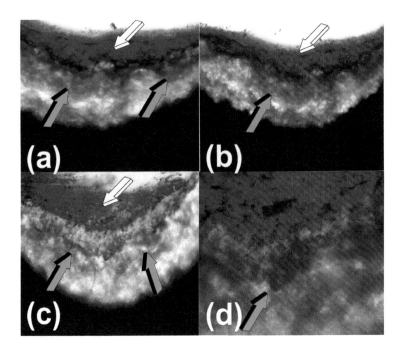

FIGURE 7.12   The critical role of the geometry of the concavity constructed along the linear surface of the hydroxyapatite-coated titania bioreactors *per se* spontaneously initiating the induction of bone formation. Titania constructs were implanted in the *rectus abdominis* muscle of *Papio ursinus* without osteogenic soluble molecular signals of the transforming growth factor-β (TGF-β) supergene family. (a,b) Induction of mineralized bone along the sintered crystalline coating (*light blue* arrows) and deposition of osteoid (*white* arrows), yet to be mineralized bone onto the mineralized newly formed bone matrix. (c,d) Details of newly formed and mineralized bone (*light blue* arrows) deposited onto the sintered crystalline hydroxyapatite. Note the tight junction of the newly formed bone by induction with the crystalline sintered hydroxyapatite (*light blue* arrows).

7.4   THE GEOMETRIC INDUCTION OF BONE FORMATION

To end this chapter reporting the induction of bone formation by osteoinductive-hydroxyapatite coated-titanium bioreactors, classic images of the effect of the geometry of the inductor in regulating the induction of bone formation by demineralized teeth and bone are re-proposed to the readership of this volume. This work highlights "the geometric induction of bone formation" (Ripamonti et al. 1999). Iconography is re-proposed to stress once again the fundamental work of Charles B. Huggins and Hari A. Reddi showing the critical role of the geometry of the inductor in controlling and modulating the induction of bone formation (Figs. 7.15, 7.16) (Reddi and Huggins 1973; Reddi 1974; Sampath and Reddi 1984).

Incisor teeth of adult rodents were demineralized in 0.5 N hydrochloric acid, washed in water 2 h.; absolute ethanol, 1 h.; diethyl ether, 0.5 h. and dried overnight at 37°C (Reddi and Huggins 1973). The apices of half of the demineralized teeth

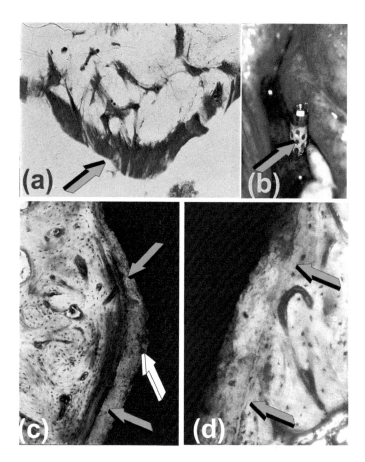

FIGURE 7.13 Translational research from *in vitro* studies to pre-clinical non-human primate studies for translation in clinical contexts. (a) Optimal polarization and alignment of MC 3T3-E1 cells within a concavity of a coral-derived calcium phosphate construct grown *in vitro* (Ripamonti et al. 2012a). (b) Geometric bioreactor ready for insertion into edentulized mandibular ridges of the non-human primate *Papio ursinus*. Note how significantly the concavity adsorbs plasma and plasma products (*light blue* arrow) within the geometries prepared along the hydroxyapatite-coated bioreactor. (c,d) High-power views of undecalcified sections detailing the concavity *motif* along the implanted bioreactor 90 days after implantation of the geometric bioreactors into edentulized mandibular ridges (Ripamonti et al. 2012a). (c) Optimal attachment and integration of the newly formed bone against the sintered crystalline hydroxyapatite (*light blue* arrows). Note in d the tight bonding of the newly formed bone against the plasma-sprayed sintered crystalline hydroxyapatite coating (*light blue* arrows). Loops of newly formed capillaries in the induced bone almost touch the sintered coated crystalline hydroxyapatite. Undecalcified mandibular blocks were processed in ascending concentrations of Technovit 7200 (Heraeus Kulzer GmbH, Wehrheim, Germany) and embedded in a fresh solution of the same resin. Undecalcified sections were ground, polished to 40–60 μm, and stained with a modified Goldner's trichrome (Ripamonti et al. 2012a). All sample preparation was performed using the Exakt precision cutting and grinding system (EXAKT Apparatebau, Nordestedt, Hamburg, Germany).

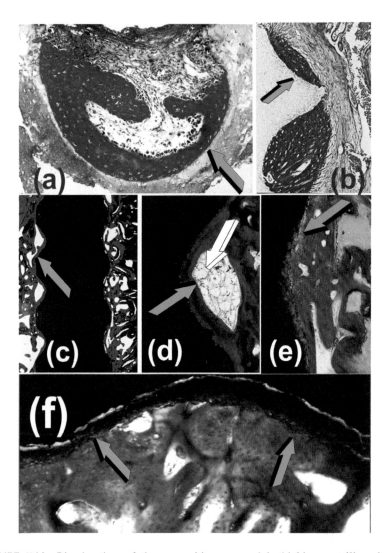

FIGURE 7.14 Biomimetism of the concavities prepared in highly crystalline sintered hydroxyapatites (a,b) vs. concavities prepared along the surface of titanium linear bioreactors later plasma-sprayed by crystalline sintered hydroxyapatite (c,d,e,f). (c) Osteointegration (*light blue* arrow) along the intra-osseous titanium bioreactor. (d) Tight osteointegration with marrow formation (*white* arrow) facing newly deposited osteoid (*light blue* arrow). (e,f) Compact newly formed bone remodelled within concavities of the substratum.

were amputated to form tooth tubes with open ends (Reddi and Huggins 1973). On day 28, in demineralized incisor transplants, bone with bone marrow formed in the pulp chamber situated near to the aperture of the tooth (Fig. 7.15a). The closed apex showed on the other hand the induction of chondrogenesis attached to the dentine walls (Fig. 7.15a *light blue* arrow c).

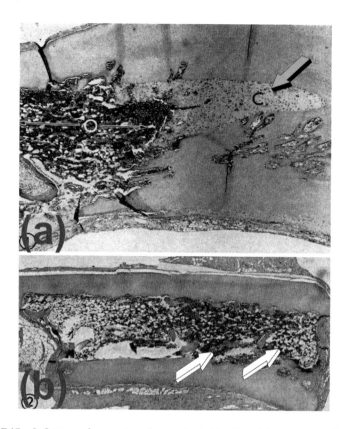

FIGURE 7.15 Influence of geometry of transplanted tooth and bone on transformation of fibroblasts (Reddi A.H., Huggins C.B. *Proceedings of the Society for Experimental Biology and Medicine*, 143, 634–637, 1973). Demineralized incisors of the rat were transplanted subcutaneously under the chest of the rat. Transplants were whole demineralized incisor teeth as well as incisors from which the apex has been removed, making the tooth a demineralized dentinal tube, open at both ends. (a) Transplanted whole incisors showed the formation of an ossicle (O) whereas the apex is populated with cartilage (*light blue* arrow c). (b) Transplants of demineralized incisor tooth tubes on day 28 show the induction of bone formation with marrow formation. The temporal sequence of fibroblast-chondroblast-osteoblast transformation was profoundly influenced by the geometry of the transformant (Reddi and Huggins 1973). (Images courtesy of Hari A. Reddi, University of California, Davis).

In contrast, in incisor-demineralized teeth with amputated root apices to form a demineralized tooth tube, the transplants were filled with bone and marrow (Fig. 7.15b *white* arrows); cartilage however was absent (Fig. 7.15b). The experiment provided evidence of the critical role of the geometry of the inductor in controlling tissue induction and morphogenesis. Demineralized transplants with closed apices showed the induction of chondrogenesis lacking vascular invasion and capillary sprouting (Fig. 7.15a *light blue* arrow c). The paper by Reddi and Huggins (1973) is the first paper to describe the effect of the geometry of the inductor in controlling tissue induction and morphogenesis (Reddi and Huggins 1973).

A further classic experiment reported by Reddi and Huggins (1973) is on the role of the geometry of demineralized bone matrices in the differentiation and transformation of fibroblasts. The influence of geometry on fibroblast differentiation was further reported by Reddi whose experiments showed that there was a strong geometric influence on the transformation of fibroblasts' differentiation. This was done by using different sizes, or geometries, of powdered demineralized bone matrices (Reddi and Huggins 1973; Reddi 1974; Sampath and Reddi 1984). Rat dry powdered acid-insoluble matrices were sieved to fine powders (44–74 μm) and coarse powders (420–850 μm) and transplanted in symmetrical contralateral sites in the subcutaneous space of the rat (Reddi and Huggins 1973; Reddi 1974; Sampath and Reddi 1984). The studies showed that alkaline phosphatase activity and ^{35}S incorporation were greater in tissues prepared from coarse powders compared to fine powders (Reddi and Huggins 1973; Reddi 1974). Implants of coarse powders also showed extensive bone and marrow formation (Fig. 7.16).

Some of the mechanistic insights into the effect of geometry of the substratum and/or the inductor have only been recently highlighted by a number of studies, which have provided partial understanding of the induction of bone formation by geometrically different calcium phosphate-based bioreactors (Ripamonti et al. 1999; Klar et al. 2013; Ripamonti 2017; Ripamonti 2018; Othman et al. 2009).

As stated in previous communications (Ripamonti 2017; Ripamonti 2018), the theme of geometry regulating cell differentiation and transformation with the induction of the osteogenic phenotype was established in the last century by the seminal papers of Reddi's group (Reddi and Huggins 1973; Reddi 1974; Sampath and Reddi 1984). We believe, however, that the fundamental contribution of the role of geometry to the induction of tissue formation is still poorly understood (Ripamonti 2017; Ripamonti 2018). It is also, and regretfully so, seldom cited in spite of the by now plethora of communications on the effect of geometry on multiple pathways. Pathways encompass the induction of angiogenesis and capillary architecture (Sun et al. 2014). Geometric cues regulate and direct stem cell differentiation (Vlacic-Zischke et al. 2011), including branching morphogenesis (Nelson et al. 2006; Gjorevski and Nelson 2009) or in general, topographically controlling cell induction and differentiation (Gospodarowicz et al. 1978; Brunette 1988; Ahn et al. 2014; Zhang et al. 2018; McNamara et al. 2010; Clark et al. 1987; Clark et al. 1990; Curtis and Wilkinson 1997; Miyoshi and Adachi 2014; Lamers et al. 2010; Kim et al. 2012; Bettinger et al. 2009; Yang et al. 2017; Zhang et al. 2017; Fiedler et al. 2013; Yoon et al. 2016; Metavarayuth et al. 2016; Karageorgiu and Kaplan 2005; Killian et al. 2010; Sammons et al. 2005; Curran et al. 2006; Muller et al. 2008; Yang et al. 2010; Gittens et al. 2011; Wilkinson et al. 2011; McNamara et al. 2011; Costa-Rodrigues et al. 2012; Zhang et al. 2018).

The recurrent theme of the role of geometry in the gene expression of a variety of genes and gene products of the TGF-β supergene family, including both bone morphogenetic proteins and the mammalian transforming growth factor-β, is articulated by describing differentiating surfaces with nanotopographical or topographical cues controlling gene expression pathways and the secretion of a number of proteins controlling cell differentiation (McNamara et al. 2010; Ahn et al. 2014; Wathari et al. 2012;

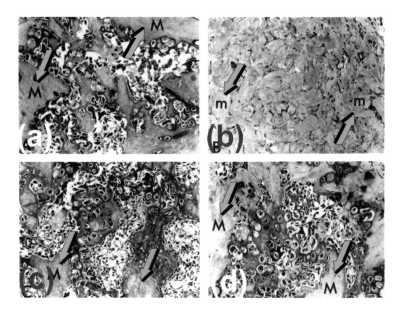

FIGURE 7.16 Importance of the geometry of the extracellular matrix in endochondral bone differentiation (Sampath T.K., Reddi A.H. *The Journal of Cell Biology* 98, 2192–2197, 1984). Coarse (74–420 µm particle size) and fine (44–74 µm particle size) powders of demineralized extracellular bone matrices were implanted in bilateral sites under the skin of the chest in rodents (Sampath and Reddi 1984). (a) Coarse demineralized bone powders (*light blue* arrows) consistently induced bone formation with vascular invasion. (b) Fine matrix powders (*light blue* arrows), on the other hand, consistently showed the complete absence of bone formation. (c) Fine matrix particles were extracted in 4 M guanidinium hydrochloride (Gdn-HCl) and extracts were reconstituted with coarse demineralized particle size. Note the complete restoration of the osteoinductive activity. (d) Reconstitution of the Gdn-HCl coarse residue after gel filtration chromatography to eliminate high molecular weight contaminants including high molecular weight collagenous proteins also restores the induction of bone formation (Sampath and Reddi 1991). The experiments showed that although fine matrix contains osteoinductive proteins, matrix geometry (size) overrules the induction of bone formation (Sampath and Reddi 1984).

McNamara et al. 2011; Liu et al. 2016). Topographical cues are nothing but the geometry of the substratum directly affecting cell differentiation (Ripamonti 2017; Ripamonti 2018). A more recent paper reported that nanopatterned titanium implants accelerate bone formation in a mouse model (Greer et al. 2020). The study showed a 20-fold increase in osteogenic gene induction of nanopatterned substrata (Greer et al. 2020), concluding that the novel nanopatterning method on titania substrata may offer critical insights into faster osteointegration and the induction of bone formation (Greer et al. 2020). Once again, however, the contribution lacked the quotation of critical studies on the induction of bone formation regulated by the geometry of the inductor.

In the published landscape of cell differentiation by geometric topographically altered substrata, a number of communications deserve a separate discussion. The studies present some deep understanding of the effect of geometry on the induction of tissue formation (Engler et al. 2006; Discher et al. 2009; Fu et al. 2010; Buxboim and Discher 2010; Buxboim et al. 2010; Liu et al. 2016). Discher et al.'s (2005) review in *Science*, "Tissue cells feel and respond to the stiffness of their substrate", elaborates on how contraction forces exerted by cells tend to increase with the stiffness of the substratum upon which cells attach and spread (Discher et al. 2005). We have thus learned that "matrix elasticity directs stem cell lineage specification", as Engler et al. (2006) titled their *Cell* paper (Engler et al. 2006). The work reports that cytoskeletal motors, one or all of the non-muscle myosin II isoforms (NMM IIA, B and C), are responsible for the cells' feeling and "sensing" the matrix upon which stem cells attach and spread (Engler et al. 2006).

How does the attached cell sense – or feel – the substratum upon which stem cells are attached? We previously reported that the work of Discher and co-workers on matrix elasticity (Discher et al. 2005; Engler et al. 2006; Buxboim et al. 2010) directing cell linage specification, cytoskeletal forces and how stem cells "feel the difference" (Buxboim and Disher 2010) is an incisive contribution that has emerged in the 21st century on how the geometry and the stiffness of the matrix control and direct cell specification and differentiation (Engler et al. 2006; Fu et al. 2010; Buxboim and Discher 2010; Buxboim et al. 2010; Liu et al. 2016). To this wealth of unique research data, Fu et al. further add the critical role of the mechanical regulation of cell function by geometrically modulated substrata (Fu et al. 2010). The published experiments in *Nature* show how micropost rigidity impacts cell morphology, focal adhesion, cytoskeletal contractility and stem cell differentiation (Fu et al. 2010).

Perhaps, however, the greatest statement of all is Discher's summary in his *Cell* paper (Engler et al. 2006) where it is stated that mesenchymal stem cells commit to different phenotypes with extreme sensitivity to matrix elasticity, i.e. "soft matrices that mimic brain are neurogenic, stiffer matrices that mimic muscle are myogenic, and comparatively rigid matrices that mimic collagenous bone prove osteogenic" (Engler et al. 2006).

We believe that the above statements and research results are the most important contribution so far on the control of cell differentiation and linage specification (Engler et al. 2006; Fu et al. 2010; Buxboim and Discher 2010; Buxboim et al. 2010; Liu et al. 2016). The research results have significant implications for our understanding of physical and geometric forces regulating cell induction and differentiation, the construction of self-inductive surface geometries for bone differentiation from the bench top to pre-clinical and clinical studies in *Homo sapiens*.

7.5 INDUCIVE MORPHOGENETIC GRADIENTS OF THE CONCAVITY

We have thus learned that the concavity is a molecular signal providing inductive and conductive microenvironments, ultimately leading to the spontaneous induction of bone formation. Bone initiates even if the macroporous calcium phosphate-based bioreactors, including hydroxyapatite-coated titanium geometries, are implanted in intramuscular heterotopic sites of animal models, where there is no bone (Ripamonti

1991; Ripamonti 1996; Ripamonti 2004; Ripamonti et al. 2012a; Ripamonti et al. 2012b; Ripamonti et al. 2013; Ripamonti 2017; Ripamonti 2018).

After the *rectus abdominis* muscle attaches to and incorporates the implanted bioreactors, resident cells are differentiated and/or de-differentiated into osteo-blastic-like cells (Ripamonti et al. 2012a; Ripamonti et al. 2012b; Klar et al. 2013; Ripamonti et al. 2015; Ripamonti 2017; Ripamonti 2018). Morphogens are later expressed, secreted and embedded into the calcium phosphate-based concavities of the assembled bioreactors.

Morphogenetic soluble signals within the concavities interact with yet to be defined and characterized stem cell niches nesting within the *rectus abdominis* striated muscle, including perivascular niches rich in pericytic and endothelial stem cells. The muscle envelops the implanted bioreactors, thus amenable to releasing a number of perivascular, myoblastic and/or pericytic cells including myo-endothelial cells and pericytes. Such cells are capable of osteoblastic cell differentiation within the differentiating selective microenvironments of the concavities. The molecular microenvironment of the concavity is set by molecular signals both morphogenetic and angiogenic (Paralkar et al. 1990; Paralkar et al. 1991; Folkman et al. 1988; Reddi 2000). Angiogenic and morphogenetic proteins bound to the extracellular matrix components of the invading and sprouting capillaries construct the supramolecular assembly of the newly formed matrix (Figs. 7.3e, 7.6b) with Ca^{++} released by osteo-clastic activity within the concavity of the crystalline bioreactors (Klar et al. 2013; Ripamonti 2017; Ripamonti 2018).

A fundamental step is osteoclastogenesis as initiated by osteoclasts recruited by the high mineral phase of the calcium phosphate bioreactors implanted in the *rectus abdominis* striated muscle. Osteoclastogenesis is followed by Ca^{++} release set free by osteoclasts within the concavity microenvironment. We have thus learned that the mechanisms initiating the ripple-like cascade of molecular and cellular micro-environments within the concavity landscape are initiated by an ionic wave of Ca^{++} released by osteoclastogenesis (Klar et al. 2013; Ripamonti et al. 2015; Ripamonti 2017; Ripamonti 2018). Osteoclastic activity along the calcium phosphate-based implanted bioreactors releases Ca^{++}. The concavity is a geometric microenvironment designed to protect and store morphogenetic and ionic signals needed for the sequential induction of bone formation (Klar et al. 2013).

Par force thus a calcium phosphate-based bioreactor is *a conditio sine qua non* for the induction of bone formation by either coral-derived, sintered or titanium constructs plasma-sprayed by sintered crystalline hydroxyapatite (Ripamonti et al. 2012a; Ripamonti et al. 2012b; Ripamonti et al. 2013 Ripamonti 2017; Ripamonti 2018). Research studies exploring the mechanistic molecular biology cascades of the spontaneous induction of bone formation by coral-derived macroporous constructs have shown that the ionic gradients of Ca^{++} are critical for both the differentiation of mesenchymal cells into osteoblastic-like cells and for the induction of angiogenesis within the concavity microenvironments (Klar et al. 2013; Ripamonti 2017; Ripamonti 2018).

To close this chapter on the spontaneous osteoinductivity of hydroxyapatite-coated titanium bioreactors, we would like to re-highlight the role of subcellular geometry that regulates stem cell differentiation (Liu et al. 2016), and, once again, to focus on the multifaceted pleiotropic capacities of endothelial cells regulating not only osteogenesis (Trueta 1963; Medici et al. 2011; Sun et al. 2014; Kusumbe et al. 2014; Ramasamy et al. 2014) but tissue morphogenesis and regeneration at large (Ramasamy et al. 2015; Gomez-Salinero and Rafii 2018).

The studies of Liu et al. (Liu et al. 2016) examined whether subcellular geometry significantly influences the extent of stem cell differentiation. Using a series of micropillar arrays of poly lactide-co-glycolide, cells were grown *in vitro* with nuclei interspacing the pillars and deformed by cellular tractional forces on the pillars (Liu et al. 2016). A persistent nuclear deformation when mesenchymal stem cells were on high micropillars influenced the differentiation of mesenchymal stem cells. Osteogenic cell differentiation was enhanced on micropillared arrays with significant self-deformation of cell nuclei. The study concludes that nuclear deformation and geometry on micropillars are a new cue to regulate the lineage commitment of stem cells, ultimately controlling tissue induction and cell differentiation (Liu et al. 2016).

The critical roles of the vessels, endothelial and pericytic cells, and of basement membrane components of the invading and sprouting capillaries are essential for tissue induction, morphogenesis and the induction of bone formation, i.e. "osteogenesis in angiogenesis". The latter term was introduced in 2006 and 2007 (Ripamonti 2006; Ripamonti et al. 2007) and re-visited in 2010 (Ripamonti 2010) to highlight the very tight relationship between the two inductive morphogenetic events so masterfully described by the grand manuscript of Trueta, "The role of the vessels in osteogenesis" (Trueta 1963).

Several authors left a prominent mark on the induction of bone formation by studying and examining the associated vascular patterning during the induction of bone formation (for reviews Ripamonti et al. 2006; Ripamonti et al. 2007; Ripamonti 2009; Ripamonti 2010). Aristotle (384–322 BC), as discussed by Lanza and Vegetti (1971) and Crivellato et al. (2007), ascribed the forming blood vessels with a patterning function during organogenesis. Aristotle had the insight to see that the vessel growth functions as a "frame" and as a "model" that shape organs and body structures.

This grand Aristotelian morphological, and no doubt molecular, insight defined a patterning function of the invading blood vessels, a patterning function Aristotle defined as "organogenetic blood vessels" (Lanza and Vegetti 1971; Crivellato et al. 2007). Vascular invasion is a prerequisite for osteogenesis (Trueta et al. 1963). Vascular invasion sets into motion chondrolysis (Reddi and Huggins 1972; Reddi and Kuettner 1981) together with the appearance of osteoblasts and osteoclasts, and the early formation of hematopoietic cells with bone marrow formation (Reddi 1981).

Ionic, cellular, molecular and mechanical signals regulating the assembly of the extracellular matrix precisely regulate angiogenesis and vascular invasion (Ripamonti 2006). As previously stated, the conceptual framework of tissue induction and regeneration would not be possible without the knowledge of the binding and sequestration of both angiogenic and morphogenetic proteins that provide the conceptual framework of the supramolecular assembly of the extracellular matrix of bone (Ripamonti 2006; Ripamonti et al. 2006; Ripamonti et al. 2007).

Angiogenic and bone morphogenetic proteins, bound to type IV collagen of the invading capillaries (Paralkar et al. 1990; Paralkar et al. 1991; Folkman et al. 1988), are presented in an immobilized form to responding mesenchymal cells to initiate osteogenesis in angiogenesis (Ripamonti 2006; Ripamonti et al. 2007; Ripamonti 2010). By sequestering initiators and promoters of angiogenesis and bone morphogenesis (Paralkar et al. 1990; Paralkar et al. 1991; Folkman et al. 1988), basement membrane components are modelling bone formation by induction in angiogenesis (Ripamonti 2006; Ripamonti et al. 2007b; Ripamonti 2010).

The induction of bone formation initiates by combining soluble osteogenic molecular signals with insoluble signals or substrata that trigger the ripple-like cascade of cell differentiation (Sampath and Reddi 1981; Sampath and Reddi 1983; Reddi 2000; Ripamonti 2006). This CRC Press volume on the geometric induction of bone formation offers a novel strategy for the induction of bone formation without the exogenous applications of the osteogenic soluble molecular signals of the TGF-β supergene family (Ripamonti 2003).

Smart self-inducing geometric concavities assembled within biomimetic matrices are endowed with the striking prerogative of differentiating osteoblastic-like cells attached to the concavities secreting bone matrix without the exogenous application of the soluble osteogenic molecular signals of the TGF-β supergene family (Ripamonti et al. 1993; Ripamonti et al. 1999; Ripamonti 2004; Ripamonti 2017; Ripamonti 2018). Expression of mRNA species is followed by secretion and embedding of the expressed gene products into the concavities (Ripamonti et al. 2007a; Klar et al. 2013; Ripamonti 2017a) (Figs. 7.1b, 7.6c,d), which later result in the induction of bone formation as a secondary response within the smart self-inducing concavities of the substratum (Figs. 7.3, 7.6, 7.12).

A critical cell that prominently contributes to the induction of bone formation is the endothelial cell (Trueta 1963; Kusumbe et al. 2014; Ramasamy et al. 2014; Ramasamy et al. 2015; Gomez-Salinero and Rafii 2018). Capillary invasion and sprouting within the mesenchymal tissue that forms within the concavities of the heterotopically implanted bioreactors are prominent morphological events with molecular cross talk with surrounding cells and the extracellular matrix nested within concavities of calcium phosphate-based biomimetic matrices (Fig. 7.3e).

Whilst the plasticity of endothelial cells sustains the regulation of organ morphogenesis, maintenance and regeneration (Ramasamy et al. 2015), the perivascular multi-lineage progenitor cells (Crisan et al. 2008; Chen et al. 2009) enveloping the invading capillaries, the pericytes, assemble the "osteogenetic vessels" of Trueta's definition (Trueta 1963). The osteogenetic vessels are the initiators and organogenetic units, controlling and finely tuning the induction of bone formation (Trueta 1963; Benjamin et al. 1998; Doherty et al. 1988; Zengh et al. 2007; Crisan et al. 2008; Chen et al. 2009; Kovacic and Boehm 2009).

The plasticity of the endothelial cells together with its basement membrane components tightly regulates the osteogenesis and morphogenesis of several tissues and organs (Ramasamy et al. 2015). Endothelial cells further control and modulate the extracellular matrix by releasing bioactive cargoes by extracellular vesicles (EVs)

(Sung and Weaver 2018) to further modulate tissue patterning, homeostasis and the induction of bone formation.

Lastly, the endothelial cells of the sprouting and invading capillaries set chondrolysis along the hypertrophic chondrocytes of the cartilage anlage. Chondrolysis together with capillary sprouting and invasion initiate the induction of the osteoblastic phenotype, the induction of bone formation, whilst the remnants of the cartilage anlage are replaced by marrow development with the full complement of the hematopoietic bone marrow.

Remarkably, the preparation of this chapter on the spontaneous induction of bone formation by the concavity plasma-sprayed by crystalline hydroxyapatite has once again showed that few phylogenetically ancient circuitries masterminded the emergence of the skeleton, the bone induction principle (Urist et al. 1967), the vertebrates and thus of the *Homo* clade. *Drosophila* genes and expressed proteins, more than 800 million years before the present, were operative to construct the unique vertebrate trait of the induction of bone formation, and later of skeletogenesis (Ripamonti 2006; Ripamonti 2010). Similarly, highly conserved circuitry has been shown between primitive marine vertebrates of the genus *Leucoraja* and other vertebrates, highlighting how *Hox* gene expression controlled the evolutionary transitions between undulatory and ambulatory motor circuit connectivity programmes (Jung et al. 2014).

The conservation of the *Hox* genes between the fruit fly *Drosophila melanogaster* and vertebrates is highly remarkable and once again reflects Nature's parsimony in organizing genes and gene products responsible for the generation of the pleiotropic diversity of the multiple species of animal life on earth, key questions of evo-devo (De Robertis 2008).

Önal et al. (2012), in *The EMBO Journal*, state that the "molecular determinants of pluripotency are conserved throughout evolution and that planaria are an informative model system for human cell biology" and that gene expression of pluripotency determinants is conserved between mammalian and planarian stem cells, and that there is deep conservation of stem cell expression of pluripotency-associated genes (Önal et al. 2012).

Studying myosin-mediated morphogenesis in choanoflagellates and animals, Brunet et al. report in *Science* the evolutionary history of tissue bending (Tomancak 2019), and that newly discovered unicellular eukaryotes show hallmark features of animal morphogenesis (Brunet et al. 2019; Tomancak 2019). Inverting sheets of connected cells, because of actomyosin activity triggered by darkness or light, results in a change of morphology, i.e. opening of collar microvilli and sheet inversion as well as behaviour, from feeding to swimming (Brunet et al. 2019; Tomancak 2019). In animal embryos, local myosin-induced apical construction or tissue bending leads to epithelia tissue invagination during gastrulation (Brunet et al. 2019). Highly conserved myosin protein sequences do exist between choanoflagellates and animals; the inversion behaviour of choanoflagellates "uses the same actomyosin molecular machinery deployed by animal cells to sculpt tissue during development through apical constriction or tissue bending" (Tomancak 2019).

The highly conserved circuitries of genes, gene products and gene pathways developing the pleiotropy and ancestry of animal phyla reflect the awesome diversity and beauty of life (De Robertis 2008). The ancestry of the concavity, the "shape of life" (Ripamonti 2004; Ripamonti 2006; Ripamonti 2010), biomimetizes the ancestral repetitive multi-million years-tested designs and topographies of Nature. The concavity, as cut into biomimetic matrices or biomimetized along titania planar surfaces plasma-sprayed by sintered crystalline hydroxyapatites, is thus the geometric signal that initiates the induction of bone formation. Concavities are endowed with smart functional shape memory geometric cues in which soluble signals induce morphogenesis, and physical forces, imparted by the geometric topography of the substratum, dictate biological patterns, constructing the induction of bone formation and regulating the expression of gene products as a function of the structure.

REFERENCES

Ahn, E.H.; Kim, Y.; Kshitiz; An, S.S.; Afzal, J.; Lee, S.; Kwak, M.; Suh, K.-Y.; Kim, D.-H.; Levchenko, A. Spatial Control of Adult Stem Cell Fate Using Nanotopographic Cues. *Biomaterials* 2014, *35*(8), 2401–10.

Benjamin, L.E.; Hemo, I.; Keshet, E. A Plasticity Window for Blood Vessels Remodeling Is Defined by Pericyte Coverage of the Preformed Endothelial Network and Is Regulated by PDGF-B and VEGF. *Development* 1988, *125*, 1591–98.

Bettinger, C.; Langer, R.; Borenstein, J. Engineering Substrate Topography at the Micro- and Nanoscale to Control Cell Function. *Angew. Chem. Int. Ed. Engl.* 2009, *48*(30), 5406–15.

Brunet, T.; Larson, B.T.; Linden, T.A.; Vermeij, M.J.A.; McDonald, K.; King, N. Light-Regulated Collective Contractility in a IMulticellular Choanoflagellate. *Science* 2019, *366*(6463), 326–34.

Brunette, D.M. The Effects of Implant Surface Topography on the Behavior of Cells. *Int. J. Oral Maxillofac. Implants* 1988, *3*(4), 231–46.

Buxboim, A.; Discher, D.E. Stem Cells Feel the Difference. *Nat. Methods* 2010, *7*(5), 695–7.

Buxboim, A.; Ivanovska, I.L.; Discher, D.E. Matrix Elasticity, Cytoskeletal IForces and Physics of the Nucleous: How Deeply Do Cells "Feel" Outside and In? *J. Cell Sci.* 2010, *123*(3), 297–308.

Chen, C.-W.; Montelatici, E.; Crisan, M.; Corselli, M.; Huard, J.; Lazzari, L.; Péault, B.; Péault Perivascular Multi-Lineage Progenitor Cella in Humans Organs: Regenerative Units, Cytokine Sources or Both? *Cytokine Growth Factor Rev.* 2009, *20*(5–6), 429–34.

Clark, P.; Connolly, P.; Curtis, A.S.G.; Dow, J.A.T.; Wilkinson, D.W. Topographical Control of Cell Behaviour. II. Multiple Grooved Substrata. *Development* 1987, *99*, 493–48.

Clark, P.; Connolly, P.; Curtis, A.S.G.; Dow, J.A.T.; Wilkinson, D.W. Topographical Control of Cell Behaviour. I. Simple Step Cues. *Development* 1990, *108*, 635–44.

Costa-Rodriguez, J.; Fernandes, A.; Lopes, M.A.; Fernandes, M.H. Hydroxyapatite Surface Roughness: Complex Modulation of the Osteoclastogenesis of Human Precursor Cells. *Acta Biomater.* 2012, *8*(3), 1137–45.

Crisan, M.; Yap, S.; Casteilla, L. et al. A Perivascular Origin for Mesenchymal Stem Cells in Multiple Human Organs. *Cell Stem Cell* 2008, *3*(3), 301–13.

Crivellato, E.; Nico, B.; Ribatti, D. Contribution of Endothelial Cells to Organogenesis: A Modern Reappraisal of an Old Aristotelian Concept. *J. Anat.* 2007, *211*(4), 415–27.

Curran, J.M.; Chen, R.; Hunt, J.A. The Guidance of Human Mesenchymal Stem Cell Differentiation In Vitro by Controlled Modifications to the Cell Substrate. *Biomaterials* 2006, *27*(27), 4783–93.

Curtis, A.C.; Wilkinson, C. Topographical Control of Cells. *Biomaterials* 1997, *18*(24), 1573–83.

Discher, D.E.; Janmey, P.; Wang, Y.-L. Tissue Cells Feel and Respond to the Stiffness of Their Substrate. *Science* 2005, *310*(5751), 1139–43.

Discher, D.E.; Mooney, D.J.; Zandstra, P.W. Growth Factors, Matrices, and Forces Combine and Control Stem Cells. *Science* 2009, *324*(5935), 1673–77.

De Groot. Carriers that Concentrate Native Bone Morphogenetic Protein In Vivo. *Tissue Eng.* 1998, *4*(4): 337–41. doi: 10.1089/ten.1998.4.337.

De Robertis, E.M. Evo-Devo: Variations on Ancestral Themes. *Cell* 2008, *132*(2), 185–95, doi:10.1016/j.cell.2008.01.003.

Doherty, M.J.; Ashton, B.A.; Walsh, S.; Beresford, J.N.; Grant, M.E.; Canfield, A.E. Vascular Pericytes Express Osteogenic Potential In Vitro and In Vivo. *J. Bone Miner. Res.* 1988, *13*, 128–39.

Engler, A.J.; Sen, S.; Sweeney, H.L.; Discher, D.E. Matrix Elasticity Directs Stem Cell Lineage Specification. *Cell* 2006, *126*(4), 677–89.

Ferretti, C.; Ripamonti, U.; Tsiridis, E.; Kerawala, C.J.; Mantalaris, A.; Heliotis, M. Osteoinduction: Translating Preclinical Promises into Reality. *Br. J. Oral Maxillofac. Surg.* 2010, *48*(7), 536–39

Fiedler, J.; Özdemir, B.; Bartholomä, J.; Pletti, A.; Brenner, R.E.; Ziemann, P. The Effect of Substrate Surface Nanotopogrphy on the Behavior of Multipotent Mesenchymal Stromal Cells and Osteoblasts. *Biomaterials* 2013, *34*(35), 8851–59.

Folkman, J.; Klagsbrun; Sasse, J.; et al. A Heparin Binding Angiogenic Protein Basic Fibroblast Growth Factor Is Stored within Basement Membranes. *Am. J. Pathol.* 1988, *130*, 393–400.

Fu, J.; Wang, Y.-K.; Yang, M.T.; Desai, R.A.; Yu, X.; Liu, Z.; Chen, C.S. Mechanical Regulation of Cell Function with Geometrically Modulated Elastomeric Substrates. *Nat. Methods* 2010, 7(9), 733–736; doi:10.1033/NMETH.1487.

Fujibayashi, S.; Nakamura, T.; Nishiguchi, S.; Tamura, J.; Uchida, M.; Kim, H.-M.; Kobuko, T. Bioactive Titanium: Effect of Sodium Removal and the Bone-Bonding Ability of Bioactive Titanium Prepared by Alkali and Heat Treatment. *J. Biomed. Mater. Res.* 2001, *56*(4), 562–70.

Fujibayashi, S.; Neo, M.; Kim, H.-M.; Kobuko, T.; Nakamura, T. Osteoinduction of Porous Bioactive Titanium Metal. *Biomaterials* 2004, *25*(3), 443–50.

Gage, F.H. Adult Neurogenesis in Mammals. *Science* 2019, *364*(6443), 827–8.

Gittens, R.A.; McLachlan, T.; Olivares-Navarrete, R.; Cai, Y.; Berner, S.; Tannenbaum, R.; Schwartz,Z.; Sandhage, K.H.; Boyan, B.D. The Effects of Combined Micron-/Submicron-Scale Surface Roughness and Nanoscale Features on Cell Proliferation and Differentiation. *Biomaterials* 2011, *32*(13), 3395–403.

Gjorevski, N.; Nelson, C.M. Bidirectional Extracellular Matrix Signaling During Tissue Morphogenesis. *Cytokine Growth Factor Rev.* 2009, *20*(5–6), 459–65.

Gomez-Salinero, J.M.; Rafii, S. Endothelial Cell Adaptation in Regeneration. *Science* 2018, *362*(6419), 1116–17.

Gospodarowicz, D.; Greenburg, G.; Birdwell, C.R. Determination of Cellular Shape by the Extracellular Matrix and Its Correlation with the Control of Cell Growth. *Cancer Res.* 1978, *38*(11 Pt 2), 4155–71.

Goss, R. *Principles of Regeneration*. Academic Press, New York, 1969.

Greer, A.I.M.; Goriainov, V.; Kanczler, J.; Black, C.R.M.; Turner, L.-A.; Meek, R.M.D.; Burgess, M.K.; MacLaren, I.; Dalby, M.J.; Oreffo, R.O.C.; Gadegaard, N. Nanopatterned Titanium Implants Accelerate Bone Formation In Vivo. *ACS Appl. Mater. Interfaces* 2020, *12*, 33541–49.

Hynes, R.O. The Extracellular Matrix: Not Just Pretty Fibrils. *Science* 2009, *326*(5957), 1216–19.

Hopkinson-Wolley, J.; Hughes, D.; Gordon, S.; Martin, P. Macrophage Recruitm,ent During Limb Development and Wound Healing in the Embryonic and Foetal Mouse. *J. Cell Sci.* 1994, *107*(5), 1159–67.

Jung, H.; Baek, M.; D'Elia, K.; Boisvert, C.; Currie, P.D.; Tay, B.-H.; Venkatesh, B.; Brown, S.T.; Heguy, A.; Schoppik, D.; Dasen, J.S. The Ancient Origins of Neural Substrates for Land Walking. *Cell* 2018, *172*(4), 667–82, doi:10.1016/j.cell.2018.01.013.

Karageorgiou, V.; Kaplan, D. Porosity of 3D Biomaterials Scaffolds and Osteogenesis. *Biomaterials* 2005, *26*(27), 5474–91.

Kawai, T.; Takemoto, M.; Fujibayashi, S.; Akiyama, H.; Tanaka, M.; Yamaguchi, S.; Pattanayak, D.K.; Doi, K.; Matsushita, T.; Nakamura, T.; Kobuko, T.; Matsuda, S. Osteoinduction on Acid and Heat Treated Porous Ti Metal Samples in Canine Muscle. *PLOS ONE* 2014, *9*(2), e88366, doi:10.1371/journal.pone.0088366.

Kilian, K.A.; Bugarija, B.; Lahn, B.T.; Mrksich, M. Geometric Cues for Directing the Differentiation of Mesenchymal Stem Cells. *Proc. Natl. Acad. Sci, U.S.A.* 2010, *107*(11), 4872–877.

Kim, D.-H.; Provenzano, P.P.; Smith, C.L.; Levchenko, A. Matrix Nanotopography as a Regulator of Cell Function. *J. Cell. Biol.* 2012, *197*(3), 351–60.

Klar, R.M.; Duarte, R.; Dix-Peek, T.; Dickens, C.; Ferretti, C.; Ripamonti, U. Calcium Ions and Osteoclastogenesis Initiate the Induction of Bone Formation by Coral-Derived Macroporous Constructs. *J. Cell. Mol. Med.* 2013, *17*(11), 1444–57.

Kobuko, T.; Miyaji, F.; Kim, H.-M. Spontaneous Formation of Apatite Layer on Chemically Treated Titanium Metals. *J. Am. Ceram. Soc.* 1996, *79*(4), 1127–29.

Kopp,J.L.; Grompe, M.; Sander, M. Stem Cells versus Plasticity in Liver and Pancreas Regeneration. *Nat. Cell. Biol.* 2016, *18*(3), 238–45.

Kovacic, J.C.; Boehm, M. Resident Vascular Progenitor Cells: An Emerging Role for Non-Terminally Differentiated Vessel-Resident Cells in Vascular Biology. *Stem Cell Res.* 2009, *2*(1), 2–15.

Kusumbe, A.P.; Ramasamy, S.K.; Adams, R.H. Coupling of Angiogenesis and Osteogenesis by a Speficic Vessel Subtype in Bone. *Nature* 2014, *507*(7492), 323–28.

Lamers, E.; van Horssen, R.; te Riet, J.; van Delft, F.C.; Luttge, R.; Walboomers, X.F.; Jansen, J.A. The Influence of Nanoscale Topographical Cues on Initial Osteoblast Morphology and Migration. *Eur. Cell Mater.* 2010, *20*, 329–43.

Lanza, D.; Vegetti, M. Aristotele, A Cura di Diego Lanza e Mario Vegetti: Opere Biologiche UTET 1971.

Lenhoff, H.M.; Lenhoff, S.G. Abraham Trembley and the origins of research on regeneration in animals. In: C.E. Dinsmore (ed.) *A History of Regeneration Research: Milestones in the Evolution of a Science.* Cambridge Univerrsity Press, Canada, 1991. 7–242.

Lensch, M.W.; Daheron, L.; Schlager, T.M. Pluripotent Stem Cells and Their Niches. *Stem Cell Rev.* 2006, *2*(3), 185–202.

Lismaa, S.E.; Kaidonis, X.; Nicks, M.; Bogush, N.; Kikuchi, K.; Naqvi, N.; Harvery, R.P.; Husain, A.; Grahm, R.M. Comparative Regenerative Mechanisms across Different Mammalian Tissues. *NPJ Reg. Med.* 2018, *3*(1), 1–20, doi:10.1038/s41536-018-0044-5.

Liu, X.; Liu, R.; Cao, B.; Ye, K.; Li, S.; Gu, Y.; Pan, Z.; Ding, J. Subcellular Cell Geometry on Micropillars Regulates Stem Cell Differentiation. *Biomaterials* 2016, *111*, 27–39.

Luyten, F.P.; Cunningham, N.S.; Ma, S.; Muthukumaran, N.; Hammonds, R.G.; Nevins, W.B.; Wood, W.I.; Reddi, A.H. Purification and Partial Amino Acid Sequence of Osteogenin, a Protein Initiating Bone Differentiation. *J. Biol. Chem.* 1999, *264*(23), 13377–80.

Martin, P.; Parkhurst, S. Parallels between Tissue Repair and Embryo Morphogenesis. *Development* 2004, *131*(13), 3021–34.

Medici, D.; Shore, E.M.; Lounev, V.Y.; Kaplan, F.D.; Kalluri, R.; Olsen, B.R. Conversion of Vascular Endothelial Cells into Multipotent Stem-Like Cells. *Nat. Med.* 2011, *16*(12), 1400–06.

Metavarayuth, K.; Sitasuwan, P.; Zhao, X.; Lin, Y.; Wang, Q. Influence of Surface Topographical Cues on the Differentiation of Mesenchymal Stem Cells In Vitro. *ACS Biomater. Sci. Eng.* 2016, *2*(2), 142–51.

McNamara, L.E.; McMurray, R.J.; Biggs, M.J.P.; Kantawong, F.; Oreffo, M.J.; Dlby, M.J. Nanotopographical Control of Stem Cell Differentiation. *J. Tissue Eng.* 2010, *2010*, doi:10.4061/2010/120623.

McNamara, L.E.; Sjöström, T.; Burgess, K.E.V.; Kim, J.J.W.; Liu, E.; Gordonov, S.; Moghe, P.V.; Meek, R.M.D.; Oreffo, O.C.; Su, B.; Dalby, M.J. Skeletal Stem Cell Physiology on Functionally Distinct Titania Topographies. *Biomaterials* 2011, *32*, 7403–10.

Miyoshi, H.; Adachi, T. Topography Design Concept of a Tissue Engineering Scaffold for Controlling Cell Function and Fate through Actin Cytoskeletal Modulation. *Tissue Eng.* 2014, *20*(6), 609–27.

Müller, P.; Buinheim, U.; Diener, A.; Lüthen, F.; Teller, M.; Klinkenberg, E.-D.; Neumann, H.-G.; Nebe, B.; Liebold, A.; Steinhoff, G.; Rychly, J. Calcium Phosphate Surfaces Promote Osteogenic Differentiation of Mesenchymal Stem Cells. *J. Cell. Mol. Med.* 2008, *12*(1), 281–91.

Nelson, C.M.; Vanduijn, M.M.; Inman, J.L.; Flertcher, D.A.; Bissell, M.J. Tissue Geometry Determines Sites of Mammary Branching Morphogenesis in Organotypic Cultures. *Science* 2006, *314*(5797), 298–300.

Nishiguchi, S.; Kat, H.; Fujita, H.; Kim, H.-M.; Miyaji, F.; Kobuko, T.; Nakamura, T. Enhancement of Bone-Bonding Strengths of Titanium Alloy Implants by Alakli and Heat Treatments. *J. Biomed. Mater. Res. (Appl. Biomater.)*, 1999,*48*(5), 689–99.

Önal, P.; Grün, D.; Adamidi, C.; Rybak, A.; Solana, J.; Mastrobuoni, G.; Wang, Y.; Rahn, H.-P.; Chen, W.; Kempa, S.; Ziebold, U.; Rajewsky, N. Gene Expression of Pluripotency Determinants Is Conserved Between Mammalian and Planarian Stem Cells. *EMBO J.* 2012, *31*(12), 2755–69.

Othman, Z.; Fernandes, H.; Groot, A.J.; Luider, T.M.; Alcinesio, A.; de melo Pereira, D.; Guttenplan, A.P.M.; Yuan, H.; Hbibovic, P. *Biomaterials* 2009, *2010*, 12–24.

Paralkar, V.M.; Nandedkar, A.K.N.; Pointer, R.H.; Kleinman, H.K.; Reddi, A.H. Interaction of Osteogenin, a Heparin Binding Bone Morphogenetic Protein, with Type IV Collagen. *J. Biol. Chem.* 1990, *265*(28), 17281–84.

Paralkar, V.M.; Vukicevic, S.; Reddi, A.H. Transforming Growth Factor β Type 1 Binds to Collagen Type IV of Basement Membrane Matrix: Implications for Development. *Dev. Biol.* 1991, *143*(2), 303–10.

Ramasamy, S.K.; Kusumbe, A.P.; Wang, L.; Adams, R.H. Endothelial Notch Activity Promotes Angiogenesis and Osteogenesis in Bone. Nature 2014, 507.

Ramasamy, S.K.; Kusumbe, A.P.; Adams, R.H. Regulation of Tissue Morphogenesis by Endothelial Cell-Derived Signals. *Trends Cell Biol.* 2015, *25*(3), 148–57.

Reddi, A.H.; Huggins, C. Biochemical Sequences in the Transformation of Normal Fibroblasts in Adolescent Rats. *Proc. Natl. Acad. Sci. U.S.A.* 1972, *69*(6), 1601–5.

Reddi, A.H.; Huggins, C.B. Influence of Geometry of Transplanted Tooth and Bone on Transformation of Fibroblasts. *Proc. Soc. Exp. Biol. Med.* 1973, *143*(3), 634–37.

Reddi, A.H. Bone Matrix in the Solid State: Geometric Influence on Differentiation of Fibroblasts. *Adv. Biol. Med. Phys.* 1974, *15*(0), 1–18.

Reddi, A.H. Cell Biology and Biochemistry of Endochondral Bone Development. *Coll. Rel. Res.* 1981, *1*(2), 209–26.

Reddi, A.H.; Kuettner, K.E. Vascular Invasion of Cartilage: Correlation of Morphology with Lysozyme, Glycosaminoglycans, Protease, and Protease-Inhibitory Activity During Endochondral Bone Development. *Dev. Biol.* 1981, *82*(2), 217–23.

Reddi, A.H. Bone Morphogenesis and Modeling: Soluble Signals Sculpt Osteosomes in the Solid State. *Cell* 1997, *89*(2), 159–61.

Reddi, A.H. Morphogenesis and Tissue Engineering of Bone and Cartilage: Inductive Signals, Stem Cells, and Biomimetic Matrices. *Tissue Eng.* 2000, *6*(4), 351–59.

Reichman, O.J. Evolution of Regenerative Capabilities. *Am. Nat.* 1984, *123*(6), 752–63.

Ripamonti, U. Inductive Bone Matrix and Porous Hydroxyapatite Composites in Rodents and Nonhuman Primates. In: J. Yamamuro, L. Wilson-Hench, L. Hench (eds.) *Handbook of Bioactive Ceramics, II: Calcium Phosphate and Hydroxylapatite Ceramics*, CRC Press, Boca Raton, FL, 1990, 245–53.

Ripamonti, U. The Morphogenesis of Bone in Replicas of Porous Hydroxyapatite Obtained from Conversion of Calcium Carbonate Exoskeletons of Coral. *J. Bone Joint Surg.* [A] 1991, *73-A*, 692–703.

Ripamonti, U. Calvarial Reconstruction in Baboons with Porous Hydroxyapatite. *J. Craniofac. Surg.* 1992a, *3*(3), 149–59.

Ripamonti, U. Calvarial Regeneration in Primates with Autolyzed Antigen Extracted Allogeneic Bone. *Clin. Orth. Rel. Res.* 1992b, *282*(282), 293–303.

Ripamonti, U.; Ma, S.; van den Heever, B.; Reddi, A.H. Osteogenin, a Bone Morphogenetic Protein, Adsorbed on Porous Hydroxyapatite Substrata, Induces Rapid Bone Differentiation in Calvarial Defects of Adult Primates. *Plast. Reconstr. Surg.* 1992, *90*(3), 382–93.

Ripamonti, U.; van den Heever, B.; van Wyk, J. Expression of the Osteogenic Phenotype in Porous Hydroxyapatite Implanted Extraskeletally in Baboons. *Matrix* 1993, *13*(6), 491 –502.

Ripamonti, U. A Method for Screening a Selected Material for Its Osteoconductive and Osteoinductive Potential. South African Patent 92/3982, May 25 1994, US patent 5,355,898, October 18, 1994.

Ripamonti, U.; Kirkbride, A.N. Biomaterial and Bone Implant for Bone Repair and Replacement. PCT/NL95/00181, WO095/3200. November 30, 1995.

Ripamonti, U.; Duneas, N. Tissue Engineering of Bone by Osteoinductive Biomaterials. MRS Bulletin November 1996, 36–9.

Ripamonti, U.; Kirkbride, A.N. A Biomaterial and Bone Implant for Bone Repair and Replacement EP760687A1, March 12, 1997.

Ripamonti, U.; Crooks, J.; Kirkbride, A.N. Sintered Porous Hydroxyapatites with Intrinsic Osteoinductive Activity: Geometric Induction of Bone Formation. *S. Afr. J. Sci.* 1999, *95*, 335–43.

Ripamonti, U.; Kirkbride, A.N. Biomaterial and Bone Implant for Bone Repair and Replacement. US Patent No. 6,302,913B6,302 B1, October 16 2001.

Ripamonti,U.; Ramoshebi, L.N.; Matsaba, T.; Tasker, J.; Crooks, J.; Teare, J. Bone Induction by BMPs/OPs and Related Family Members. The Critical Role of Delivery Systems. *J. Bone Joint Surg.* [A] 2001a, 83-A(Suppl. 1), S116–127.

Ripamonti, U.; Crooks, J.; Rueger, D.C. Induction of Bone Formation by Recombinant Human Osteogenic Protein-1 (hOP-1) and Sintered Porous Hydroxyapatite in Adult Primates. *Plast. Reconstr. Surg.* 2001b, *107*(4), 977–88.

Ripamonti, U. Osteogenic Proteins of the Transforming Growth Factor-ß Superfamily. In: H.L. Henry, A.W. Norman (eds.) *Encyclopedia of Hormones*, Elsevier, San Diego, CA, 2003, 80–6.

Ripamonti, U. Soluble, Insoluble and Geometric Signals Sculpt the Architecture of Mineralized Tissues. *J. Cell. Mol. Med.* 2004, *8*(2), 169–80.

Ripamonti, U.; Ramoshebi, L.N.; Patton, J.; Matsaba, T.; Teare, J.; Renton, L., Soluble Signals and Insoluble Substrata: Novel Molecular Cues Instructing the Induction of Bone. In: E.J. Massaro, J.M. Rogers (eds.) *The Skeleton*, Humana Press, Totowa, NJ, 2004, Chapter 15, 217–27.

Ripamonti, U. Soluble Osteogenic Molecular Signals and the Induction of Bone Formation. *Biomaterials* Leading Opinion Paper 2006, *27*(6), 807–22.

Ripamonti, U.; Ferretti, C.; Heliotis, M. Soluble and Insoluble Signals and the Induction of Bone Formation: Molecular Therapeutics Recapitulating Development. *J. Anat.* 2006, *209*(4), 447–68.

Riapmonti, U.; Richter, P.W.; Thomas, M.E.; Shape, Self-Inducing Memory Geomteric Cues Embedded within Smart Hydroxyapatite-Nased Biomimetic Matrices. *Plast. Reconstr. Surg.* 2007a, *120*, 1796–07.

Ripamonti, U.; Heliotis, M.; Ferretti, C. Bone Morphogenetic Proteins and the Induction of Bone Formation: From Laboratory to Patients. *Oral Maxillofac. Surg. Clin. North Am.* 2007b, *19*(4), 575–89.

Ripamonti, U. Biomimetism, Biomimetic Matrices and the Induction of Bone Formation. *J. Cell. Mol. Med.* 2009, *13*(9B), 2953–72.

Ripamonti, U.; Crooks, J.; Khoali, L.; Roden, L. The Induction of Bone Formation by Coral-Derived Calcium Carbonate/Hydroxyapatite Constructs. *Biomaterials* 2009, *30*(7), 1428–39.

Ripamonti, U. Soluble and Insoluble Signals Sculpt Osteogenesis in Angiogenesis. *World J. Biol. Chem.* 2010, *26*(5), 109–32.

Ripamonti, U.; Klar, R.M.; Renton, L.F.; Ferretti, C. Synergistic Induction of Bone Formation by hOP- 1, hTGF-β_3 and Inhibition by Zoledronate in Macroporous Coral-Derived Hydroxyapatites. *Biomaterials* 2010, *31*(25), 6400–410.

Ripamonti, U. The Concavity: The "Shape of Life"and the Control of Bone Differentiation – Feature Paper – *Science in Africa* May 2012.

Ripamonti, U.; Roden, L.C.; Renton, L.F. Osteoinductive Hydroxyapatite-Coated Titanium Implants. *Biomaterials* 2012a, *33*(15), 3813–23.

Ripamonti, U.; Roden, L.; Renton, L.; Klar, R.M.; Petit, J.-C. The Influence of Geometry on Bone: Formation by Autoinduction. *Science in Africa* 2012b. http://www.scienceinafric a.co.za/2012/Ripamonti_bone.htm.

Ripamonti, U.; Renton, L.; Petit, J.-C. Bioinspired Titanium Implants: The Concavity - The Shape of Life. In: M. Ramalingam, P. Vallitu, U. Ripamonti, W.-J. Li (eds.) *CRC Press Taylor & Francis, Boca Raton USA;* Tissue Engineering and Regenerative Medicine. A Nano Approach, 2013, CRC Press, Chapter 6, 105–23.

Ripamonti, U.; Duarte, R.; Ferretti, C. Re-Evaluating the Induction of Bone Formation in Primates. *Biomaterials* 2014, *35*(35), 9407–22.

Ripamonti, U.; Dix-Peek, T.; Parak, R.; Milner, B.; Duarte, R. Profiling Bone Morphogenetic Proteins and Transforming Growth Factor-βs by hTGF-β_3 Pre-treated Coral-Derived Macroporous Constructs: The Power of One. *Biomaterials* 2015, *49*, 90–102.

Ripamonti, U. Biomimetic Functionalized Surfaces and the Induction of Bone Formation. *Tissue Eng.* 2017, *23*(21,22), 1197–209.

Ripamonti, U. Functionalized Surface Geometries Induce: "Bone: Formation by Autoinduction". *Front. Physiol.* 2018, *8*, 1084, doi:10.3389/fphys.2017.01084.

Sammons, R.L.; Lumbikanoda, N.; Gross, M.; Cantzler, P. Comparison of Osteoblast Spreading on Microstructured Dental Implant Surfaces and Cell Behaviour in an Explant Model of Osteointegration. A Scanning Electron Microscopre Study. *Clin. Oral Impl. Res.* 2005, *16*, 657–66, doi:10.1111/j.1600-0501.2005.01168.x.

Sampath, T.K.; Reddi, A.H. Dissociative Extraction and Reconstitution of Extracellular Matrix Components Involved in Local Bone Differentiation. *Proc. Natl. Acad. Sci. U.S.A.* 1981, *78*(12), 7599–603.

Sampath, T.K.; Reddi, A.H. Homology of Bone-Inductive Proteins from Human, Monkey, Bovine, and Rat Extracellular Matrix. *Proc. Natl. Acad. Sci. U.S.A.* 1983, *80*(21), 6591–95.

Sampath, T.K.; Reddi, A.H. Importance of Geometry of the Extracellular Matrix in Endochondral Bone Differentiation. *J. Cell Biol.* 1984, *98*(6), 2192–97.

Sampath, T.K.; Rashka, K.E.; Doctor, J.S.; Tucker, R.F.; Hoffmann, F.M. Drosophila Transforming Growth Factor-β Superfamily of Proteins Induce Endochondral Bone Formation in Mammals. *Proc. Natl. Acad. Sci. U.S.A.* 1993, *90*(13), 6004–08.

Sun, J.; Jamilpour, N.; Wang, F.-Y.; Wong, P.K. Geometric Control of Capillary Architecture via Cell-Matrix Mechanical Interactions. *Biomaterials* 2014, *35*(10), 3273–80.

Sung, B.H.; Weaver, A. Direct Migration: Cells Navigate by Extracellular Vesicles. *J. Cell Biol.* 2018, *217*(8), 2613–14.

Takemoto, M.; Fujjibayadshi, S.; Neo, M.; Suzuki, J.; Matsushita, T.; Kobuko, T.; Nakamura, T. Osteoinductive Porous Titanium Implants: Effects of Sodium Removal by Dilute HCl Treatment. *Biomaterials* 2006, *27*(13), 2682–91.

Tanaka, E.M.; Reddien, P.W. The Cellular Basis for Animal Regeneration. *Dev. Cell* 2011, *21*(1), 172–85.

Tomancak, P. Evolutionary History of Tissue Bending. *Science* 2019, *366*(6463), 300–01, doi:10.1126/science.aaz1289.

Trueta, J. The Role of the Vessels in Osteogenesis. *J. Bone Joint Surg.* 1963, *45B*, 402–18.

Uchida, M.; Kim, H.-M.; Kobuko, T.; Fujibayashi, S.; Nakamura, T. Effect of Water Treatment on the Apatite-Forming Ability of NaOH-Treated Titanium Metal. *J. Biomed. Mater. Res. (Appl. Biomater.)* 2002, *63*(5), 522–30.

Uchida, M.; Kim, H.-M.; Kobuko, T.; Fujibayashi, S.; Nakamura, T. Structural Dependence of Apatite Formation on Titania Gels in a Simulated Body Fluid. *J. Biomed. Mater. Res.* 2003, *64A*, 164–70.

Urist, M.R. Bone: Formation by Autoinduction. *Science* 1965, *220*(3698), 893–99, doi:10.1126/science.150.3698.893.

Urist, M.R.; Silverman, F.; Büring, K.; Dubuc, F.L.; Rosenberg, J.M. The Bone Induction Principle. *Clin. Orthop. Rel. Res.* 1967, *53*, 243–83.

Urist, M.R.; Hou, Y.K.; Brownell, A.G.; Hohl, W.; Buyske, J.; Lietze, A.; Tempst, P.; Hunkapiller, M.; DeLange, R.J. Purification of Bovine Bone Morphogenetic Protein by Hydroxyapatite Chromatography. *Proc. Natl. Acad. Sci. U.S.A.* 1984, *81*(2), 371–5.

van Eeden, S.; Ripamonti, U. Bone Differentiation in Porous Hydroxyapatite Is Regulated by the Geometry of the Substratum: Implications for Reconstructive Craniofacial Surgery. *Plast. Reconstr. Surg.* 1994, *93*(5), 959–66.

Vlacic-Zischke, J.; Hamlet, S.M.; Friis, et al The Influence of Surface Microroughness and Hydrophilicity of Titanium on the Up-Regulation of TGFβ/BMP Signalling in Osteoblasts. *Biomaterials* 2011, *32*, 7403–10.

Watari, S.; Hayashi, K.; Wood, J.A.; Russell, P.; Nealey, P.F.; Murphy, C.J.; Genetos, D.C. Modulation of Osteogenic Differentiation in hMSCs Cells by Submicron Topographically-Patterned Ridges and Grooves. *Biomaterials* 2012, *33*(1), 128–36.

Wilkinson, A.; Hewitt, R.N.; McNamara, L.E.; McCloy, D.; Meek, D.R.M.; Dalby, M.J. Biomimetic Microtopography to Enhance Osteogenesis *In Vitro*. *Acta Biomater.* 2011, *7*(7), 2919–25.

Yang, G.-L.; He, F.-M.; Song, E.; Hu, J.-A.; Wang, X.-X.; Zhao, S.-F. In Vivo Comparison of Bone Formation on Titanium Implant Surfaces Coated with Biomimetically Deposited Calcium Phosphate or Electrochemically Deposited Hydroxyapatite. *Int. J. Oral Maxillofac. Implants* 2010, *25*(4), 669–80.

Yang, Y.; Wang, K.; Gu, X.; Leong, K.W. Biophysical Regulation of Cell Behaviour-Cross Talk between Substrate Stiffness and Nanotopography. *Engineering* 2017, *3*(1), 36–54.

Yoon, J.-L.; Kim, H.N.; Bhang, S.H.; Shin, J.-Y.; Han, J.; La, W.-G.; Jeong, G.-J.; Kang, S.; Lee, J.-R.; Oh, J.; Kim, M.S.; Jeon, N.L.; Kim, B.-S. Enhanced Bone Repair by Guided Osteoblast Recruitment Using Topographically Defined Implant. Tissue Eng. 2016, 22(7), 654.

Yuan, H.; Fernandes, H.; Habibovic, P.; de Boer, P.; Barradas, A.M.; de Ruiter, A.; Walsh, W.R.; van Blitterswijk, C.A.; de Bruijnij, J.D. Osteoinductive Ceramics as Synthetic Alternative to Autologous Bone Grafting. *Proc. Natl. Acad. Sci. U.S.A.* 2010, *107*(31), 13614–19.

Zhang, J.; Sun, L.; Luo, X.; Barbieri, D.; de Bruijn, J.D.; van Blitterswijk, C.A.; Moroni, L.; Yuan, Y. Cells Responding to Surface Structure of Calcium Phosphate Ceramics for Bone Regeneration. *J. Tissue Eng. Reg. Med.* 2017, doi:10.1002/term.2236.

Zhang, Y.; Chen, S.E.; Shao, J.; van den Beucken, J.J.J.P. Combinatorial Surface Roughness Effects on Osteoclastogenesis and Osteogenesis. *ACS Appl. Mater. Interfaces* 2018, doi:10, 36652-663.

Zheng, B.; Cao, B.; Crisan, M.; Sun, B.; Li, G.; Logar, A.; Yap, S.; Pollett, J.B.; Drowley, L.; Cassino, T.; Gharaibeh, B.; Deasy, B.; Huard, J.; Péault, B. Prospective Identification of Myogenic Endothelial Cells in Human Skeletal Muscle. *Nat. Biotechnol.* 2007, *25*(9), 10125–34.

Index